Designed Maps

A Sourcebook for GIS Users

Cynthia A. Brewer

ESRI Press
REDLANDS, CALIFORNIA

ESRI Press, 380 New York Street, Redlands, California 92373-8100

 First edition 2008
10 09 08 1 2 3 4 5 6 7 8 9 10

Printed in the United States of America

Library of Congress Cataloging-in-Publication Data
Brewer, Cynthia A., 1960–
Designed maps : a sourcebook for GIS users / Cynthia A. Brewer.—1st ed.
p. cm.
Includes bibliographical references and index.
ISBN 978-1-58948-160-2 (alk. paper)
1. Cartography. 2. Geographic information systems. I. Title.
GA105.3.B73 2008
526.0285—dc22 2007051571

Ask for ESRI Press titles at your local bookstore or order by calling 1-800-447-9778. You can also shop online at www.esri.com/esripress. Outside the United States, contact your local ESRI distributor.

ESRI Press titles are distributed to the trade by the following:

In North America:
Ingram Publisher Services
Toll-free telephone: (800) 648-3104
Toll-free fax: (800) 838-1149
E-mail: customerservice@ingrampublisherservices.com

In the United Kingdom, Europe, and the Middle East:
Transatlantic Publishers Group Ltd.
Telephone: 44 20 7373 2515
Fax: 44 20 7244 1018
E-mail: richard@tpgltd.co.uk

Cover design and production *Stefanie Tieman*
Interior design and production *Jennifer Hasselbeck*
Image editing *Michael Law and Jennifer Hasselbeck*
Editing *Amy Collins*
Copyediting and proofreading *Tiffany Wilkerson*
Permissions *Kathleen Morgan*
Printing coordination *Cliff Crabbe and Lilia Arias*

On the cover
GIS for Gettysburg Battlefield Planning and Rehabilitation map courtesy of National Park Service.
Tauranga Harbour Tidal Movements Map courtesy of Environment Bay of Plenty (The Bay of Plenty Regional Council).

CONTENTS

PREFACE

Successful mapmakers study well-designed maps throughout their careers, building a store of visual solutions for representing geographic data. The goal of this book is to present a broad collection of excellent map designs to inspire those who want to create attractive maps that communicate information effectively.

HOW THE BOOK BEGAN

At the 2005 conference of the North American Cartographic Information Society (NACIS), Professor Barbara Buttenfield (University of Colorado–Boulder) organized a panel discussion called "What Goes On Before You Make the Map?" During the initial presentations, four expert cartographers—Alex Tait (International Mapping Associates), Stuart Allan (Allan Cartography, Raven Maps & Images, Benchmark Maps), Dennis McClendon (Chicago Cartographics), and Tom Patterson (National Park Service)—talked about how they start a mapping project. Each reflected that, when they begin designing a map, they think back on other maps they have created or have seen that solved a similar design challenge. Calling images to mind from a store of previous observations was an essential part of their map design work.

As I sat in the audience, I thought about my students who are just starting out with map design. Many of them have not seen a variety of maps. Or they have seen a great many maps, but all with a similar design. The range of maps a young person sees today is, I think, quite limited. Even the distinctive font and careful styling of a National Geographic Society map, the icon of excellent map design from my youth, seems to no longer be immediately recognized by students. Likewise, professionals starting out with a new mapping task unfamiliar to them may not have seen maps of the general sort they set out to make. Mapping textbooks show a variety of example maps, but they rely on schematized designs that make conceptual points, and they do not offer many complete maps designed for use in navigation or data interpretation.

The NACIS panel also brought to mind repeated advice from Professor Judy Olson (Michigan State University) to not show our students poorly designed maps. These maps are fun to critique, but they fill the mind's eye with flawed solutions that compete with the good designs students need to call on to make their own maps. Fundamental to this book is a desire to show readers good map examples that inspire them to likewise make good maps. The collection of maps in this book are all positive examples that I would use to guide my students. Not only students, but professional cartographers, GIS technicians, and amateur mapmakers can absorb good practices from this collection.

The maps you'll see in the following pages were selected for various reasons: some were maps I purchased out of interest and for travel, some are favorite maps of my colleagues, others are maps shown at professional conferences, published in atlases, exhibited in the International Map Exhibition at International Cartographic Association (ICA) meetings, and displayed in the Map Gallery at the ESRI International User Conference and in the resulting map books published by ESRI Press. I did not attempt to divine the best maps in the world (don't worry if I left you out), and I do not cover all possible map types. But the maps in this collection offer wide ranging ideas for overall look and treatment of graphic details that will help people making all types of maps. The discussion that accompanies each map is intended to highlight design options novice mapmakers might not notice when looking at the maps. This book is an exercise in learning to see.

HOW THE BOOK IS ORGANIZED

Each chapter begins with a brief introduction, followed by a map that I have redesigned three different ways. The intent of the redesigns is not to improve on the original map, but simply to show different approaches one could take depending on audience and the medium in which the map will be displayed. These redesigns emphasize different features by using alternative color palettes and symbols. With each map design I've pointed out the specific symbols, fonts, fills, and colors I used, which will help you create similar effects. The "ArcMap Tips" feature at the back of this book provides guidance for creating particular effects using the ArcMap application in ArcGIS Desktop. Each chapter closes with a collection of nine to thirteen maps that are good examples of maps in the category described by the chapter title.

When I started developing the book with the help of Michael Law, the ESRI Press cartographer, it was an inductive process. Rather than finding maps to fit predetermined categories, we started by pulling apart recent volumes of the *ESRI Map Book* and sorting the maps into piles of related topics. We started with about twenty groups and then aggregated groups when we had too many maps that could fit into multiple categories. We further collapsed categories as we attempted to name our groups, and finally ended up with the six chapters in this book.

Two standard categories of maps are general reference and thematic maps. We found many maps that did not fit readily into either category, and thus developed two transitional categories (chapters 3 and 4). We dubbed these special-purpose maps. These sit between the reference maps in chapters 1 and 2 and the thematic maps in chapters 5 and 6, and they contain aspects of both categories. For example, a bicycling map has a general reference base and a theme specific to bike routes. It could be used for recreational cycling or for navigation of the road network by bicycle commuters. Similarly, a hiking map that emphasizes terrain could be in either the topographic or recreation chapters. It does not fit easily into a single category, and this is the case with many of the examples in the book, so don't be distracted if you feel a map is incorrectly categorized—the point is simply to show map design samples in an easy-to-see manner.

CARTOGRAPHIC PRINCIPLES

Label placement is time consuming, but it is the heart of readable maps. Doing it right means taking time to sort out classes, styles, and priority settings for automatic labeling tools, then to correct and adjust the results. Area labels have quite a wide variety of treatments on the maps here, with some centered and spread across their polygons and others labeling areas on either side of a boundary line by placing labels along that line. The particular fonts, sizes, and styles used for the redesigned maps at the start of each chapter are listed. In the map descriptions, you'll encounter a basic font distinction: serif versus sans serif typefaces. Serifs are small curls and tips that sit perpendicular to the strokes that build characters (the type on this page has serifs). Sans serif fonts have strokes with blunt ends; they have no serifs.

Another distinction I make repeatedly throughout the book is whether map elements are part of the foreground or background. Elements that provide supporting background information are designed differently than when the same elements are part of the main message of the map. The foreground elements have stronger contrasts, larger sizes, and more distinct categorizations. Map designs may be dissected into a more detailed hierarchy with many levels, but I emphasize simply putting visual emphasis on the main message of the map, and pushing supporting information into the background with careful design decisions. I repeat this guidance with a variety of maps because I think it is fundamental to good map design.

Another fundamental principle of cartography that I emphasize is systematic use of visual variables. These are the basic categories of marks on the page that can be used to symbolize data. They were first systematized for cartography by Jacques Bertin (see the "Resources" section) and many other authors have since added to his set. The visual variables I emphasize in the book are color hue, color lightness, color saturation, size, shape, pattern angle, and pattern spacing. Three of these visual variables are perceptual dimensions of color. Hue is described with color names such as blue and orange (and is akin to dominant wavelength in the color spectrum). The colors of the rainbow—red, orange, yellow, green, blue—can be arranged in a hue circle and joined with nonspectral purples and magentas to provide a circular continuum from which many sequences of map colors are selected. Lightness and saturation of a color are perceived separately from hue, though many color names evoke particular hue-lightness-saturation combinations (yellow hues are often used as light saturated colors). Any hue may be seen as darker or lighter (lightness) and as vivid or grayish (saturation). The best maps use the three perceptual dimensions of color in a structured manner to create effective map symbols.

The goal of this book is to invite you to see design elements and to prompt you to combine and customize them for your own mapping. The goal is not to suggest that you create these maps exactly. Principles of mapmaking are discussed in an abbreviated manner throughout the book, and the "Resources" section at the end of the book will point you to more in-depth reference books and journals with the detail and theory underpinning these descriptions.

MAPMAKING AND GIS

Not all maps shown in this book were made using GIS. Some include techniques that would still be hard to produce using GIS tools, such as Alex Tait's curving flow lines with color gradients on the Census Bureau's migration map (page 159). Perhaps examples of these effects may give GIS vendors a push toward more complete graphic toolboxes and symbol types. The redesigned maps with full specifications at the start of each chapter are limited to effects that can be produced with GIS tools. The reality of getting data from the original mapmakers meant that they sent us AI or PDF files that I continued to edit in Adobe Illustrator CS2. It was not appropriate or practical for the authors to send their original projects and datasets. As I edited the original maps, I limited my choices to effects that were available in ArcGIS 9.2 software, though you may see a few things that behave a bit differently between AI and ArcMap. Furthermore, the maps emphasize designs suited to the resolution and color capabilities of print cartography, since the reader will evaluate them on the printed pages of the book.

In the years that I have been observing and evaluating maps created with GIS, they have improved quite a bit. Most of the basic graphic design tools are now included with GIS software. For example, improved transparency and hillshading capabilities allow the mapmaker to better show the form of terrain along with the mapped topic of interest. Map legends have also improved. I show the legends for most of the example maps in this book, and many are laid out to organize and group related elements to assist map use. Natural legend designs that show elements together as they are logically arranged in the landscape are included. For example, a small section of coastline in a legend may show symbols for low and high water marks, a tidal flat area, jetty, and bathymetric contours as they are typically arranged within the map. Example legends that show the results of combining hillshading with transparent data colors are also included. So you can see that, while there are still some limitations, GIS is geared toward the cartographer more than ever before.

ACKNOWLEDGMENTS

Michael Law and Amy Collins of ESRI Press were key to completing this project. The three of us talked through final map choices and their groupings into chapters. Amy was the book editor and project manager. She kept me working when I got distracted by other responsibilities and made sure what I wrote made sense. At the core of this book are many real examples from mapmakers around the world, and Michael organized the map files and listings for over a hundred maps and their status as we sought permissions. Michael also reworked the map files I edited by building layers that made them more editable graphics. His assistance has been vital to this project. Jennifer Hasselbeck designed the look and layout of the book. She has an expert eye for working with geographic content, and she customized page structures to best showcase the maps and legends. Tiffany Wilkerson's copyediting improved the text. Kathleen Morgan and Kelley Heider assisted with all those permissions. Thanks also to Judy Hawkins for her enthusiasm for *Designing Better Maps* and encouragement to continue as an author for the Press with its sequel.

ESRI President Jack Dangermond advances good cartographic design through publication of the *Map Book* for all participants at each ESRI User Conference. He consistently highlights excellence in mapmaking and improvements in GIS cartography tools in his talks to tens of thousands of people at these events. Clint Brown promotes basemap innovation and ArcGIS usability for mapping as he directs the software products department at ESRI. Charlie Frye leads cartography research at ESRI and assisted me with many tips on using ESRI software for mapmaking, reflected in the design options shown in the book.

At Penn State, Roger Downs, head of the Department of Geography from 1994 to 2007, encouraged and valued the applied side of my work. My husband, David DiBiase, leads Geography's online certificate and master's programs in GIS, and this book is intended for his adult students and the many GIS professionals like them.

Finally, thank you to all of the map authors and staff at mapping companies and agencies who answered our requests for permission and map files. Obviously, the book would not have come to fruition without you.

Cindy Brewer
October 2007
State College, Pennsylvania

1

REFERENCE MAPS

TOPOGRAPHIC

Reference maps present a wide range of themes together on a map without strong emphasis on one over another. The goal of reference maps is to provide locational details for both casual map users and professionals. The largest scale reference maps often include details of landforms, usually represented with contour lines. Topographic maps are produced as series at a variety of scales covering entire countries across the globe. In the United States, topographic series created by the U.S. Geological Survey (USGS) are available at 1:24,000, 1:100,000, and 1:250,000 scales (these may be downloaded from USGS at www.store.usgs.gov/locator/ or other Web services). Topographic maps from other countries have quite different designs. The Netherlands and Switzerland, for example, have distinctively different terrains, so their topographic maps emphasize different landforms. Land cover and water features are more prominent on the Dutch map, and mountain detail dominates the Swiss map.

Many contemporary reference maps combine hillshading and elevation tints, rather than relying on contour lines to symbolize landforms. Hillshading techniques model the surface by rendering the hypothetical reflection of sunlight (often coming from the northwest) that illuminates a homogeneous surface (no rock colors or vegetation textures are displayed). Hillshading may be calculated using a digital elevation model (DEM), or hand drawn by interpreting contour lines. It produces a portrait that emphasizes the shape of the landforms. The effect has a variety of names with very similar meanings, such as relief shading, analytical shaded relief, and terrain shading. Elevation tints, also called hypsometric tints, are often added to hillshading. They may be presented as continuous gradation through a series of colors or split into specific elevation ranges with each class bordered by contours and assigned a color. The combination of land shape and elevation gives a complete sense of the landscape. Flat landforms in a hillshade layer may be colored as a lowland or high plateau by the elevation tints, providing additional information to map readers. The general depth of a gorge and height of a mountain range are communicated through elevation tints.

Terrain and physical features are a key element of some of the maps highlighted in this chapter, with complete labeling for hydrography, glaciers, valleys, ranges, and peaks. On others, physical features function more as supporting information to the political and cultural elements in the foreground, such as boundaries, transportation, cities, and points of interest. Features are generally categorized with serif fonts for physical features and sans serif fonts for human features in most of the maps presented in this chapter.

The chapter progresses from large-scale topographic maps, through country reference maps, to small-scale continent representations, and contains a wide variety of reference map styles that emphasize topography. Together these maps offer design ideas for a common map genre.

1.0 USGS DIGITAL TOPOGRAPHIC MAP

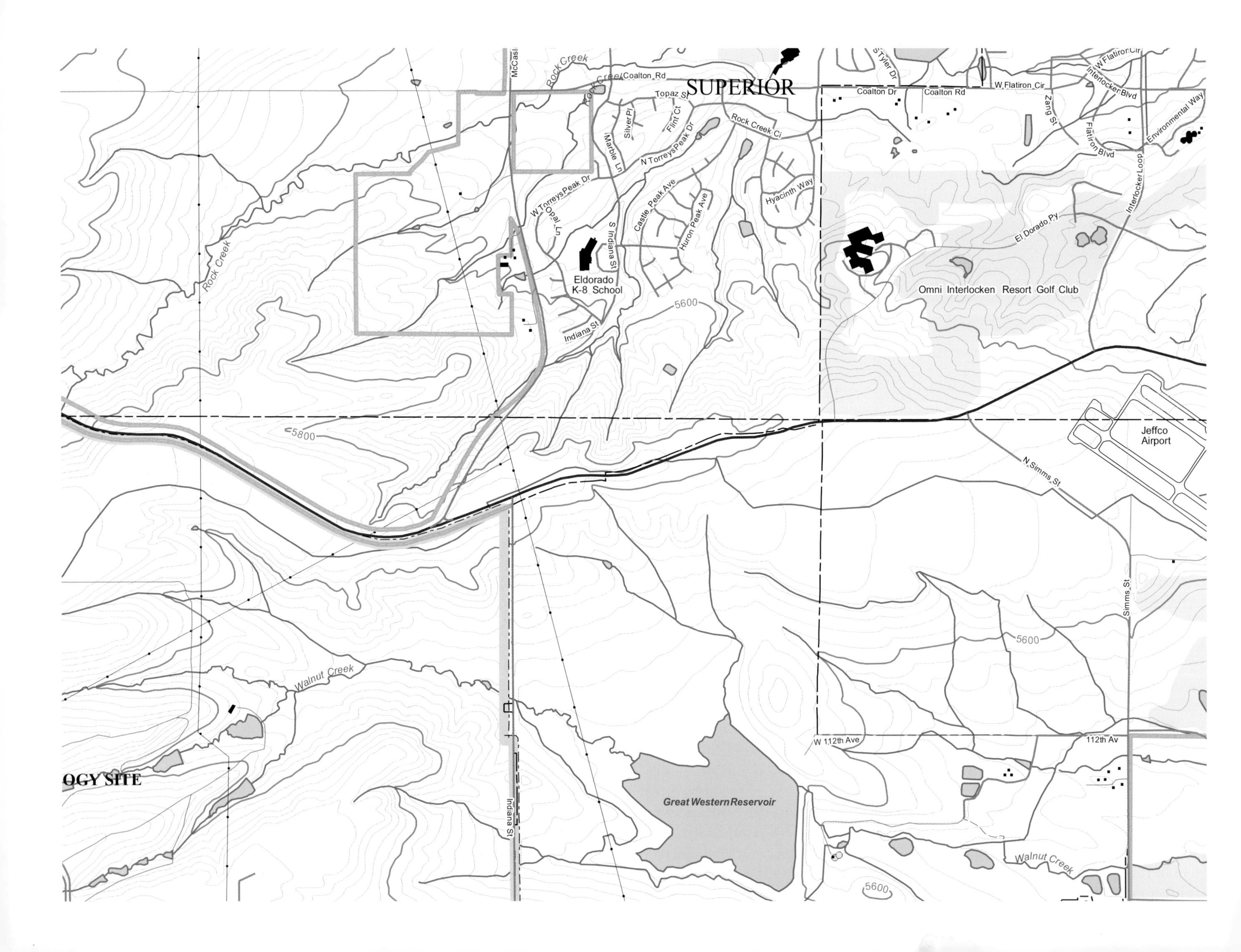

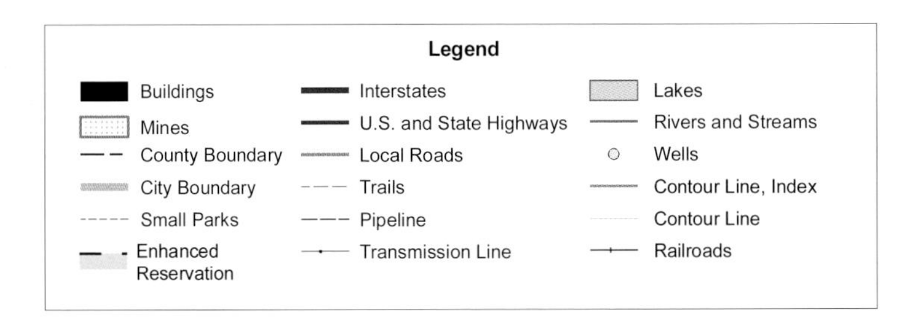

Topographic map data at a scale of 1:24,000 for an area near Superior, Colorado, is shown with different visual emphases in four designs on the following pages. The original and black-and-white designs balance emphasis among features, and the other two designs place more emphasis on human boundaries or natural features.

Courtesy of Parallel Incorporated and U.S. Geological Survey.

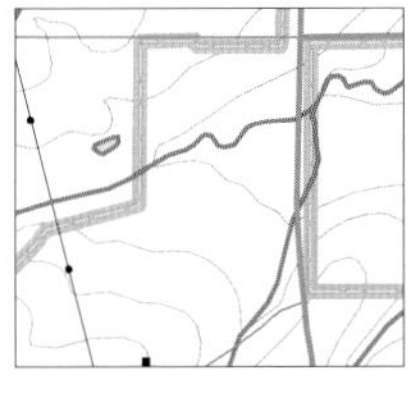

ORIGINAL DESIGN (Pages 2 and 4)

This map shows a prototype design by the U.S. Geological Survey (USGS) for *The National Map* using a limited set of digital data for contours, hydrography, transportation, boundaries, structures, and geographic names (prepared using data from local, state, and federal agencies). Index contours, streams, and pond outlines are bold elements in this design. Different dashed black line patterns distinguish line symbols for county, park, and enhanced reservation boundaries and for trails and pipelines. The gray band under the yellow dashed line of the city boundaries provides contrast for easy visibility and preserves a clean dash pattern when the line overlaps another dashed boundary. Gray local road lines are frequently broken by white halos on black labels, ensuring that labels are readable.

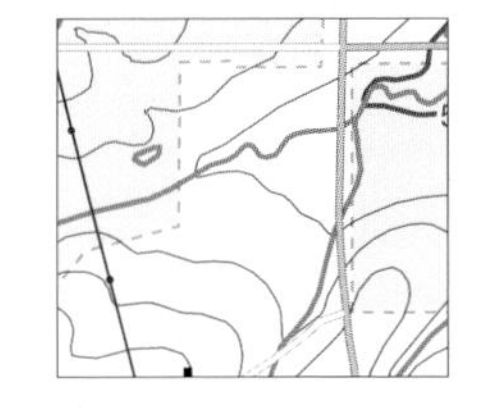

EMPHASIS ON NATURAL FEATURES (Page 5)

In this design, contours are wider, more saturated brown lines that are labeled with greater frequency, and bold hydrography from the original design is repeated. This combination emphasizes natural features. The road network becomes a background element with a progression of purple to white cased lines for major roads to small rural lanes. These light roads allow labels to be placed on roads, but the tradeoff is that the reader needs to look closely at the map to extract information about roads. The city boundary is a fine gray dash that would be barely visible, but it is assisted by a light fill outside urban areas to establish city limits behind the landscape information.

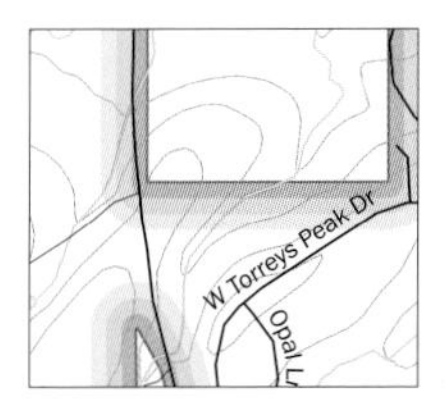

EMPHASIS ON BOUNDARIES (Page 4)

A wider variety of boundary styles is shown with this design. Green dashes for county lines separate them from other boundaries. A fine red line details city boundaries and reddish tint bands distinguish city from noncity sides, adding clarity to an irregular boundary. A band of progressively finer diamond textures outline the interior side of the Rocky Flats Environmental Technology Site boundary at the lower left (in the enhanced reservation category). Road symbols are simplified to a hierarchy of widths of black lines and are broken less often by using tighter label halos. Water features are pushed into the background with lighter colors and thinner lines.

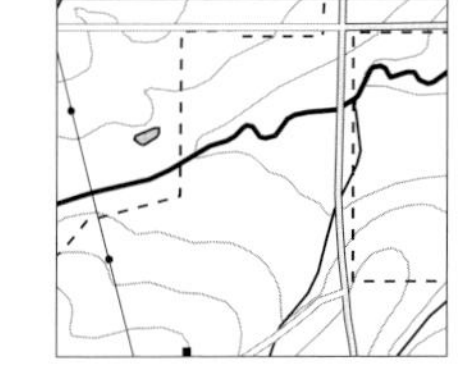

BLACK-AND-WHITE DESIGN (Page 5)

Road, contour, and boundary line weights are balanced in this design so one does not dominate the map. The streams fit better into the landscape because they are tapered; in other words, small tributaries are indicated by the thinnest lines, slightly thicker lines are used after tributaries join, and this progression continues, building wider lines with increases in stream flow. Gray is used for water body areas; gray inside cased roads distinguishes them from other fine black lines; large buildings are black; and contours lines are gray with wider index contours. Patterns are used for line dashes, but notice there are no patterned area fills, thereby preserving the readability of contour lines.

REDESIGNS

1A ORIGINAL DESIGN

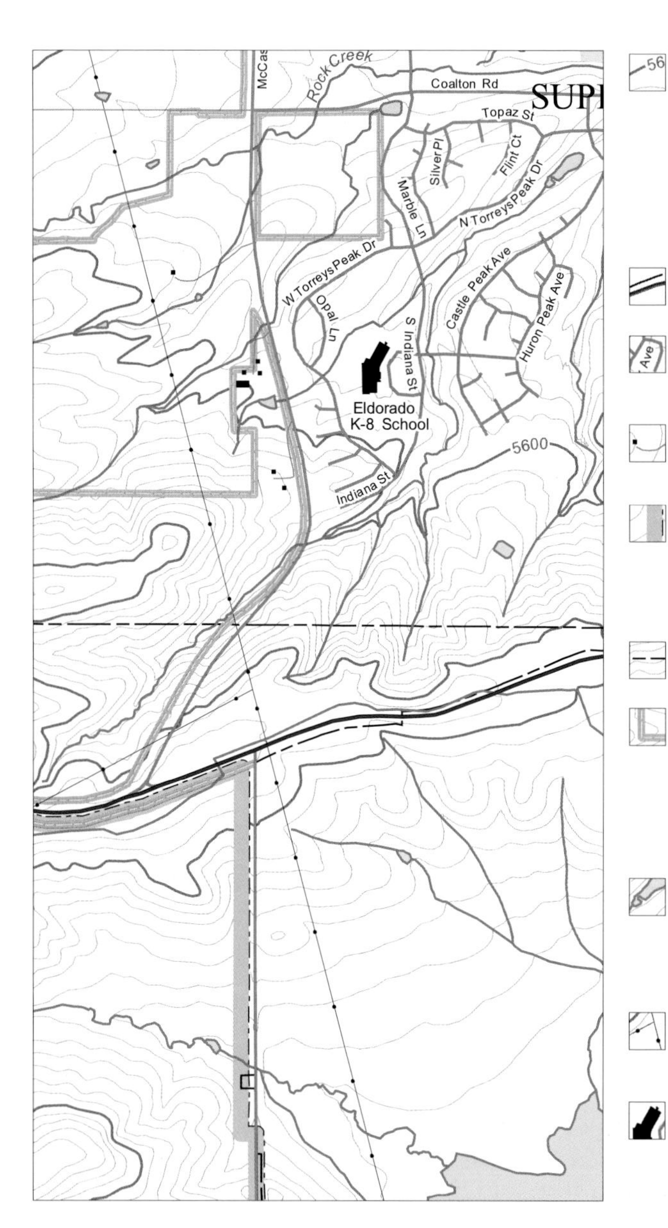

CONTOURS
Index: 1 pt, 33C 48M 100Y 12K, round join
Intermediate: 0.25 pt, 12C 34M 88Y, round join
Label: 6 pt, Arial, 33C 48M 100Y 12K
Halo: 1 pt, white

ROADS
Highways
Line: 1 pt, 15C 100M 100Y
Casing: 1.5 pt, 100K
Major
Line: 1.2 pt, 50K
Label: 5 pt, Arial, 100K
Halo: 1 pt, white
Minor
Line: 0.5 pt, 50K

BOUNDARIES
Enhanced Reservation
Line: 0.5 pt, 100K, dash pattern 6/2/2/2 *(dash/gap in pts)*
Offset band: 6 pt, 31M 15Y, offset -3 pt
County
Line: 0.75 pt, 100K, dash pattern 17/2/5/2
City
Line: 0.5 pt, 6C 96Y, dash pattern 6/1/3/1
Casing: 3 pt, 30K
Label: 14 pt, Times Roman, 100K

OTHER
Water
Fill: 34C 4Y
Line: 1 pt, 70C 40M, round join
Label: 6 pt, Arial Italic, 80C 57M
Transmission line
Line: 0.25 pt, 100K, circle marker line pattern 1.5/28 *(symbol size/spacing in pts)*
Buildings
Fill: 100K
Label: 6 pt, Arial, 100K

ArcMap Tips (see pages 161 to 167)

2 Multilayer line	5 Marker line
6 Offset line	40 Layer order

1B EMPHASIS ON BOUNDARIES

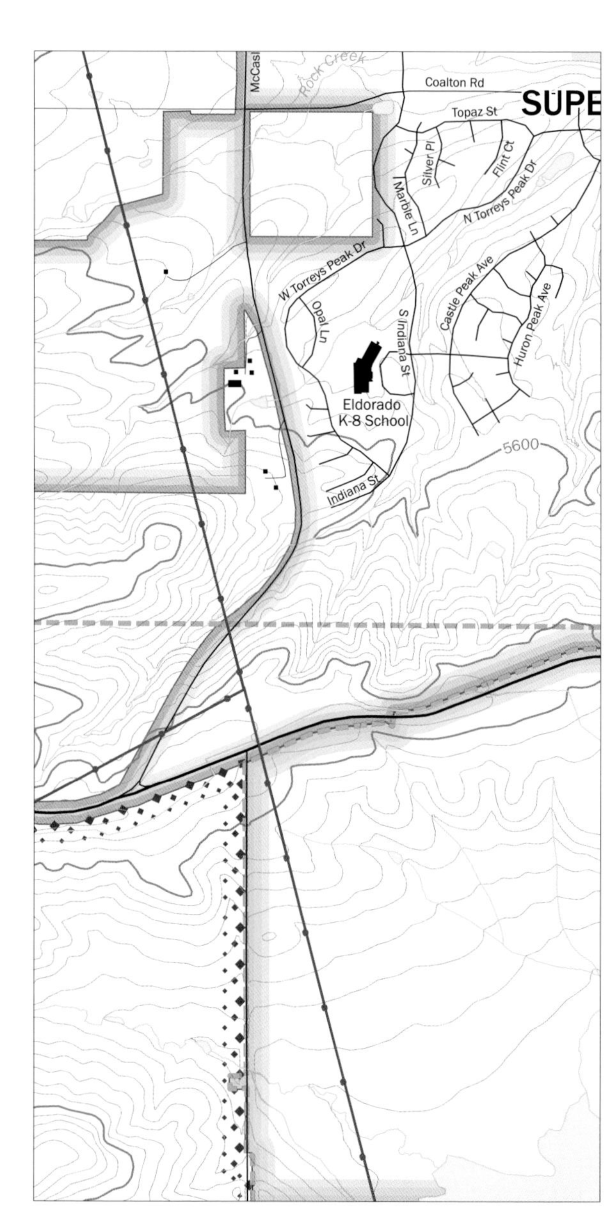

CONTOURS
Index: 0.75 pt, 30C 50M 70Y, round join
Intermediate: 0.25 pt, 20C 40M 60Y, round join
Label: 6 pt, Arial, 30C 50M 70Y
Halo: 1 pt, white

ROADS
All lines: round cap/join
Highways
Line: 1 pt, 100K
Major
Line: 0.5 pt, 100K
Label: 5 pt, Franklin Gothic Book, 100K
Halo: 0.5 pt, white
Minor
Line: 0.25 pt, 100K

BOUNDARIES
Enhanced Reservation
Line: 1 pt, 100M 100Y, three diamond marker line patterns: 3/14; 2/14 offset 2.5 pt; 1.5/7 offset 6.5 pt
County
Line: 2 pt, 60C 60Y, dash pattern 5/2
City
Line: 0.5 pt, 10C 80M 80Y, round join
Offset bands: 3 pt, 40M 40Y; 6 pt, 20M 20Y; 9 pt, 10M 10Y
Offset bands: 3 pt, 20M 55Y; 6 pt, 15M 40Y; 9 pt, 10M 25Y
Label: 14 pt, Franklin Gothic Medium, 100K
Halo: 1 pt, white

OTHER
Water
Fill: 15C
Line: 0.5 pt, 40C, round join
Label: 6 pt, Franklin Gothic Book Italic, 60C 10M
Transmission line
Line: 1 pt, 50C 80M, circle marker line pattern 2.5/28
Buildings
Fill: 100K
Label: 6 pt, Franklin Gothic Book, 100K

ArcMap Tips

6 Offset line	10 Tint bands
12 Index contours	34 Maplex settings

1C EMPHASIS ON NATURAL FEATURES

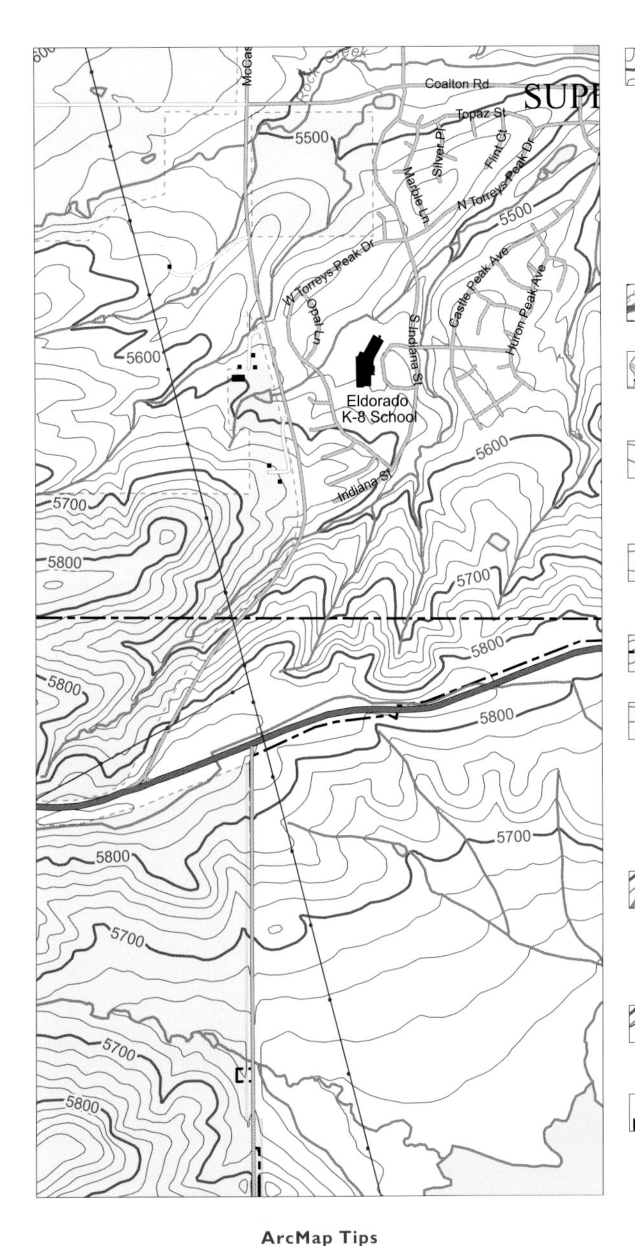

CONTOURS
Index: 1 pt, 25C 85M 100Y, round join
Intermediate: 0.4 pt, 20C 75M 90Y, round join
Label: 6 pt, Arial, 25C 85M 100Y
Halo: 1 pt

ROADS
All lines: round cap/join

Highways
Line: 2 pt, 40C 70M
Casing: 2.5 pt, 100K

Major
Line: 1 pt, 20C 20M
Casing: 1.5 pt, 50K
Label: 5 pt, Arial, 100K

Minor
Line: 1 pt, white
Casing: 1.5 pt, 30K

BOUNDARIES
Enhanced Reservation
Line: 1 pt, 4C 90Y
Offset band: 8 pt, 40Y, offset -4 pt

County
Line: 1 pt, 100K, dash pattern 10/2/3/2

City
Fill: 3M 20Y (outside city)
Line: 0.5 pt, 35K, dash pattern 2/2
Label: 14 pt, Times Roman, 100K

OTHER
Water
Fill: 15C
Line: 0.5 pt, 65C 35M, round join
Label: 6 pt, Arial Italic, 65C 35M

Transmission line
Line: 0.5 pt, 60C 100M, circle marker line pattern 1/28

Buildings
Fill: 100K
Label: 6 pt, Arial, 100K

ArcMap Tips

1 Dashed line	3 Cased line
10 Tint bands	28 Variable-depth masking

1D BLACK-AND-WHITE DESIGN

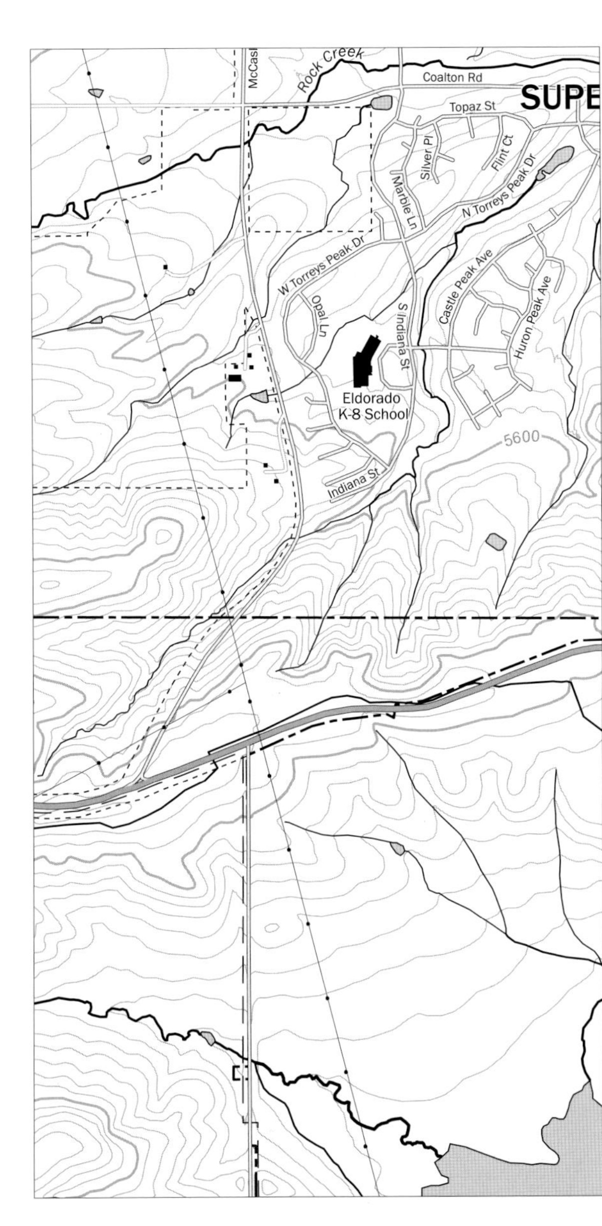

CONTOURS
Index: 1 pt, 35K, round join
Intermediate: 0.4 pt, 35K, round join
Label: 6 pt, Arial, 50K
Halo: 1 pt, white

ROADS
All lines: round cap/join

Highways
Line: 2 pt, 35K
Casing: 2.5 pt, 100K

Major
Line: 1 pt, 5K, round cap/join
Casing: 1.5 pt, 100K, round cap/join
Label: 5 pt, Franklin Gothic Book, 100K
Halo: 0.5 pt, white

Minor
Line: 1 pt, white, round cap/join
Casing: 1.5 pt, 50K

BOUNDARIES
Enhanced Reservation
Line: 0.5 pt, 100K, dash pattern 10/5

County
Line: 1 pt, 100K, dash pattern 10/2/3/2

City
Line: 0.5 pt, 100K, dash pattern 2/2
Label: 14 pt, Franklin Gothic Medium, 100K

OTHER
Water
Fill: 20K
Lines: 0.5 pt; 0.75 pt; 1 pt; 1.25 pt; 100K
Label: 6 pt, Franklin Gothic Book Italic, 100K

Transmission line
Line: 0.25 pt, 100K, circle marker line pattern 2/28

Buildings
Fill: 100K
Label: 6 pt, Franklin Gothic Book, 100K

ArcMap Tips

4 Merge/overpass	14 Tapered lines
34 Maplex settings	37 Curved labels

1.1 USGS TOPOGRAPHIC MAPS, 1980 AND 1990

USGS topographic maps have varied in design over the years. This example shows most of Dayton, Tennessee, with the southeast corner of the Morgan Springs quadrangle positioned above the northeast corner of the Graysville quadrangle. The maps show content field checked in 1972, with a 1980 photorevision for the upper half and 1990 limited updating on the lower half. These are the most current printed topo sheets at 1:24,000 for this area. Printed USGS topo quads are on average over twenty years old, and data currency efforts now focus on digital products. The fonts USGS uses for newer topographic print products (lower half) are bolder with a large x-height to aid reproduction. Other symbols are also more simply designed. For example, roads rely less on cased lines and cross-hatch patterns are not used in large buildings. U.S. topographic maps may be printed with five spot colors (green, blue, red, black, brown inks) plus purple for the photorevisions. This simplicity of production maintains the detail of fine contours, for example (because brown is not produced by registering cyan, magenta, and yellow inks during the press run), but it requires a five- or six-color press, which may increase costs.

BOUNDARIES
- National
- State or territorial
- County or equivalent
- Civil township or equivalent
- Incorporated city or equivalent
- Federally administered park, reservation, or monument (external)
- Federally administered park, reservation, or monument (internal)
- State forest, park, reservation, or monument and large county park
- Forest Service administrative area*
- Forest Service ranger district*
- National Forest System land status, Forest Service lands*
- National Forest System land status, non-Forest Service lands*
- Small park (county or city)

BUILDINGS AND RELATED FEATURES
- Building
- School; house of worship
- Athletic field
- Built-up area
- Forest headquarters*
- Guard station or work center*

CONTOURS

Topographic
- Index
- Approximate or indefinite
- Intermediate
- Approximate or indefinite
- Supplementary
- Depression
- Cut
- Fill
- Continental divide

Bathymetric
- Index***
- Intermediate***
- Index primary***
- Primary***
- Supplementary***

RAILROADS AND RELATED FEATURES
- Standard guage railroad, single track
- Standard guage railroad, multiple track
- Narrow guage railroad, single track
- Railroad siding
- Railroad in highway

RIVERS, LAKES, AND CANALS
- Perennial stream
- Perennial river
- Intermittent stream
- Intermittent river
- Disappearing stream
- Falls, small
- Falls, large
- Rapids, small
- Perennial lake/pond
- Intermittent lake/pond
- Dry lake/pond
- Narrow wash
- Wide wash
- Canal, flume, or aqueduct with lock
- Elevated aqueduct, flume, or conduit
- Aqueduct tunnel
- Water well, geyser, fumarole, or mud pot
- Spring or seep

ROADS AND RELATED FEATURES

Please note: Roads on Provisional-edition maps are not classified as primary, secondary, or light duty. These roads are all classified as improved roads and are symbolized the same as light duty roads.
- Primary highway
- Secondary highway
- Light duty road
- Light duty road, paved*
- Light duty road, gravel*
- Light duty road, dirt*
- Light duty road, unspecified*
- Unimproved road
- Unimproved road*
- 4WD road
- 4WD road*
- Trail
- Highway or road with median strip
- Highway or road under construction
- Highway or road underpass; overpass
- Highway or road bridge; drawbridge
- Highway or road tunnel
- Road block, berm, or barrier*
- Gate on road*
- Trailhead*

SUBMERGED AREAS AND BOGS
- Marsh or swamp
- Submerged marsh or swamp
- Wooded marsh or swamp
- Submerged wooded marsh or swamp
- Land subject to inundation

SURFACE FEATURES
- Levee
- Sand or mud
- Disturbed surface
- Gravel beach or glacial moraine
- Tailings pond

TRANSMISSION LINES AND PIPELINES
- Power transmission line; pole; tower
- Telephone line
- Aboveground pipeline
- Underground pipeline

VEGETATION
- Woodland
- Shrubland
- Orchard
- Vineyard
- Mangrove

* USGS-USDA Forest Service Single-Edition Quadrangle maps only.
In August 1993, the U.S. Geological Survey and the U.S. Department of Agriculture's Forest Service signed an Interagency Agreement to begin a single-edition joint mapping program. This agreement established the coordination for producing and maintaining single-edition primary series topographic maps for quadrangles containing National Forest System lands. The joint mapping program eliminates duplication of effort by the agencies and results in a more frequent revision cycle for quadrangles containing National Forests. Maps are revised on the basis of jointly developed standards and contain normal features mapped by the USGS, as well as additional features required for effi

See the full legend at erg.usgs.gov/isb/pubs/booklets/symbols/topomapsymbols.pdf.

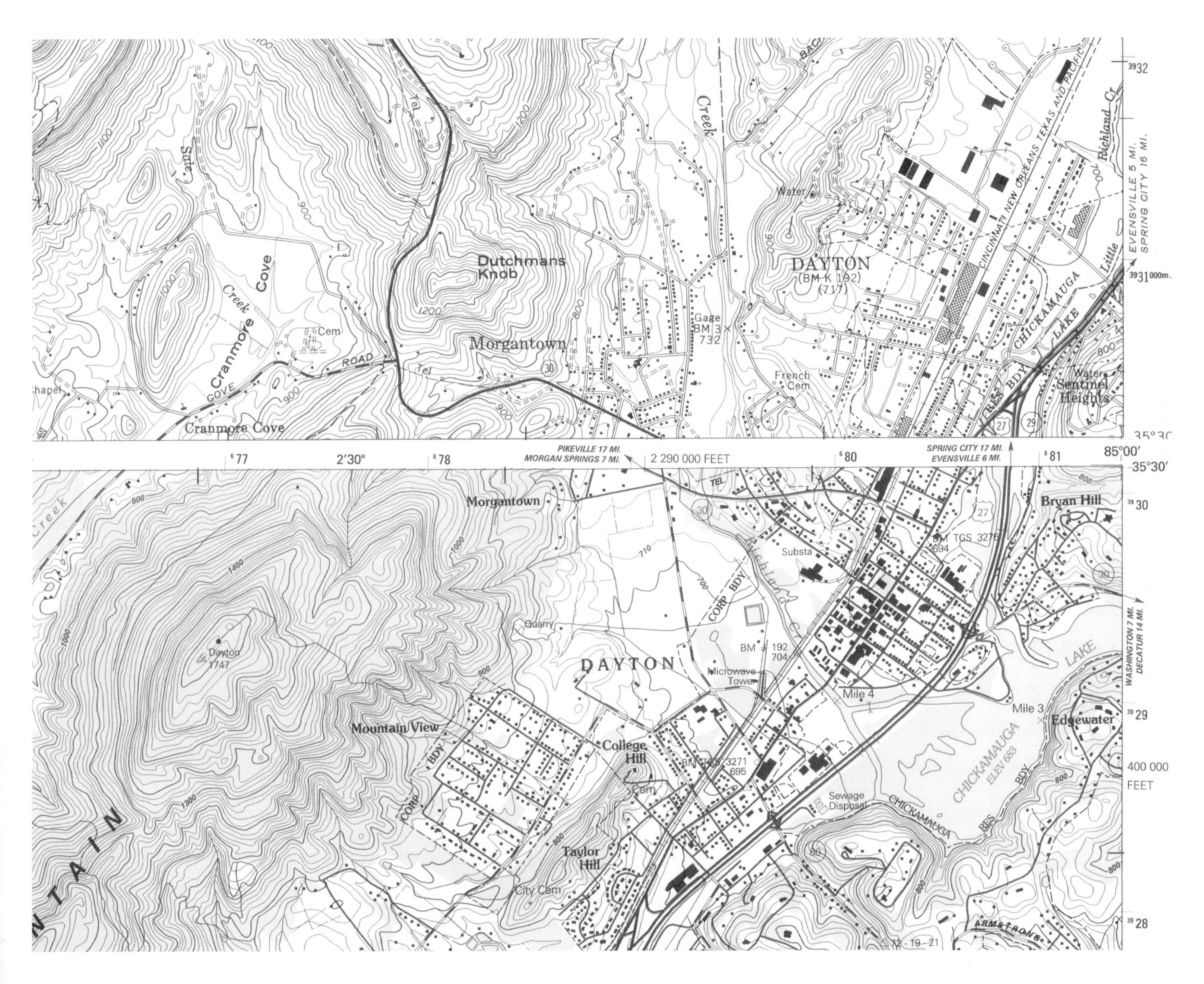
Dutchmans Knob
Morgantown
Cranmore Cove
Cranmore Cove
DAYTON
(BM K 192)
(717)
Gage BM 3 732
French Cem
Water
CINCINNATI NEW ORLEANS TEXAS AND PACIFIC
CHICKAMAUGA LAKE
Sentinel Heights
Little Richland Cr
EVENSVILLE 5 MI.
SPRING CITY 16 MI.
PIKEVILLE 17 MI.
MORGAN SPRINGS 7 MI.
2 290 000 FEET
SPRING CITY 17 MI.
EVENSVILLE 6 MI.
85°00′
35°30′
2′30″
Morgantown
Bryan Hill
Substa
Quarry
Dayton 1747
DAYTON
Microwave Tower
Mile 4
Mile 3
Edgewater
Mountain View
College Hill
Taylor Hill
City Cem
Sewage Disposal
CHICKAMAUGA ELEV 683
WASHINGTON 7 MI.
DECATUR 14 MI.
400 000 FEET
ARMSTRONG
NTAIN

1.2 NETHERLANDS TOPOGRAPHIC MAP

Topographic maps for the Netherlands use a bold and carefully structured color scheme. Roads range from purple highways through a series of red, orange, yellow, and white cased lines for the progression from main roads to residential streets. Vegetation information is also detailed in this map series, showing agricultural land, forest types, and hedges. Built-up areas are symbolized with a variety of reds, with pink for residential blocks and vivid red for houses and larger buildings. Detailed mapping of dikes, banks, and slopes augments the limited terrain information provided by contouring. Many cultural features are symbolized with black and blue shapes but few are labeled, allowing the area and line symbols to structure the map. The distinction between built-up areas in reds, vegetated land in greens, and water features in blue are the primary visual partitions achieved with this design.

Legenda	*Legend*
bebouwd gebied	***built-up area***
a huizenblok b huizen	*residential block houses*
c wandelgebied	*pedestrian precinct*
d muur	*wall*
e groot gebouw	*large building*
f hoogbouw	*high-rise building*
g kassen	*greenhouses*
h industriegebied	*industrial area*
wegen	***roads***
autosnelweg	*motorway*
hoofdweg:	*main road:*
met gescheiden rijbanen	*dual carriageway*
7 m of breder	*7 m wide or over*
4-7 m breed	*4-7 m wide*
smaller dan 4 m	*less than 4 m wide*
regionale weg:	*regional road:*
met gescheiden rijbanen	*dual carriageway*
7 m of breder	*7 m wide or over*
4-7 m breed	*4-7 m wide*
smaller dan 4 m	*less than 4 m wide*
lokale weg:	*local road:*
met gescheiden rijbanen	*dual carriageway*
7 m of breder	*7 m wide or over*
4-7 m breed	*4-7 m wide*
smaller dan 4 m	*less than 4 m wide*
straat	*street*
overige weg	*other road*
weg met losse of slechte verharding	*loose or light surface road*
onverharde weg	*unmetalled road*
fietspad	*cycle-track*
pad, voetpad	*path, footpath*
weg in aanleg	*road under construction*
weg in ontwerp	*planned road*
viaduct	*viaduct*
mini-rotonde	*small roundabout*
tunnel	*tunnel*
beweegbare brug	*movable bridge*
brug op pijlers	*bridge on piers*
spoorwegen	***railways***
spoorweg: enkelspoor	*railway: single track*
spoorweg: dubbelspoor	*railway: double track*
spoorweg: driesporig	*railway: three tracks*
spoorweg: viersporig	*railway: four tracks*
a station b laadperron	*station loading-bay*
tram	*tramway*
metro a metrostation	*underground station*
hydrografie	***hydrography***
waterloop:	*watercourse:*
smaller dan 3 m	*less than 3 m wide*
3-6 m breed	*3-6 m wide*
breder dan 6 m	*6 m wide or over*
a schutsluis b brug	*lock bridge*
c vonder	*foot-bridge*
a grondduiker b stuw	*culvert siphon weir*
c duiker d sluis	*culvert sluice*
a pontveer	*ferry*
b voetveer	*ferry for pedestrians*
c peilschaal	*water-level gauge*
d kilometerraaibord	*kilometre sign*
e stroomrichting	*direction of flow*
f baak g dok	*beacon dock*
h lichtopstand	*light beacon*
i aanlegsteigers	*landing-stages*
j versterkt talud	*reinforced slope*
k eb/vloed aanduiding	*indication of tides*
l dieptegetal	*sounding*
m hoogwaterlijn	*high water mark*
n laagwaterlijn	*low water mark*
o dieptelijnen	*bathymetric contours*
p droogvallende grond	*tidal flat*
q krib, golfbreker	*jetty, breakwater*
getrianguleerde punten	***triangulation points***
a RD-steen	*state survey stone*
b GPS-kernnetpunt	*GPS point*
c toren, hoge koepel	*tower, high dome*
d kerk met toren	*church with tower*
e markant object	*landmark*
f watertoren g vuurtoren	*water tower lighthouse*
overige symbolen	***other symbols***
a kerk zonder toren	*church without tower*
b toren, hoge koepel	*tower, high dome*
c kerk met toren	*church with tower*
d markant object	*landmark*
e watertoren f vuurtoren	*water tower lighthouse*
a gemeentehuis	*town hall*
b postkantoor	*post office*
c politiebureau	*police-station*
a kapel	*chapel*
b kruis	*cross*
c begraafplaats	*cemetery*
a vlampijp	*flare pipe*
b wegwijzer	*signpost*
c telescoop	*telescope*
a windmolen	*windmill*
b watermolen	*watermill*
c windmolentje	*windpump*
d windturbine	*windturbine*
a oliepompinstallatie	*oil-pumping unit*
b seinmast	*signalpost*
c zendmast	*wireless mast*
a hunebed	*cairn*
b monument	*monument*
c gemaal	*pumping-station*
a paal b markante boom	*pole conspicuous tree*
c boom d opslagtank	*tree tank*
a kampeerterrein	*camp site*
b sportcomplex	*sports-ground or hall*
c ziekenhuis	*hospital*
schietbaan	*firing range*
afrastering	*wire fence*
hoogspanningsleiding	*high tension line*
geluidswering	*sound-proof barrier*
wegen-informatie	***road-information***
wegnummering	*road numbering*
a tankstation	*filling-station*
b parkeerplaats	*parking*
c carpoolplaats	*carpool*
d afritnummer	*number of exit*
a aantal rijstroken	*lane-information*
b kilometerpaal	*kilometre post*
c wegafsluiting	*road closing*
grenzen	***boundaries***
rijksgrens	*national boundary*
provinciegrens	*provincial boundary*
gemeentegrens	*municipal boundary*
grens nationaal park	*boundary National Park*
reliëf	***relief***
a dijk: 2,5 m of hoger	*dike: 2.5 m high or over*
b dijk: 1-2,5 m hoog	*dike: 1-2.5 m high*
kade, wal: 0,5-1 m hoog	*earth bank: 0.5-1 m high*
a berijdbare dijk	*dike with road*
b ingraving	*cutting*
hoogtelijnen	*contours*
hoogtepunt	*spot height*
a steile rand	*escarpment*
b helling	*slope*
bodemgebruik	***vegetation***
a weide met sloten	*meadow with ditches*
b bouwland met greppels	*arable land with trenches*
c boomgaard	*orchard*
d fruitkwekerij	*orchard (low)*
e boomkwekerij	*tree nursery*
f weide met populieren	*meadow with poplar*
g loofbos	*deciduous forest*
h naaldbos	*coniferous forest*
i gemengd bos	*mixed forest*
j griend	*osier*
k heide	*heath*
l zand	*sand*
m dras en riet	*marsh and reed*
n heg en houtwal	*hedge and hedge-bank*

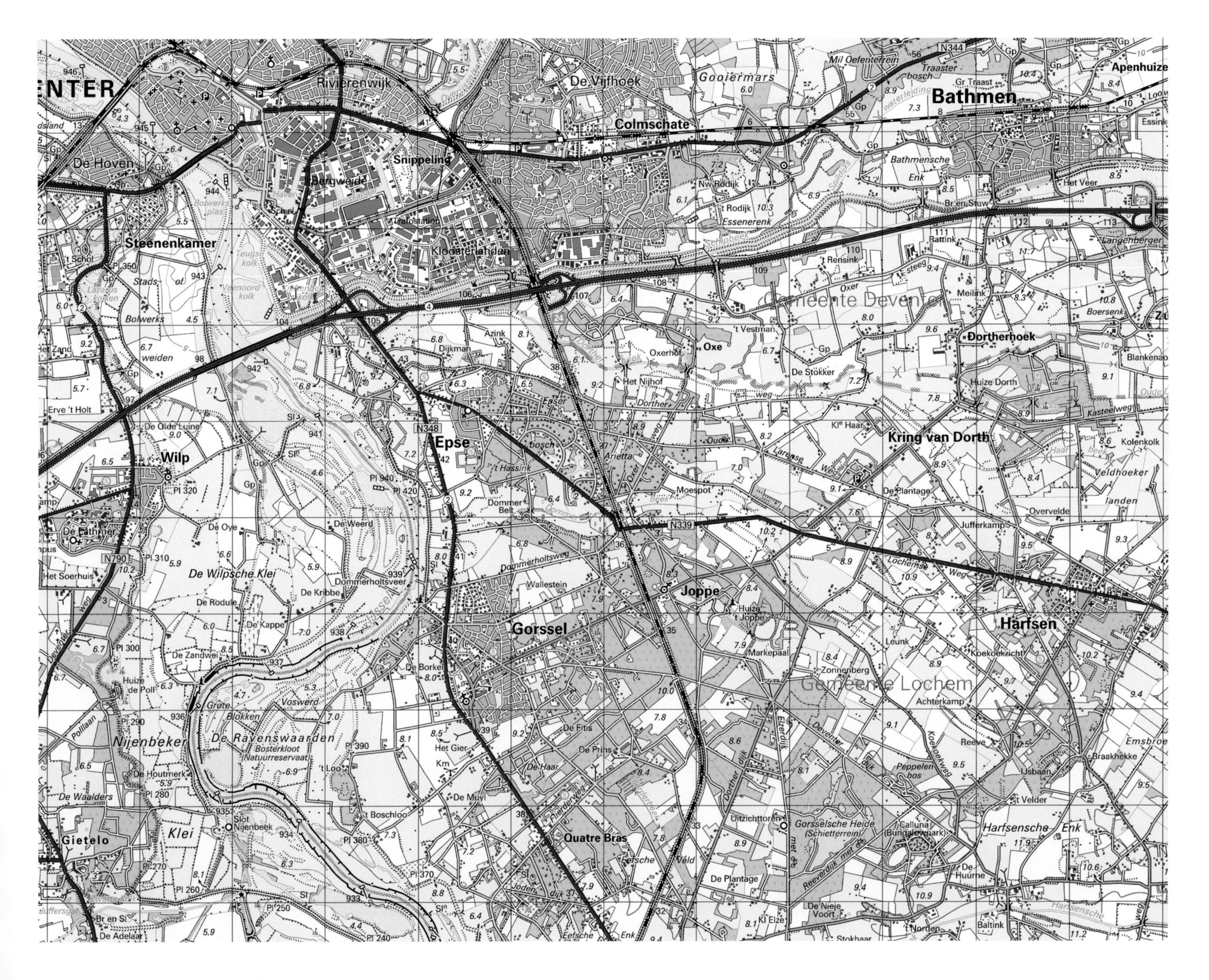
NTER
Rivierenwijk
De Vijfhoek
Gooiermars
Bathmen
Apenhuize
Colmschate
De Hoven
Snippeling
Bergweide
Steenenkamer
Kloosterlanden
Gemeente Deventer
Oxe
Dortherhoek
Huize Dorth
Kring van Dorth
Epse
Wilp
De Lathmer
De Wilpsche Klei
Gorssel
Joppe
Harfsen
Gemeente Lochem
Nijenbeker
De Ravenswaarden
Bosterkloot (Natuurreservaat)
Gietelo
Klei
Slot Nijenbeek
Quatre Bras
Gorsselsche Heide (Schietterrein)
Harfsensche Enk
Mil Oefenterrein
Traaster bosch
Gr Traast
Bathmensche Enk
Het Veer
Langenberger
Meilink
Boersenk
Blankenao
Kolenkolk
Veldhoeker landen
Overvelde
Jufferkamp
Achterkamp
Emsbroe
Braakhekke
Ijsbaan
Calluna (Bungalowpark)
De Huurne
Baltink
Norden
Stokhaar
De Nieje Voort
De Plantage
Uitzichttoren
Markepaal
Leunk
Koekoekzicht
Zonnenberg
Peppelen bos
Wallestein
Dommerholtsweg
Dommerholtsveer
De Kribbe
De Rodulo
De Kappe
De Weerd
De Oye
Het Soerhuis
De Olde Luine
Erve 't Holt
Bolwerks
Lazarus kolken
Teugse kolk
Veenoord kolk
't Schol
Het Zand
De Zandwei
Huize de Poll
Grote Blokken
Voswerd
Pollaan
De Houtmerk
De Waalders
Juffersgat
Br en Sl
De Adelaar
't Boschloo
De Muyl
Het Gier
De Haar
De Prins
De Fitis
De Borkel
Fliederweg
Jodendijk
Eefsche Veld
Eefsche Enk
't Loo
Dommer Belt
't Hassink
Arietta
Moespot
Het Nijhof
Oxerhof
De Stokker
't Vestman
Azink
Dijkman
't Rensink
Rattink
't Rodijk
Nw Rodijk
Essenerenk
Br en Stuw
Kasteelweg
Lochemse Weg
Laren Weg
De Plantage
Reeve
't Velder
Reeverdijk
Dortherdijk
Elzerdijk
Deventerdijk
Koekoekweg
N344
N348
N339
N790

1.3 SWISS TOPOGRAPHIC MAP

Swiss topographic maps present the country's high-relief landscape with beautifully rendered detail. Hand-drawn cliffs and rocks are placed where slopes are too rugged to describe with contours. Light hillshading is augmented with nuanced use of blue hues on shadowed slopes and warm yellow and peach hues highlighting sunlit slopes. The forms of glaciated surfaces are represented with blue contours and crevasse drawings, striking against a cold white background. Fine, mostly italic labels lie on the surface, elegantly spaced across features. Similarly, buildings, roads, and trails are characterized by thin or small black symbols that do not compete with the terrain details, which are the main message of the map. Labels are carefully fit to the terrain with attention to the placement of each letter. For example, the letters of the Stechelberg label running vertically up the valley are each placed between building clusters and labels for towns and spot height.

Roads, tracks

1:50 000

Highway (divided lanes) Junction	under construction	Adorf
Rest area Parking		Ried, A1
2nd cl. highway (undivided lanes) Exit / Access	under construction	Wil
Trunk road	Main connecting road	7, 3
1st cl. road (at least 6 m wide)	conspicuous bridge	
2nd cl. road (at least 4 m wide)	conspicuous bridge	
Suburban road (at least 4 m wide)	conspicuous bridge	
3rd cl. road (at least 2.8 m wide)	covered bridge	
4th cl., narrow road (at least 1.8 m)	Bridge	
5th cl., path, trail, bicycle path	Footbridge, catwalk	
6th cl., footpath	Passenger ferry, attached free	
Traces, mountain passage	on glacier	
Barrier, traffic ban		
conspicuous roundabout	Parking lot	
Level crossings		
Underpasses		
Overpasses		
Tunnels	Ventilation shaft	
Gallerie		
Parklane	Tank road	
Traces of historic road		
Airport, hard surface runway Airfield, grass strip		

Boundaries

- National boundary with numbered markers
- Cantonal boundary with markers
- District boundary with markers
- Municipal boundary with markers
- Boundary for National Park or protected area

Railways

1:50 000

Railway station, tracks	Platform roof
Stop with separate track	
Stop without separate track	
Normal gauge railway: multiple tracks	Bridge
Normal gauge railway: single track	Bridge
Narrow gauge railway: multiple tracks	Bridge
Narrow gauge, rack, cable railway: single track	Bridge
Freight or nostalgic railway Railway out of service	Bridge
Intercommunal tramway with stop	Bridge
Industrial track	Bridge
Tunnels	
Galleries	
Aerial cable way, chairlift with intermediate station	Pylon
Goods lift	Pylon
Skilift	

Topography

Contour lines	earth, scree / shingle, ice / lake	eau) 20 m
Index contours	earth, scree / shingle, ice / lake	200 m
Intermediate contours	earth, scree / shingle, ice / lake	10 m
Small depression	Doline	
Escarpment, earth	Escarpment, stone	
Cutting	Embankment	
Earth slip	Gravel pit	
Clay pit	Quarry	
Rock	Scree	
Glacier	Moraine	

Individual symbols

1:50 000

House	Ruin
Remote inn	Tower
Greenhouse	Storage tank
Allotment (garden)	Monument
Church	Chapel
Cemetery	Shrine, cross
Cooling tower	Wind power station
Chimney-stack	Castle
Lookout tower	Radio transmitter
large antenna	small antenna
Camp site	Summer toboggan-run
Sports ground	Stadium
Rifle range	
Race course (horses)	
Border of an area	Golf course
Ski jump	dry wall
Wall	Avalanche barricade
Cave, grotto	erratic bloc

Trigonometric points, spot heights

Trigonometric points 1st to 3rd order and LV95		2127.6
Spot height		1587, 713
Index contour		800
Lake level	Spot height at lake bottom	419, 387

Vegetation

Forest, defined outline	undefined outline
Scattered forest	isolated tree / group of trees
Scrub	Hedge
Orchard	Tree nursery
Vineyard	

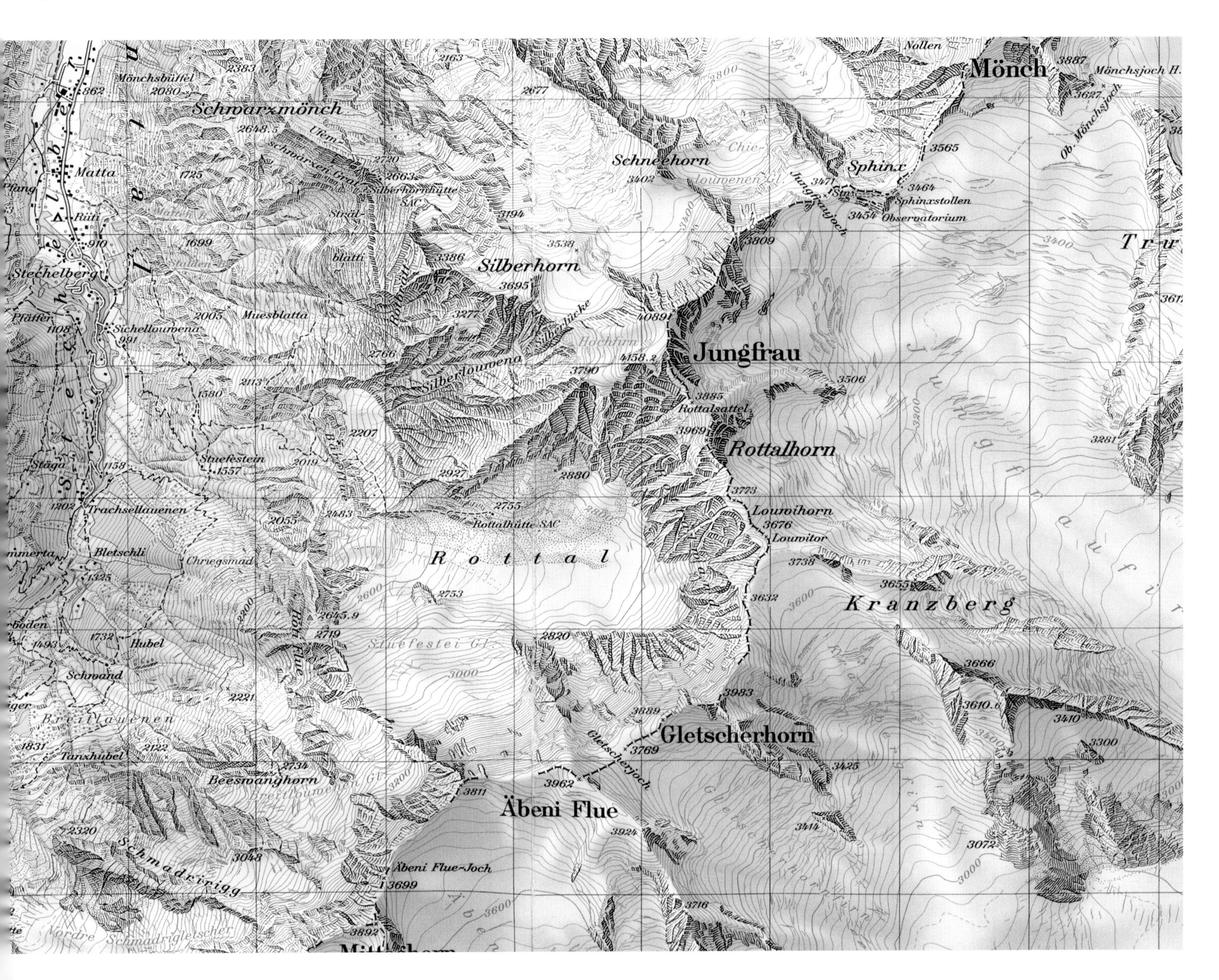
Mönch
Mönchsjoch H.
Nollen
Sphinx
Sphinxstollen
Observatorium
Jungfraujoch
Schneehorn
Silberhorn
Silberhornhütte SAC
Schwarzmönch
Mönchsbüffel
Matta
Rüti
Stechelberg
Pfäffer
Sichellouwena
Muesblatta
Jungfrau
Hochfirn
Silberlouwena
Silberlücke
Rottalsattel
Rottalhorn
Louwihorn
Louwitor
Kranzberg
Rottal
Rottalhütte SAC
Trachsellauenen
Bletschli
Chriegsmad
Stuefestein
Stäga
Stuefestei Gl.
Hubel
Schmand
Gletscherhorn
Gletscherjoch
Äbeni Flue
Äbeni Flue-Joch
Beeswanghorn
Tanzhubel
Schmadririgg
Vordre Schmadrigletscher
4158.2
3969
3885
3773
3676
3738
3655
3632
3666
3610.6
3983
3889
3769
3962
3811
3924
3699
3892
3716
3414
3425
3410
3300
3072
3281
3506
3790
4089
3695
3538
3386
3277
2766
3194
3402
3471
3464
3454
3565
3887
3627
3809
2383
2163
2677
2648.5
2720
2663
2080
862
1725
910
1699
2005
1108
991
2113
1580
2207
2019
1557
1158
1202
2927
2880
2755
2483
2055
1325
2753
2645.9
2719
2820
1732
1493
2221
2122
2734
1831
2320
3048

1.4 CORDILLERA HUAYHUASH CLIMBER'S MAP

Cordillera Huayhuash, a range in the Peruvian Andes, offers challenging high-elevation trekking. Routes are published by the Alpine Mapping Guild on their climber's map with the goal of providing terrain and trek information at a useful scale. The map improves on imperfect contour information available for map production by using terrain shading and classed Landsat imagery rendered in smooth translucent colors. The glaciated area shown in this example area is differentiated from the land areas with an abrupt shift to cyan contours, a bluish tint on the hillshading, and the indefinite ice boundary marked with a fine dashed line. Glacier labels are printed in the darker blue used for hydrographic features so they contrast with the contours. A slightly darker blue tint band within lakes gives them depth. Additional elevations for lakes, passes, and peaks plus a detailed grid add to the utility for circuit hikers, the target audience for the map.

Courtesy of Martin Gamache, Alpine Mapping Guild.

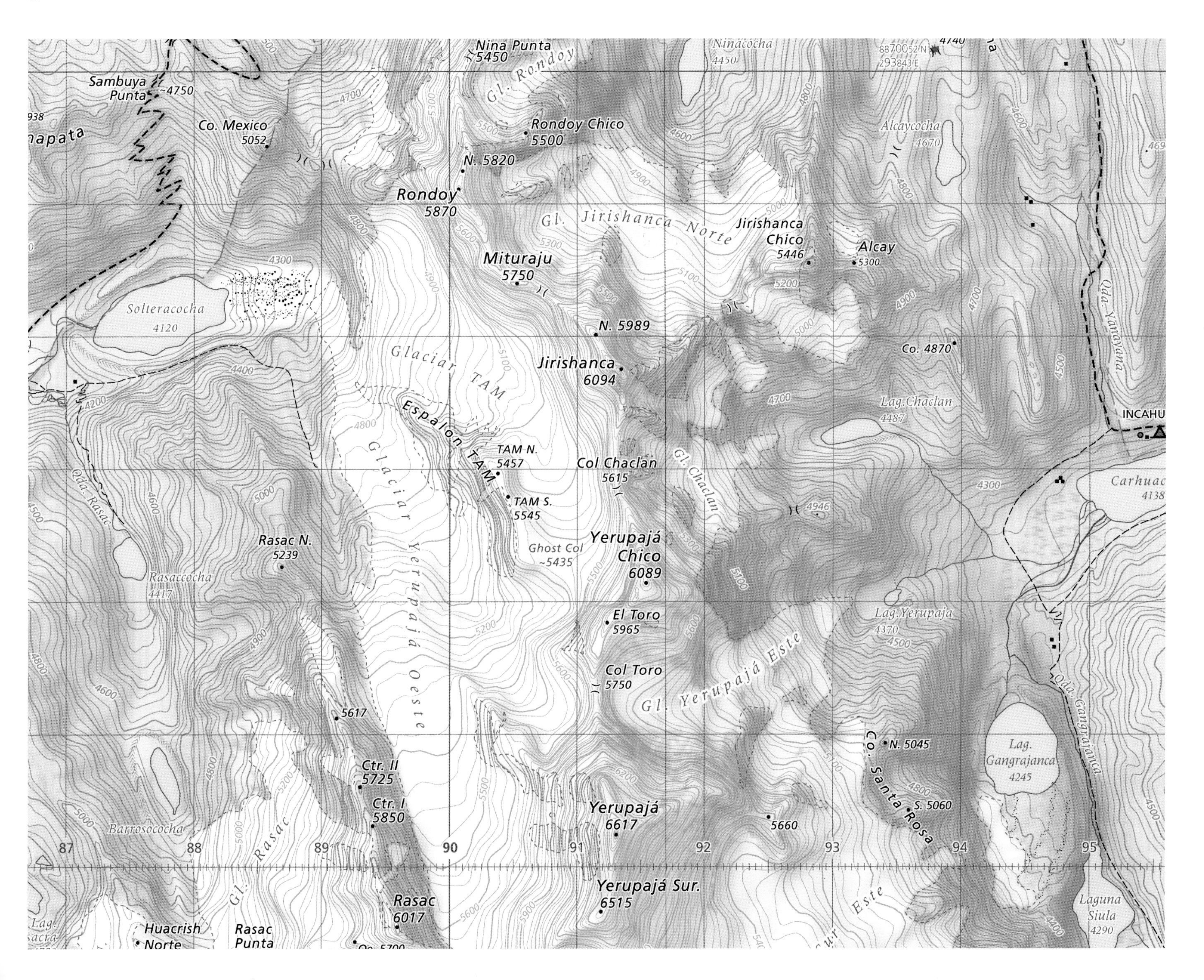

Nina Punta
5450
Gl. Rondoy
Ninacocha
4450
8870052 N
293843 E
4740
Sambuya Punta
~4750
Co. Mexico
5052
Rondoy Chico
5500
N. 5820
Rondoy
5870
Alcaycocha
4670
Gl. Jirishanca Norte
Jirishanca Chico
5446
Alcay
5300
Mituraju
5750
Solteracocha
4120
N. 5989
Jirishanca
6094
Co. 4870
Qda. Yanayana
Glaciar TAM
Espalon TAM
TAM N.
5457
TAM S.
5545
Col Chaclan
5615
Gl. Chaclan
Lag. Chaclan
4487
INCAHU
Carhuac
4138
Qda. Rasac
Glaciar Yerupajá Oeste
4946
Rasac N.
5239
Rasaccocha
4417
Ghost Col
~5435
Yerupajá Chico
6089
El Toro
5965
Lag. Yerupaja
4370
Col Toro
5750
Gl. Yerupajá Este
Qda. Gangrajanca
5617
Co. Santa Rosa
N. 5045
Lag. Gangrajanca
4245
Ctr. II
5725
Ctr. I
5850
S. 5060
Yerupajá
6617
5660
Barrosococha
Gl. Rasac
87
88
89
90
91
92
93
94
95
Yerupajá Sur.
6515
Rasac
6017
Laguna Siula
4290
Huacrish Norte
Rasac Punta

1.5 CENTRAL OREGON PROTECTION DISTRICT MAP

The Oregon Department of Forestry's fire protection program prepares maps that serve both internal and public use. The mesh of boundaries associated with state and national districts (portrayed by a collection of dashed lines) overlay landownership areas (in light hues) for administrative detail specific to forest protection and firefighting. Labels for smaller administrative areas are centered with words spaced across the area, and labels for larger areas parallel each other on corresponding sides of boundary lines. A detailed township and range grid in red facilitates communication of fire locations and recreation sites. This map uses hillshading and elevation labels at particular locations to show terrain. The lack of contour lines avoids the visual clutter that would result from their presence along with the dense network of dirt roads through portions of the forests. The transparent landownership hues combine with the hillshading, and accurate renditions of these mixtures are shown in the legend to assist identification of map colors.

Courtesy of Oregon Department of Forestry.

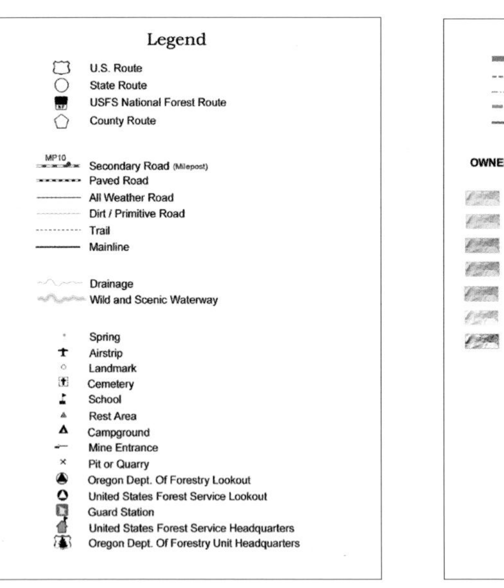

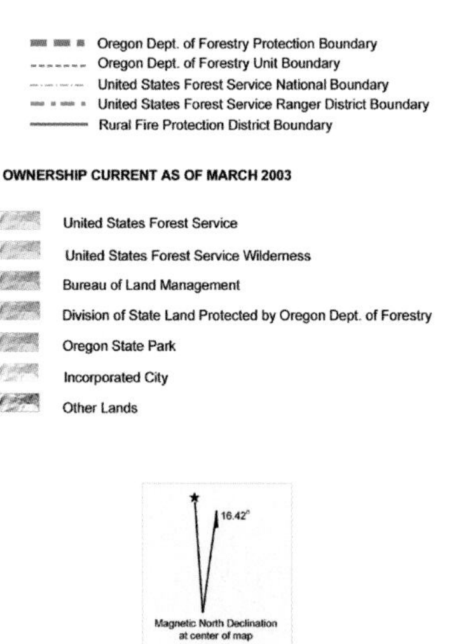

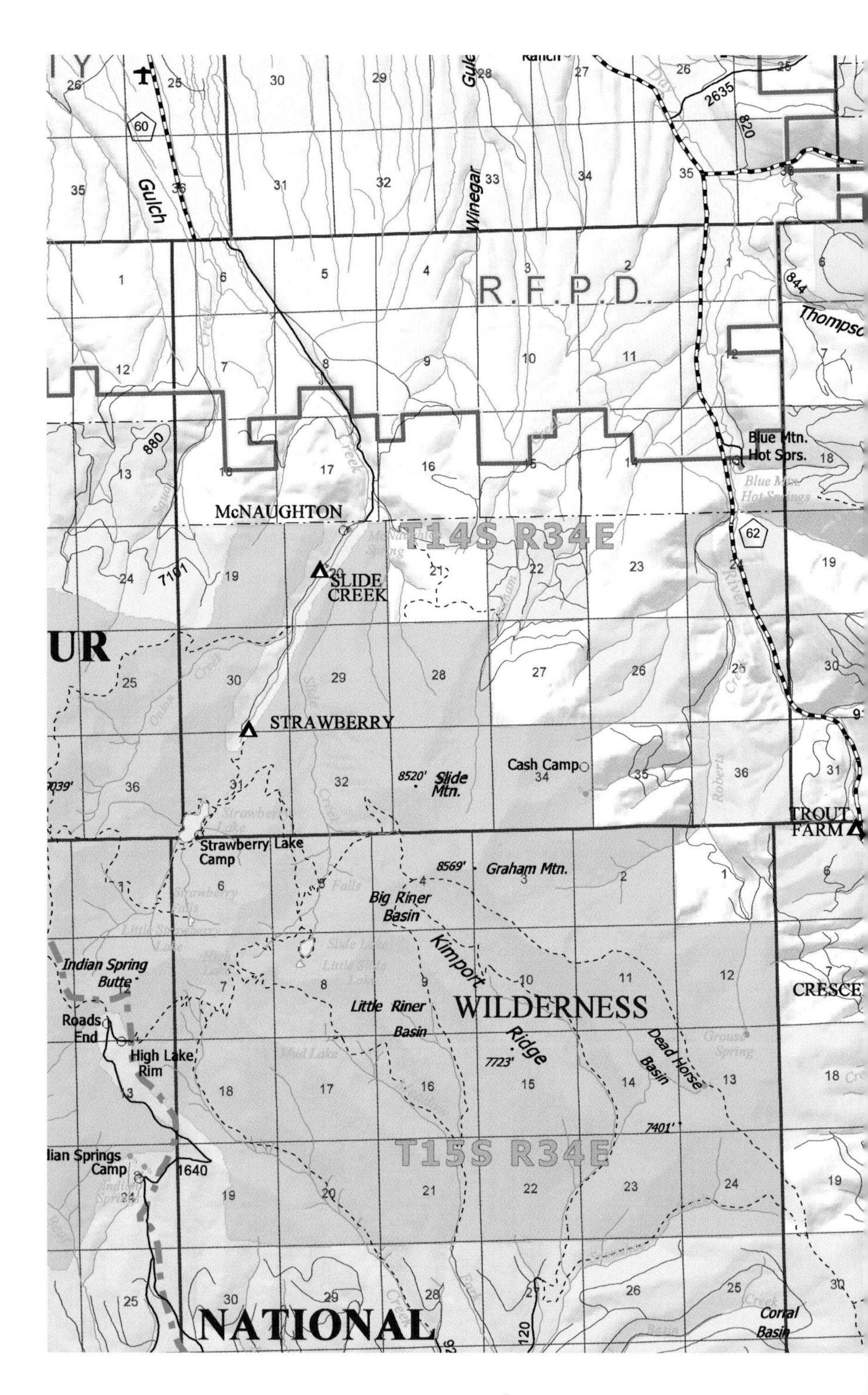

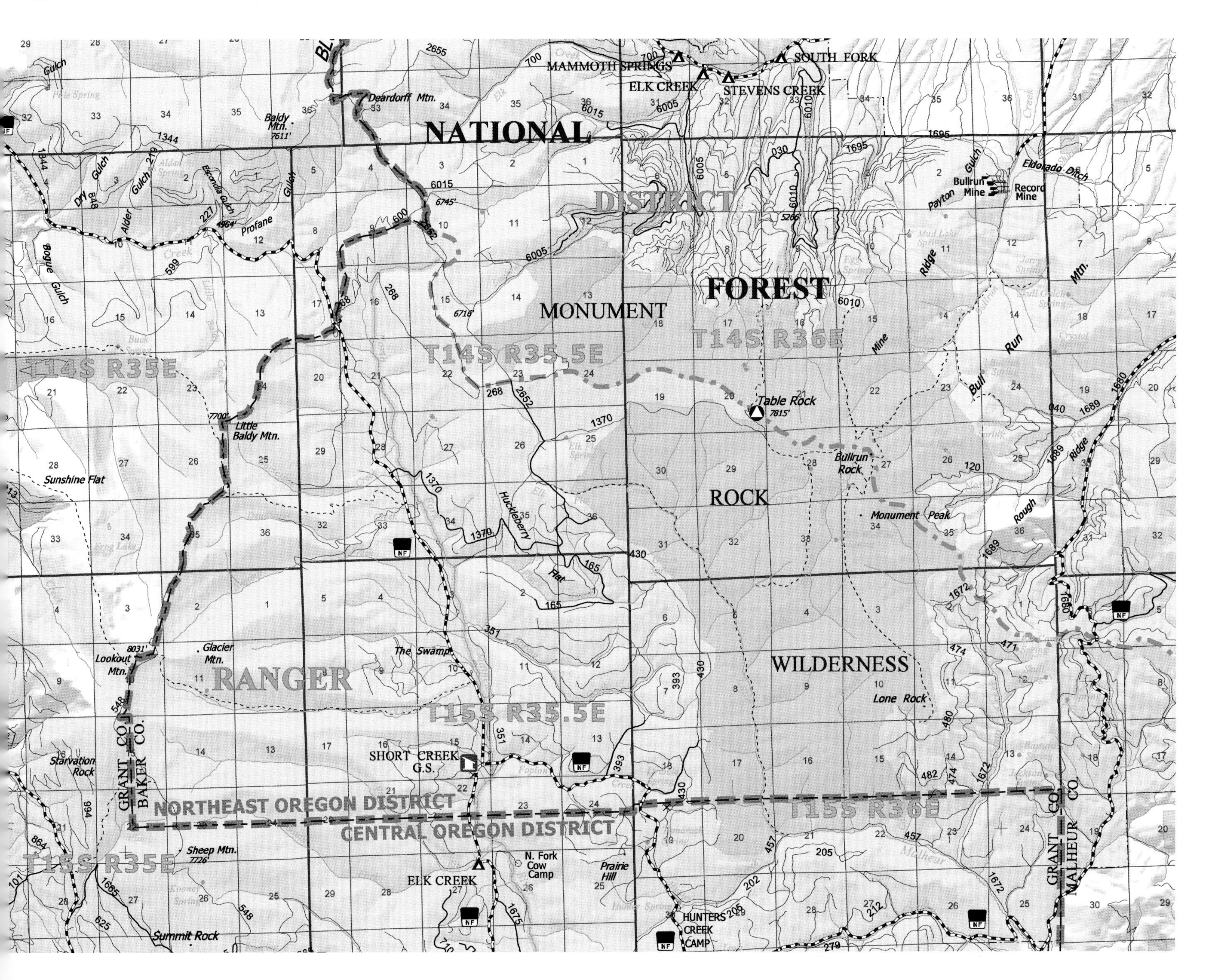

NATIONAL
DISTRICT
FOREST
MONUMENT
ROCK
WILDERNESS
RANGER
T14S R35E
T14S R35.5E
T14S R36E
T15S R35E
T15S R35.5E
T15S R36E
NORTHEAST OREGON DISTRICT
CENTRAL OREGON DISTRICT
MAMMOTH SPRINGS
ELK CREEK
STEVENS CREEK
SOUTH FORK
Deardorff Mtn.
Baldy Mtn. 7611'
Little Baldy Mtn.
Table Rock 7815'
Bullrun Rock
Monument Peak
Lone Rock
Glacier Mtn.
Lookout Mtn.
Sheep Mtn. 7726'
Starvation Rock
Summit Rock
Sunshine Flat
The Swamp
SHORT CREEK G.S.
N. Fork Cow Camp
Prairie Hill
HUNTERS CREEK CAMP
Bullrun Mine
Record Mine
Eldorado Ditch
GRANT CO.
BAKER CO.
MALHEUR CO.

1.6 TWO KOREAS REFERENCE MAP

The National Geographic Society (NGS) presents North and South Korea with this general reference map. Landforms are indicated by muted green-to-orange elevation tints and a simple hillshading scheme with abrupt transitions between lit and shaded surfaces. River lines taper from thick to thin at headwaters. Reddish and purple two-level tint bands define the boundaries of the two countries and fill island features to specify ownership. Provincial boundaries continue the darker of the tint band colors through the interiors, centered on very fine dotted lines that define the boundaries.

The distinctive NGS font is captured nicely in this map. Its slightly flared strokes give it a unique style that suggests serifs but does not invite the attendant problems with maintaining tiny serif tips against complex map backgrounds. Italics label physical features such as hydrography in blue and islands and peaks in black. Uppercase letters are used effectively at both small and large type sizes. They emphasize the importance of large city labels, classify provincial areas as large (but gray letters push these labels into the background), and provide a distinctive categorization for small red labels for infrastructure (such as airports and expressways). The relative importance of cities is established with size and boldness of labels and size and shape of city point symbols. The NGS font generally has round letterforms, but the larger city labels on this map are slightly condensed so they do not extend too far beyond city positions.

Map used with permission of the National Geographic Society.

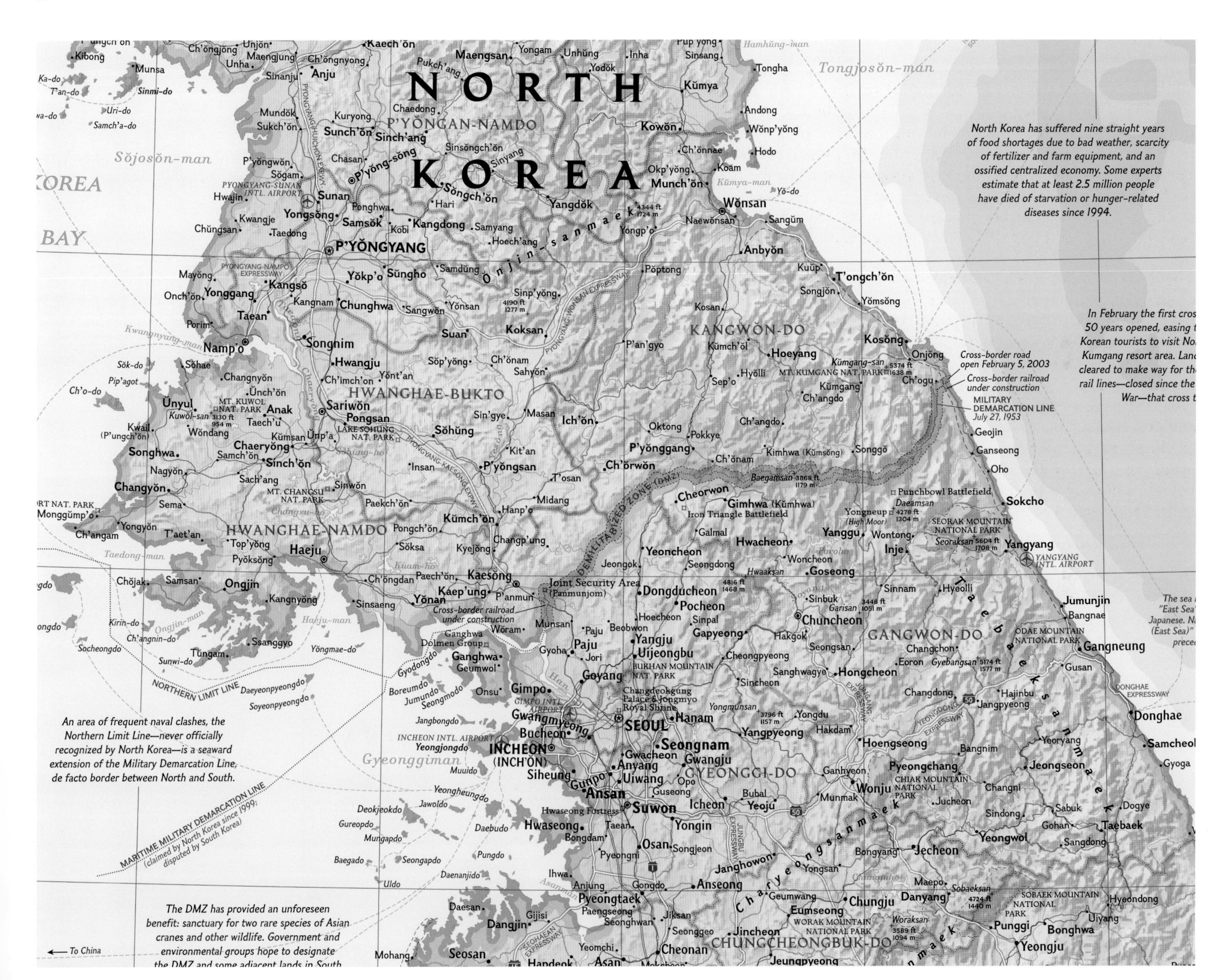
NORTH KOREA
North Korea has suffered nine straight years of food shortages due to bad weather, scarcity of fertilizer and farm equipment, and an ossified centralized economy. Some experts estimate that at least 2.5 million people have died of starvation or hunger-related diseases since 1994.
In February the first cros
50 years opened, easing t
Korean tourists to visit No
Kumgang resort area. Lan
cleared to make way for th
rail lines—closed since the
War—that cross t
An area of frequent naval clashes, the Northern Limit Line—never officially recognized by North Korea—is a seaward extension of the Military Demarcation Line, de facto border between North and South.
The DMZ has provided an unforeseen benefit: sanctuary for two rare species of Asian cranes and other wildlife. Government and environmental groups hope to designate the DMZ and some adjacent lands in South
The sea
"East Sea"
Japanese. N
(East Sea)"
prece
Cross-border road open February 5, 2003
Cross-border railroad under construction
MILITARY DEMARCATION LINE July 27, 1953
NORTHERN LIMIT LINE
MARITIME MILITARY DEMARCATION LINE (claimed by North Korea since 1999; disputed by South Korea)
DEMILITARIZED ZONE (DMZ)
Cross-border railroad under construction
Joint Security Area (Panmunjom)
To China
KOREA
BAY
Sŏjosŏn-man
Tongjosŏn-man
Hamhŭng-man
Kŭmya-man
Taedong-man
Kwangnyang-man
Ongjin-man
Haeju-man
Gyeonggiman
P'YŎNGAN-NAMDO
HWANGHAE-BUKTO
HWANGHAE-NAMDO
KANGWŎN-DO
GANGWON-DO
GYEONGGI-DO
CHUNGCHEONGBUK-DO
Ŏnjin-sanmaek
Taebaeksanmaek
Charyeongsanmaek
P'YŎNGYANG
SEOUL
INCHEON (INCH'ŎN)
Kaesŏng
Haeju
Nampo
Sariwŏn
Wŏnsan
Chuncheon
Gangneung
Suwon
Seongnam
Goyang
Uijeongbu
Bucheon
Anyang
Ansan
Gimpo
Gwangmyeong
Hanam
Wonju
Jecheon
Chungju
Cheonan
Pyeongtaek
Osan
Icheon
Donghae
Samcheok
Sokcho
Yangyang
Kosŏng
T'ongch'ŏn
Anbyŏn
Munch'ŏn
Kowŏn
Anju
Sunan
Sunch'ŏn
Kaech'ŏn
Kangsŏ
Songnim
Hwangju
Ongjin
Yŏnan
Kaep'ung
Ch'ŏrwŏn
Cheorwon
Dongducheon
Pocheon
Yangju
Paju
Hongcheon
Hoengseong
Pyeongchang
Jeongseon
Taebaek
Yeongwol
Danyang
Yeongju
Punggi
Jincheon
Anseong
Yongin
Hwaseong
Siheung
Gunpo
Uiwang
Gwacheon
Gwangju
Yangpyeong
Gapyeong
Inje
Yanggu
Hwacheon
Goseong
Jumunjin
PYONGYANG-SUNAN INTL. AIRPORT
INCHEON INTL. AIRPORT
GIMPO INTL. AIRPORT
YANGYANG INTL. AIRPORT
MT. KUMGANG NAT. PARK
SEORAK MOUNTAIN NATIONAL PARK
ODAE MOUNTAIN NATIONAL PARK
CHIAK MOUNTAIN NATIONAL PARK
SOBAEK MOUNTAIN NATIONAL PARK
WORAK MOUNTAIN NATIONAL PARK
BUKHAN MOUNTAIN NAT. PARK
Punchbowl Battlefield
Iron Triangle Battlefield
Changdeokgung Palace & Jongmyo Royal Shrine
Hwaseong Fortress
Ganghwa Dolmen Group
PYONGYANG-KAESONG EXPWY
PYONGYANG-WONSAN EXPRESSWAY
PYONGYANG-NAMPO EXPRESSWAY
YEONGDONG EXPRESSWAY
JUNGANG EXPRESSWAY
JUNGBU EXPRESSWAY
SEOHAEAN EXPRESSWAY
DONGHAE EXPRESSWAY

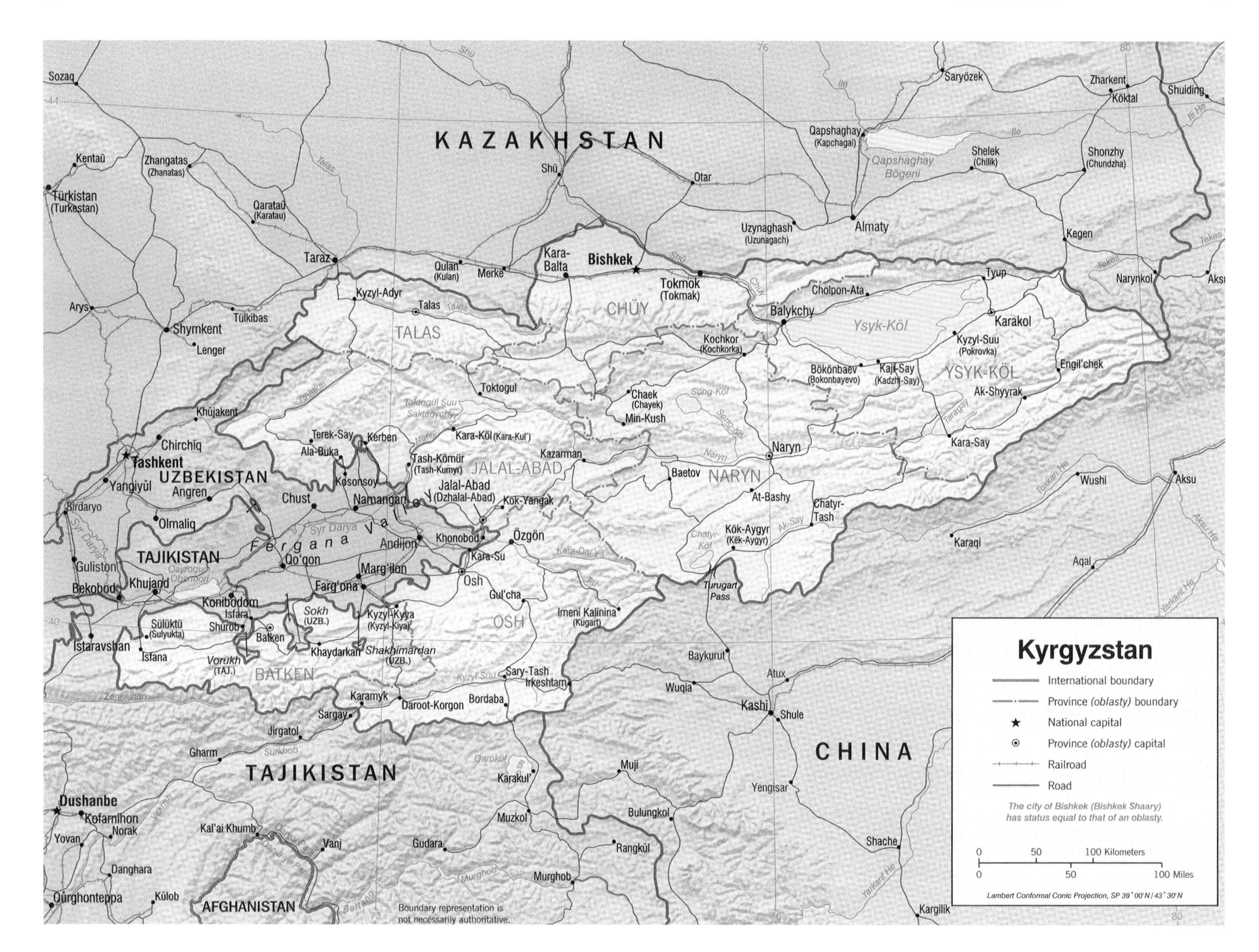
KAZAKHSTAN
Sozaq
Saryözek
Zharkent
Köktal
Shuiding
Kentaū
Zhangatas
(Zhanatas)
Qapshaghay
(Kapchagai)
Qapshaghay Bögeni
Shelek
(Chilik)
Shonzhy
(Chundzha)
Shū
Otar
Türkistan
(Turkestan)
Qaratau
(Karatau)
Uzynaghash
(Uzunagach)
Almaty
Kegen
Taraz
Qulan
(Kulan)
Merke
Kara-Balta
Bishkek
Tokmok
(Tokmak)
Tyup
Narynkol
Aksu
Kyzyl-Adyr
Talas
Chüy
Chüy
Cholpon-Ata
Balykchy
Ysyk-Köl
Karakol
Arys
Tülkibas
Shymkent
Lenger
TALAS
Kochkor
(Kochkorka)
Kyzyl-Suu
(Pokrovka)
Bökönbaev
(Bokonbayevo)
Kaji-Say
(Kadzhi-Say)
YSYK-KÖL
Engil'chek
Toktogul
Chaek
(Chayek)
Min-Kush
Song-Köl
Ak-Shyyrak
Toktogul Suu Saktagychy
Khŭjakent
Chirchiq
Tashkent
Terek-Say
Kerben
Ala-Buka
Kara-Köl (Kara-Kul')
Tash-Kömür
(Tash-Kumyr)
Kazarman
Naryn
Kara-Say
UZBEKISTAN
Yangiyŭl
Angren
Kosonsoy
JALAL-ABAD
Baetov
NARYN
Wushi
Aksu
Chust
Namangan
Jalal-Abad
(Dzhalal-Abad)
Kok-Yangak
At-Bashy
Chatyr-Tash
Sirdaryo
Olmaliq
Fergana Valley
Syr Darya
Khonobod
Özgön
Kök-Aygyr
(Kök-Aygyr)
Chatyr-Köl
Ak-Say
Karaqi
TAJIKISTAN
Andijon
Qo'qon
Kara-Su
Kara-Dar'ya
Guliston
Marg'ilon
Osh
Aqal
Bekobod
Khujand
Qayroqqum Obanbori
Farg'ona
Torugart Pass
Konibodom
Gul'cha
Isfara
Sokh
(UZB.)
Kyzyl-Kyya
(Kyzyl-Kiya)
Imeni Kalinina
(Kugart)
Sülüktü
(Sulyukta)
Shŭrob
Batken
OSH
Istaravshan
Khaydarkan
Shakhimardan
(UZB.)
Baykurut
Isfana
Vorukh
(TAJ.)
BATKEN
Kyzyl-Suu
Sary-Tash
Irkeshtam
Atux
Wuqia
Zeravshan
Karamyk
Daroot-Korgon
Bordaba
Kashi
Shule
Sargay
Jirgatol
CHINA
Gharm
Surkhob
Qarokŭl
TAJIKISTAN
Karakul'
Muji
Yengisar
Dushanbe
Kofarnihon
Norak
Muzkol
Bulungkol
Yovan
Kal'ai Khumb
Vanj
Gudara
Rangkŭl
Shache
Danghara
Murghob
Murghob
Qŭrghonteppa
Kŭlob
AFGHANISTAN
Bartang
Boundary representation is not necessarily authoritative.
Yarkant He
Kargilik
Ili He
Tekes
Toxkan He
Aksu He
Kyrgyzstan
International boundary
Province (oblasty) boundary
National capital
Province (oblasty) capital
Railroad
Road
The city of Bishkek (Bishkek Shaary) has status equal to that of an oblasty.
0
50
100 Kilometers
0
50
100 Miles
Lambert Conformal Conic Projection, SP 39° 00' N / 43° 30' N

1.7 KYRGYZSTAN REFERENCE MAP

This country reference map is part of a long-running series produced by the U.S. Central Intelligence Agency. The maps are easy-to-read basic references for foreign countries. All labels are sans serif. City labels and symbols are represented with a size hierarchy with the largest cities in bold. Alternative spellings for place names are parenthetically listed below labels. Administrative region names, such as country and oblasty names, are uppercase. Oblasty labels are green, echoing the green dash-dot line marking their boundaries. Complexities of inholdings are detailed with country abbreviations below labels, and areas are linked to the larger unit with "fish hook" symbols (for example, Sokh and Shakhimardan areas are part of Uzbekistan). Fish hooks are a common cartographic annotation used with multipart features when color hue is not used to group all members of an area. Physical features and areas are labeled in italics with blue lettering for lakes and rivers to match the color of the water features. Definite identification of country extent is a main goal of these reference maps, so country boundaries are bold and the focal country terrain is lighter than surrounding countries. Kyrgyzstan's shape stands out well on the map. River lines are thinner than country boundaries so both lines are seen where they are coincident. A subtle elevation tint, from green to brown, pulls mountainous areas forward, but use of brown rather than grays separates the hillshading from other map elements to ensure readability of labels and small symbols. Roads are a background element with no hierarchy in width, unlike maps with navigation purposes.

Courtesy of CIA—The World Factbook.

1.8 UNITED STATES PHYSICAL MAP

This small-scale (1:4,000,000) map of coterminous United States terrain is produced from space shuttle elevation data. The mapmaker constructed "plan oblique relief" for a strong three-dimensional appearance. Looking closely at small features, readers see volcanic mountains as cones, tilted slightly back to see the south side of the mountain, rather than the bulls-eye forms they have on regular planimetric views. Northerly canyon walls are also visible, giving a sense of canyon depth, seen along the Columbia River, for example. Boundary lines that appear straight on planimetric maps actually trace irregular routes as they move up and down the oblique terrain visible on this map. Landforms are also emphasized with elevation tints that range from low-elevation dark greens through pale yellows to oranges and up through desaturated browns to light gray. Because readers often misinterpret greens that represent low elevation as a vegetation symbol, arid lowlands are represented with a parallel gray green to reduce the verdant appearance of these dry areas while still distinguishing them as low elevation. The mapmaker describes this technique as "cross-blended hypsometric tints," and the bifurcation is seen at the bottom of the map legend. For more information, go to www.shadedrelief.com/physical.

Courtesy of Tom Patterson, www.shadedrelief.com.

Elevation

Meters

Feet

14,000

4,000

13,000

12,000

3,500

11,000

3,000

10,000

9,000

2,500

8,000

7,000

2,000

6,000

1,500

5,000

4,000

1,000

3,000

2,000

500

1,000

0 (sea level)

0 (sea level)

Arid lowland

Humid lowland

Strait of Juan de Fuca
CAPE FLATTERY
Cape Alava
SAN JUAN ISLANDS
Mt Baker 10785
Skagit
Iron Mt 4793
Mt Olympus 7969
OLYMPIC MTS
Quinault
Puget Sd
SEATTLE
Glacier Pk 10541
Oval Pk 8795
Okanogan Range
Lake Chelan
Stevens Pass 4061
Cashmere Mt 8501
Stormy Mt 7198
Snoqualmie Pass 3022
Grays Harbor
OLYMPIA
Willapa Bay
Leadbetter Pt
Mt Rainier 14410
Storm King Mt 4752
CAPE DISAPPOINTMENT
Willapa Hills
Cowlitz
White Pass 4469
Tillamook Head
Mt St Helens 8363
Mt Adams 12276
Bare Mt 4360
Cape Lookout
PORTLAND
Columbia River Gorge
Mt Hood 11239
Yaquina Head
SALEM
Marys Peak 4097
Willamette
Mt Jefferson 10497
EUGENE
McKenzie Pass 5325
South Sister 10358
Deschutes
Umpqua
Calapooya Mts
Paulina Pk 7984
Mt Thielsen 9182
Christmas
COAST RANGES
CASCADE RANGE
Wenatchee Mts
Grand Coulee
Franklin D Roosevelt Lake
Kettle River Ra
Columbia
Selkirk Mts
Priest Lake
Purcell Mts
7539 Robinson Mt
Whitefish Ra
Lake Koocanusa
Lake Pend Oreille
Cabinet Mts
Clark
SPOKANE
Channelled Scablands
Coeur d'Alene Lake
Flathead Lake
Marmot Pk 7208
Stevens Pk 6838
COLUMBIA BASIN
Palouse Hills
Rattlesnake Hills
Yakima
Snake
Monumental Buttes 6722
Illinois Pk 7690
Horse Heaven Hills
Columbia
Pot Mt 7139
Lolo Pk 9139
MISS
Oregon Butte 6387
COLUMBIA PLATEAU
BITTERROOT RANGE
CLEARWATER MTS
John Day
Black Mt 5932
Umatilla Range
BLUE MTS
Sacajawea Pk 9839
WALLOWA MTS
HELLS CANYON
Pilot Knob 7135
Buffalo Hump 8938
Salmon
Cannon Ball Mt 7197
Bitterroot
Sapphire Mts
Lost Trail 7014
Big Hole
Ochoco Mts
Big Lookout Mt 7120
Seven Devils Mts
Sturgill Pk 7589
Rainbow Pk 9325
Mt McGuire 10082
Strawberry Mt 9038
Cascade Res
SALMON RIVER MOUNTAINS
Lem 7373

1.9 ARCTIC CONSERVATION AREA MAP

This map emphasizes the lack of land at the Arctic Pole by showing the form of the ocean floor in the region rather than the ice cap. A thin two-tone shadow line along the coasts pulls the land above the water, and green-brown hypsometric tints provide a strong hue contrast with the blue progression for bathymetric tints. The landform colors provide a dark background to contrast with white areas representing permanent ice on land. International boundaries in white also contrast with the landform colors. Shaded relief combines with the elevation and depth colors for a fuller representation of form both above and below sea level. The map provides a small-scale reference to physical features within the area of concern for the Conservation of Arctic Flora and Fauna (CAFF) working group of the Arctic Council. The boundary of this area is represented with a bold green outline and a wide gradient from white to transparent inward from the boundary line. This gradient lightens areas at the inner edge of the area CAFF addresses, producing an appearance of illumination.

Courtesy of Hugo Ahlenius, UNEP/GRID-Arendal.

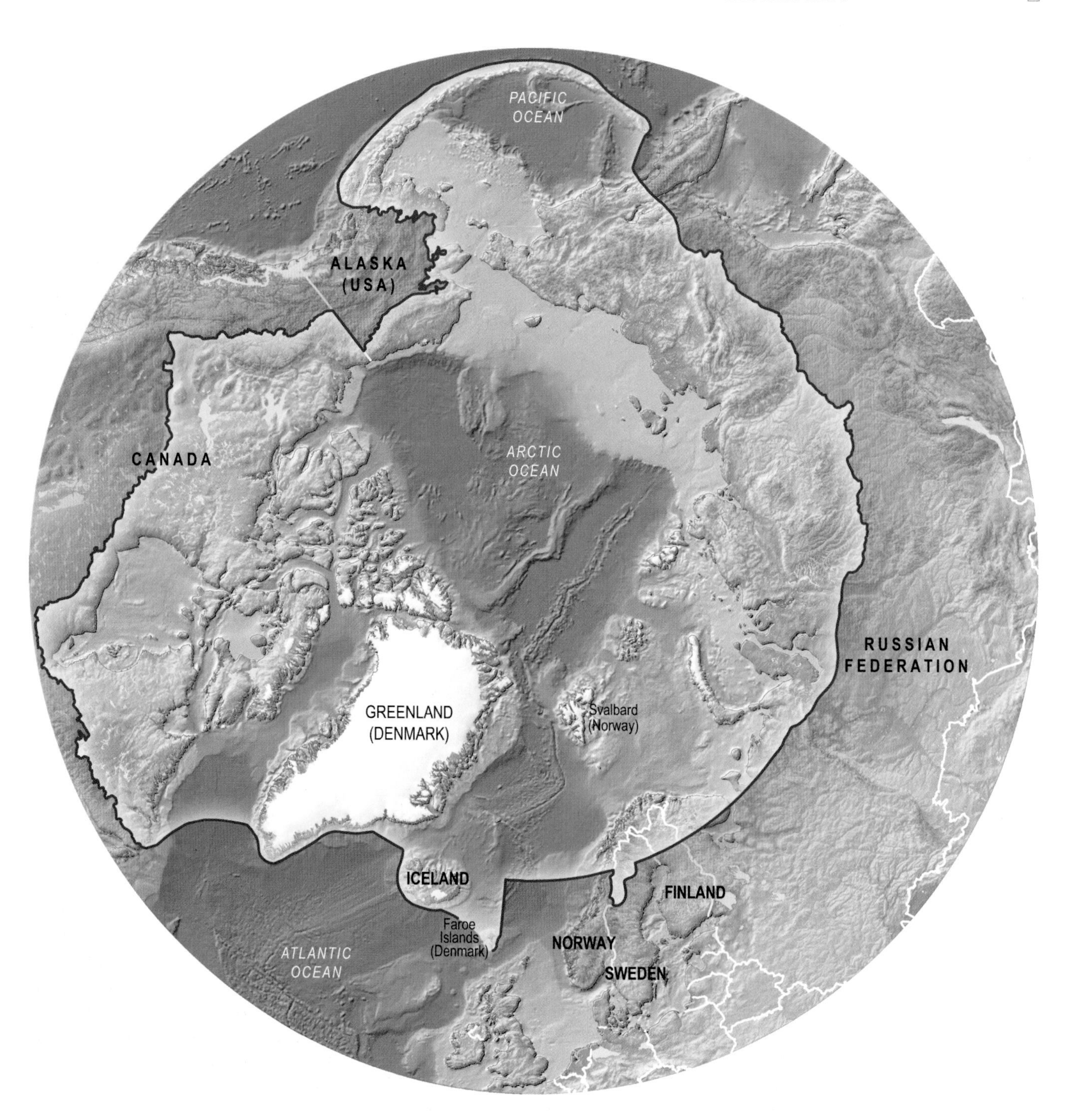

2

REFERENCE MAPS

NAVIGATION

Maps used for navigation must be more than an inventory of roads, bus stops, towns, or islands. They need to be useful to people trying to get from one place to another. Map users need to be able to search for their destinations, make route decisions, and read map labels rapidly as they travel. The best map designs respond to these needs with carefully labeled, systematically symbolized, and well-generalized data.

Care in label design improves map readability. Features are categorized by label hue and style differences, such as italic and nonitalic, and by different typefaces (fonts). Prominence and ordering of features is accomplished with differences in character size, case, lightness, and transparency. For example, an uppercase label in 18-point, black type draws attention and is a clear indicator that a feature is more important than a feature labeled with lowercase letters in 6-point, italic, light blue type. Systematic use of these distinguishing label characteristics makes a navigation map easier to search because readers can attend to just one label type and know they are being thorough. Labels are also visually linked to map features in obvious ways; for example, wide brown boundaries go with bold brown labels and red tent symbols go with red labels.

To reduce clutter, small-scale maps are generalized by omitting elements, such as bends in a road or river, or even entire cities. Features are also simplified and displaced, so the connections in a road network and its relationships to rivers and towns are clearly symbolized. These design decisions clean up the map to make it useful for navigation decisions. The most relevant aspects of a navigation map—topological relationships and completeness along major routes through the network—are maintained. Details of points of interest, such as historical sites or parking, and distances or bearings between places are some of the tools offered to navigation map users.

Navigation maps have base information much like general reference maps, though data selected for the base is customized to the distinctive features of the locale. For example, landownership hues underlay the Oregon road map but not the Washington, D.C. map, because destination planning and recreational opportunities in the western United States have a stronger link to large areas owned by the Forest Service, Bureau of Land Management, National Park Service, and other agencies. Terrain shading and contour lines are also common features on navigation maps for the obvious reason that the type of terrain you travel through affects travel time, difficulty, and scenic interest. Roads may be the main message on some maps, or they may be relegated to background information on others. They often drop out of background colors as light lines to make visual room for labels. The base information serves the dual purpose of indicating characteristics of the surrounding area and providing contrast for the navigation symbols that are the main message of the map.

2.0 SPOKANE ROAD MAP

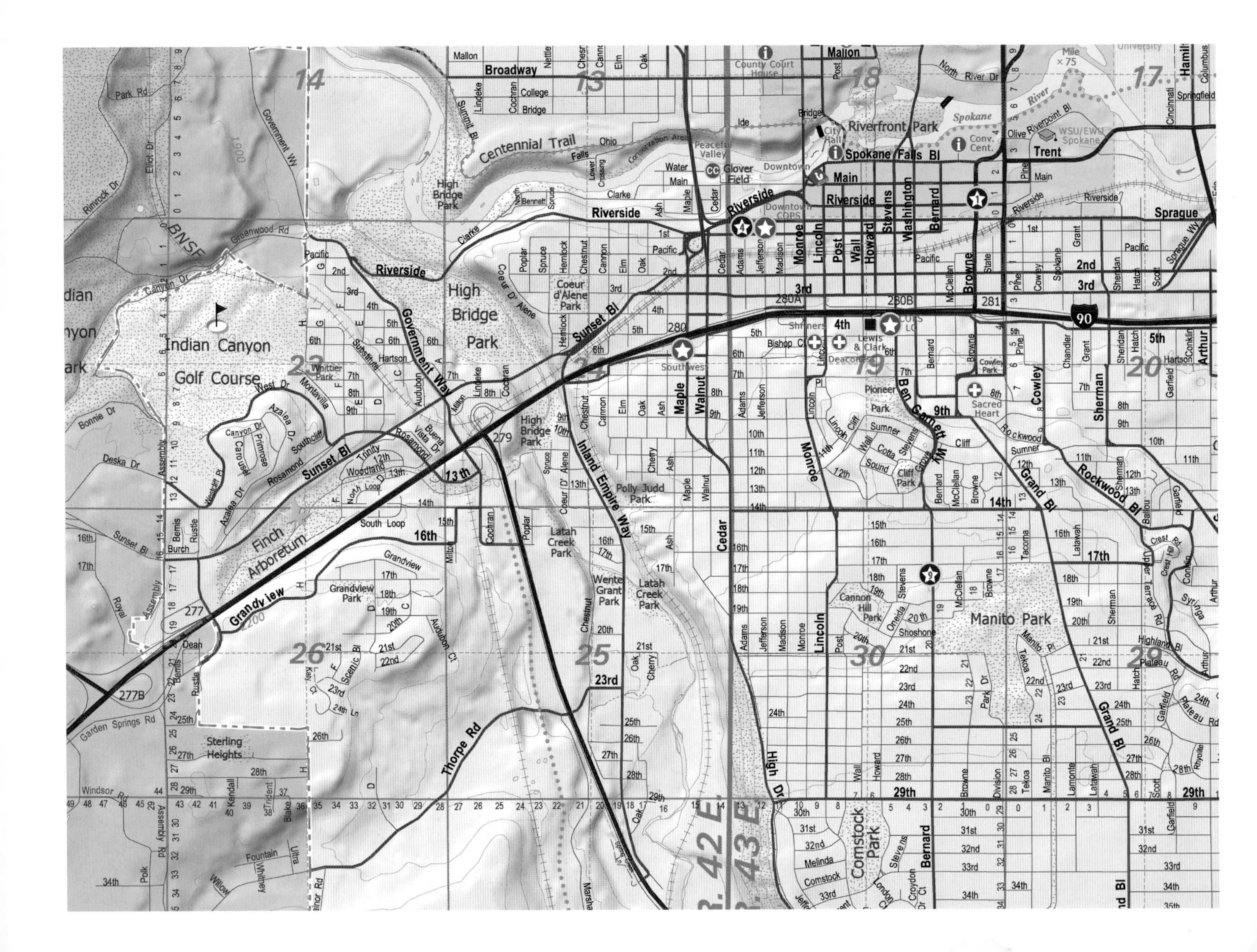

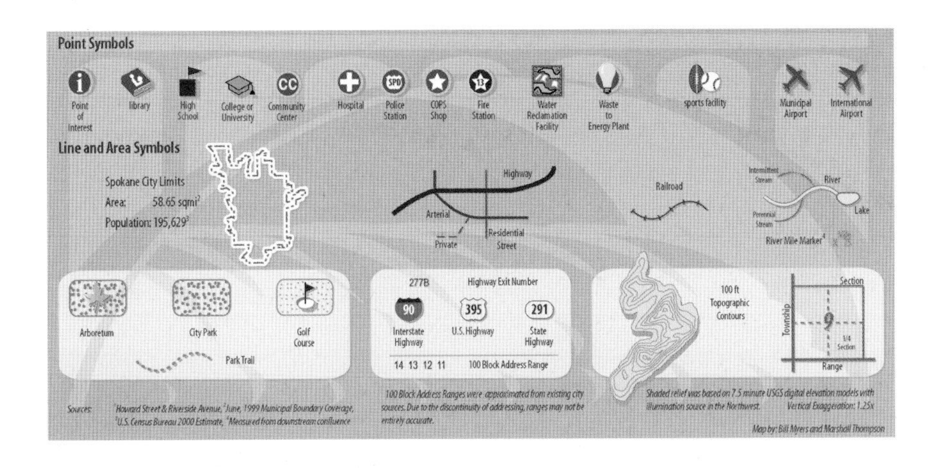

The following four designs of a road map for the city of Spokane, Washington, demonstrate different road symbols, effective in each map context: thin dark lines, light colorful wide lines, greens in three widths, and thin white lines. These line symbols are combined with area colors and textures for unique overall effects from the same data.

Courtesy of the City of Spokane.

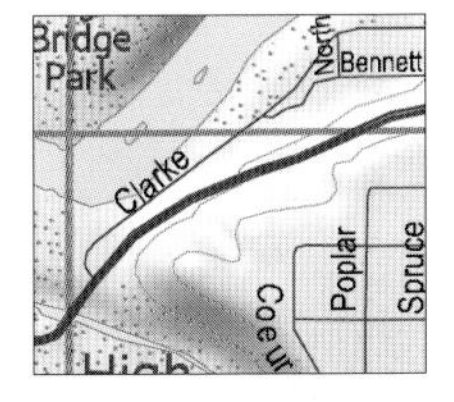

ORIGINAL DESIGN (Pages 24 and 26)

The categories of roads have strong contrasts in the original design, with fine and wide black lines for local and arterial roads, respectively, and wide red lines for limited-access highways. Trails represented by lines of dots coordinate well with the random dot pattern used for park areas. This road network sits lightly on the terrain shading, and the map also includes contour lines to further indicate landforms. Township and range lines in brown are set below the street information.

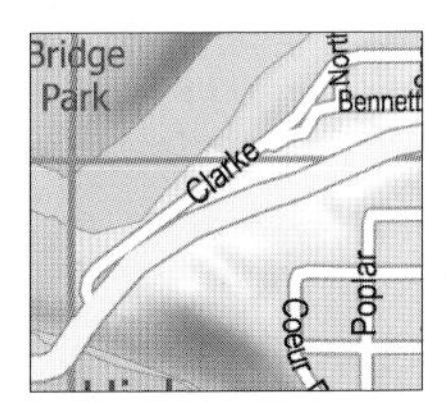

EMPHASIS ON ROAD HIERARCHY (Page 26)

Here the roads are wider so labels may run within road lines. Color lightness and hue work together to produce a strong road hierarchy, with white lines knocked out of the terrain for residential roads and wider yellow lines for arterial roads. Both of these symbols use a brown casing that does not interfere with road labels but ensures the definition of the road line when it runs over lighter areas in the terrain shading. The widest lines, in dark red-orange, establish the highways as most visually prominent. The line casings break where roads join and ramps merge onto the highways to communicate access detail.

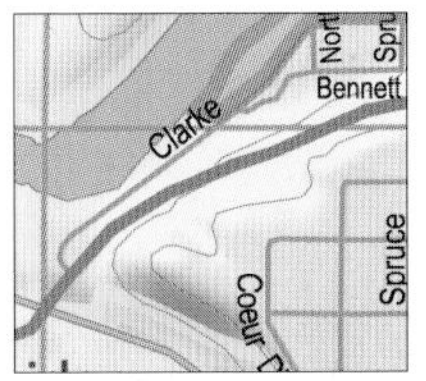

EMPHASIS ON PARKS (Page 27)

Parks are emphasized on this map: the large multipart park, High Bridge, has its own yellow hue, the Arboretum is in yellow green, and smaller parks share an orange symbol. Each is highlighted with tint bands at the edge of the park area in more saturated versions of the park hue. The road labels focus on roads en route and adjacent to parks to emphasize navigation to the parks, rather than having the map serve as a road map to the entire city. Contrast within the road network is lessened, though size and modest variation in lightness within a range of greens distinguishes three road categories. Trails are emphasized by using a saturated red hue and a lively footprint symbol to draw the map reader's eye.

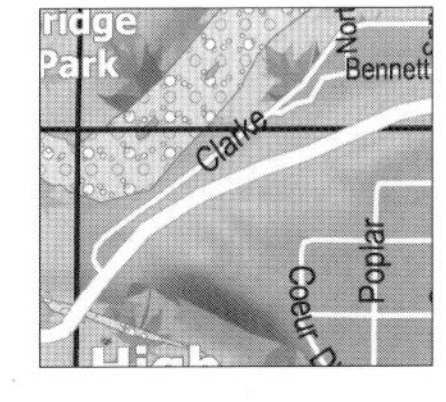

BLACK-AND-WHITE DESIGN (Page 27)

This map reverses the contrast in the original design with white roads dropped out of a darker version of the terrain shading. The major roads are wider, and black labels are positioned directly on roads rather than to the side. Multilayer textures for rivers and parks add visual interest, and some features, such as the rail lines, are omitted to reduce clutter (an important consideration here, given the limited range of contrasts available in monochrome). Categorization of labels makes use of outlines, with park labels in white with a black outline and section numbers in transparent outlined characters that keep them in the background.

REDESIGN

2A ORIGINAL DESIGN

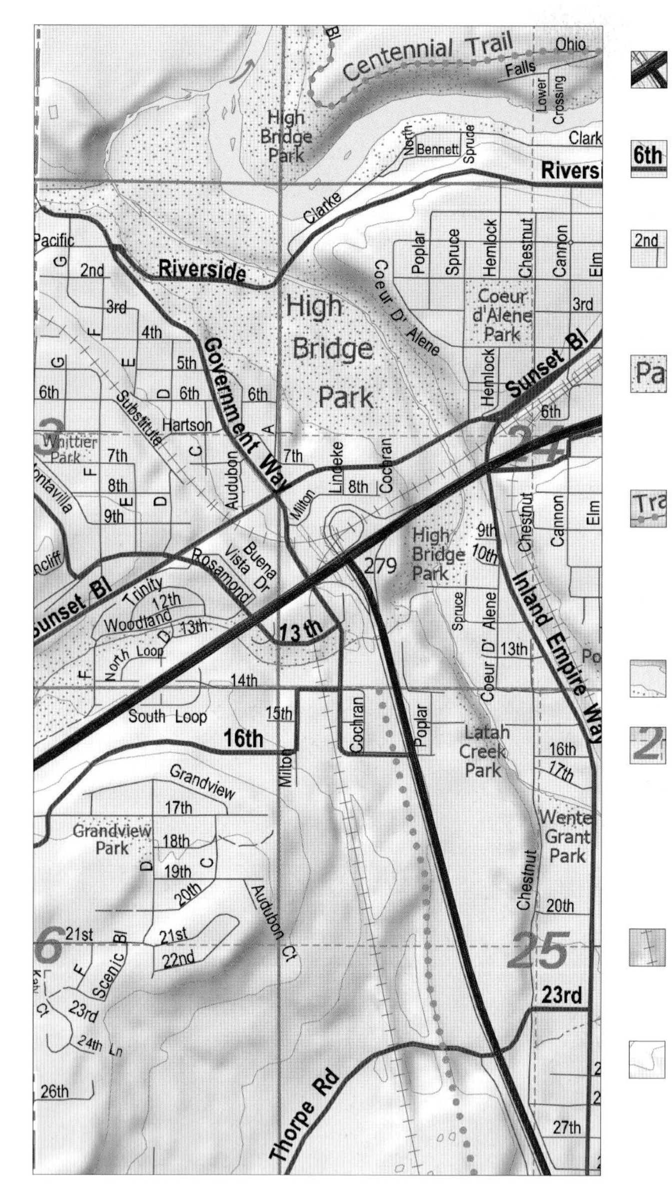

ROADS

Highways
Line: 3.5 pt, 100M 100Y
Label: 8 pt, Tahoma, 100M 100Y (exit number)

Major
Line: 2 pt, 80K
Label: 9 pt, Arial Narrow Bold, 80K

Minor
Line: 0.5 pt, 80K
Label: 6.75 pt, Arial Narrow, 80K

RECREATION

Parks
Fill: 100C 100Y, 0.7 pt, circle marker fill with random arrangement
Labels: 11 to 7 pt, Tahoma, 80K

Trails
Line: 100C 2M 40Y, marker line pattern 2.5/5.5 *(symbol size/spacing)*
Label: 9 pt, Arial Narrow, 80K

OTHER

Water
Fill: 19C 3M 3Y
Line: 0.25 pt, 100C 25M

Grid
Lines: 1.3 pt, 30C 50M 50Y (section); 0.5 pt, 30C 50M 50Y, dash pattern 3/1 (dash/gap in pts) (quarter section)
Label: 19 pt, Arial Bold Italic, 30C 50M 50Y

Railways
Lines: 0.25 pt, 80K; 4 pt, 80K, hash line pattern 0.25/5.5 *(hash width/spacing in pts)*

Contours
Line: 0.25 pt, 20C 50M 75Y

ArcMap Tips (see pages 161 to 167)

5 Marker line	7 Hash line
16 Random fill	36 Convert to annotation

2B EMPHASIS ON ROAD HIERARCHY

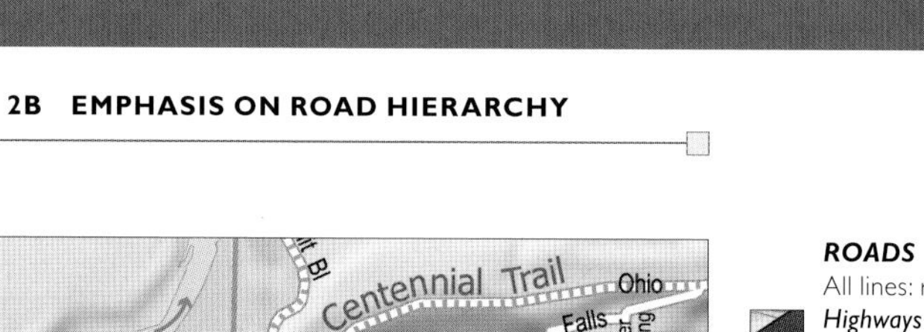

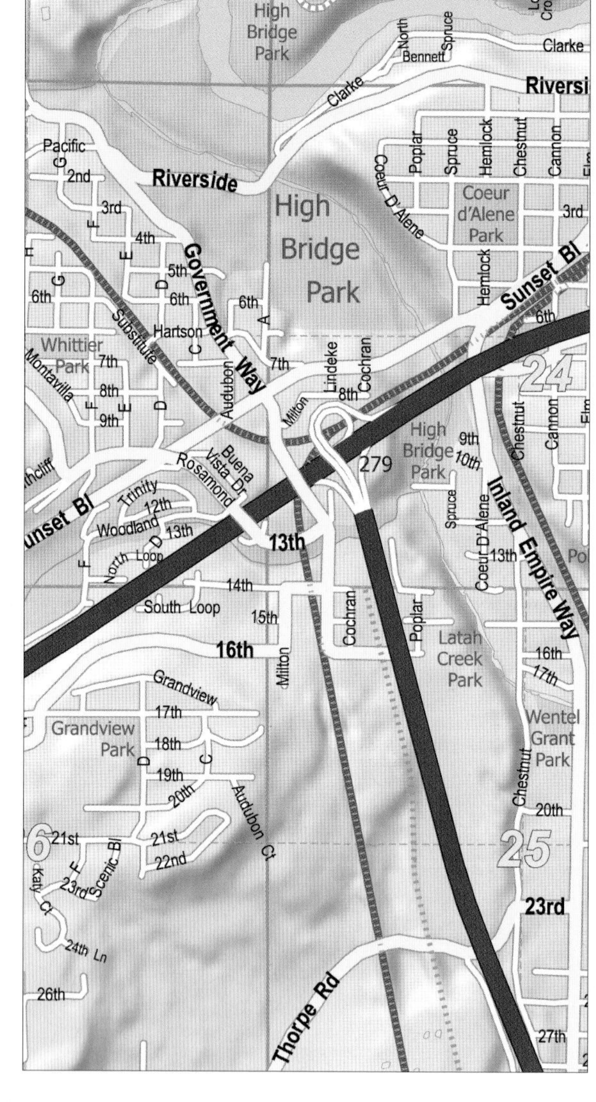

ROADS

All lines: round cap/join

Highways
Line: 7 pt, 80M 80Y 10K
Casing: 8 pt, 100M 100Y 50K

Major
Line: 4.5 pt, 40Y
Casing: 5.5 pt, 20M 40Y 30K
Label: 9 pt, Arial Narrow Bold, 100K

Minor
Line: 2 pt, white
Casing: 3 pt, 20M 40Y 30K
Label: 6.75 pt, Arial Narrow, 100K

RECREATION

Labels: 11 to 7 pt, Tahoma, 80K

Parks
Fills: 80C 100Y, 20% transparent

Trails
Line: 2 pt, 75C 90Y, hash line pattern 1/2

OTHER

Water
Fill: 20C 10M
Line: 0.25 pt, 60C 20M 10K

Grid
Lines: 1.3 pt, 10C 50M (section); 0.5 pt, 10C 50M, dash pattern 3/1 (quarter section)
Label: 19 pt, Arial Bold Italic, 5C 10M
Halo: 0.3 pt, 50C 100M

Railways
Line: 2 pt, 20M 30Y 20K, hash line pattern 0.5/2
Casing: 3 pt, 40M 30Y 50K

ArcMap Tips

2 Multilayer line	4 Merge/overpass
25 Eliminate by size	34 Maplex settings

2C EMPHASIS ON PARKS

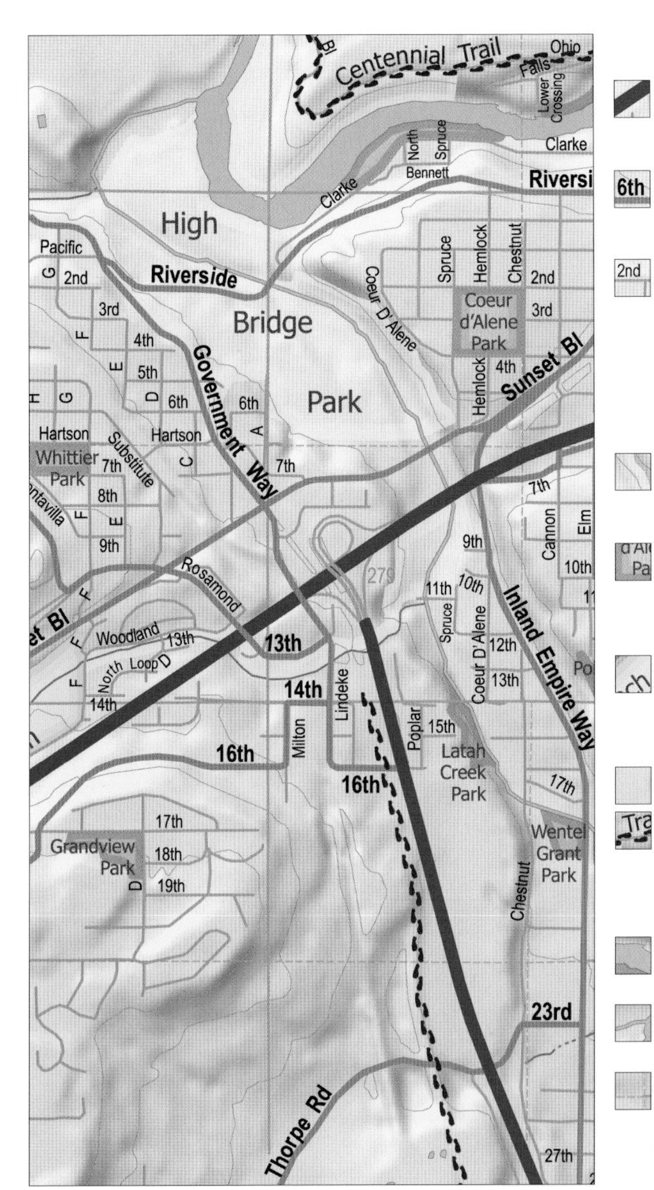

ROADS

All lines: round cap/join

Highways
Line: 5.5 pt, 100C 70Y 50K
Label: 8 pt, Arial Narrow, 100C (exit number)

Major
Line: 2.5 pt, 70C 70Y 20K,
Label: 9 pt, Arial Narrow Bold, 100K

Minor
Line: 1 pt, 60C 80Y 10K,
Label: 6.75 pt, Arial Narrow, 100K

RECREATION

Labels: 11 to 7 pt, Tahoma, 40M 80Y 60K (parks); 100M 100Y 20K (trails)

Large Parks
Fill: 50Y, 40% transparent
Offset bands: 3 pt, 60%, 100Y; 6 pt, 40%, 70Y

Small Parks
Fill: 30M 50Y, 40%
Offset bands: 3 pt, 60%, 55M 100Y; 6 pt, 40%, 45M 70Y

Arboretum
Fill: 15C 50Y
Offset bands: 3 pt, 60%, 30C 100Y; 6 pt, 40%, 25C 70Y

Golf Course
Fill: 40C 20Y, 40%

Trails
Line: 100M 100Y 20K, custom footprint drawing along picture line

OTHER

Water area
Fill: 40C 15M
Line: 0.25 pt, 70C 50M

Streams
Line: 1 pt, 40C 15M
Casing: 1.5, 70C 50M

Grid
Line: 1 pt, 30C 40M (section); 0.5 pt, 30C 40M, dash pattern 3/1 (quarter section)

ArcMap Tips

8 Picture line	10 Tint bands
21 Hillshade colors	24 Centerline

2D BLACK-AND-WHITE DESIGN

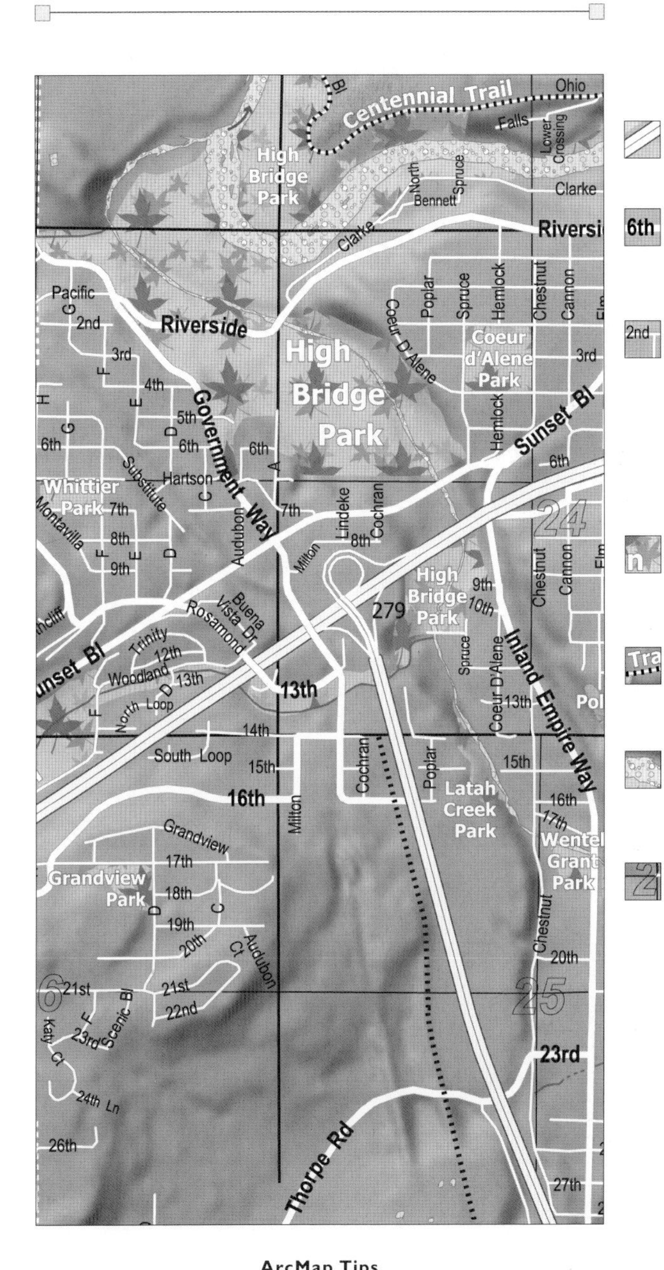

ROADS

All lines: round cap/join

Highways
Centerline: 0.25 pt, 80K
Line: 5 pt, white,
Casing: 6.5 pt, 80K, 100K

Major
Line: 2.5 pt, white,
Casing: 3 pt, 40K
Label: 9 pt, Arial Narrow, 100K

Minor
Line: 1 pt, white,
Casing: 1.5 pt, 40K,
Label: 6.75 pt, Arial Narrow, 100K

RECREATION

Labels: 11 to 7 pt, Tahoma, white
Halo: 0.25 pt, 70K

Parks
Multilayer fill: custom leaf drawings in five picture fills; white, 40% transparent base fill

Trails
Line: 2 pt, 100K, hash line pattern 1/2

OTHER

Water
Multilayer fill: custom circle drawings in two picture fills; 10K base fill
Line: 0.25 pt, 70K

Grid
Line: 0.5 pt, 100K
Label: 19 pt, Arial Bold Italic, no fill
Halo: 0.25 pt, 100K

ArcMap Tips

2 Multilayer line	7 Hash line
18 Picture fill	41 Convert grid

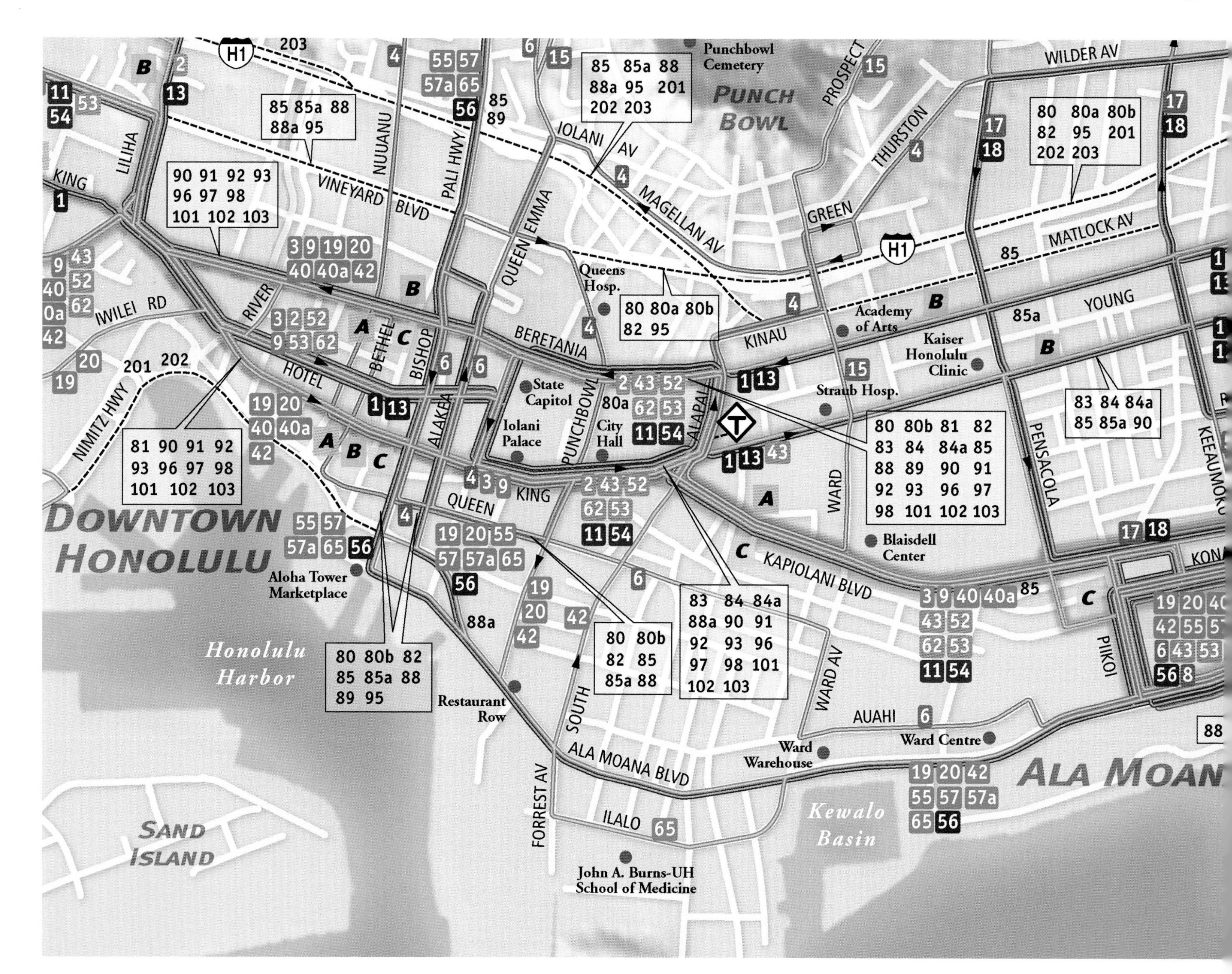
DOWNTOWN HONOLULU
PUNCH BOWL
ALA MOAN
SAND ISLAND
Honolulu Harbor
Kewalo Basin
Punchbowl Cemetery
Queens Hosp.
Academy of Arts
Kaiser Honolulu Clinic
Straub Hosp.
State Capitol
Iolani Palace
City Hall
Blaisdell Center
Aloha Tower Marketplace
Restaurant Row
Ward Warehouse
Ward Centre
John A. Burns-UH School of Medicine
WILDER AV
MATLOCK AV
YOUNG
KINAU
BERETANIA
HOTEL
KING
QUEEN
KAPIOLANI BLVD
ALA MOANA BLVD
AUAHI
ILALO
VINEYARD BLVD
IOLANI AV
MAGELLAN AV
GREEN
PROSPECT
THURSTON
NUUANU
PALI HWY
QUEEN EMMA
LILIHA
RIVER
IWILEI RD
NIMITZ HWY
BETHEL
BISHOP
ALAKEA
PUNCHBOWL
ALAPAI
WARD
WARD AV
PENSACOLA
KEEAUMOKU
PIKOI
SOUTH
FORREST AV
KONA
85 85a 88 88a 95
90 91 92 93 96 97 98 101 102 103
85 85a 88 88a 95 201 202 203
80 80a 80b 82 95 201 202 203
80 80a 80b 82 95
83 84 84a 85 85a 90
80 80b 81 82 83 84 84a 85 88 89 90 91 92 93 96 97 98 101 102 103
81 90 91 92 93 96 97 98 101 102 103
80 80b 82 85 85a 88 89 95
80 80b 82 85 85a 88
83 84 84a 88a 90 91 92 93 96 97 98 101 102 103

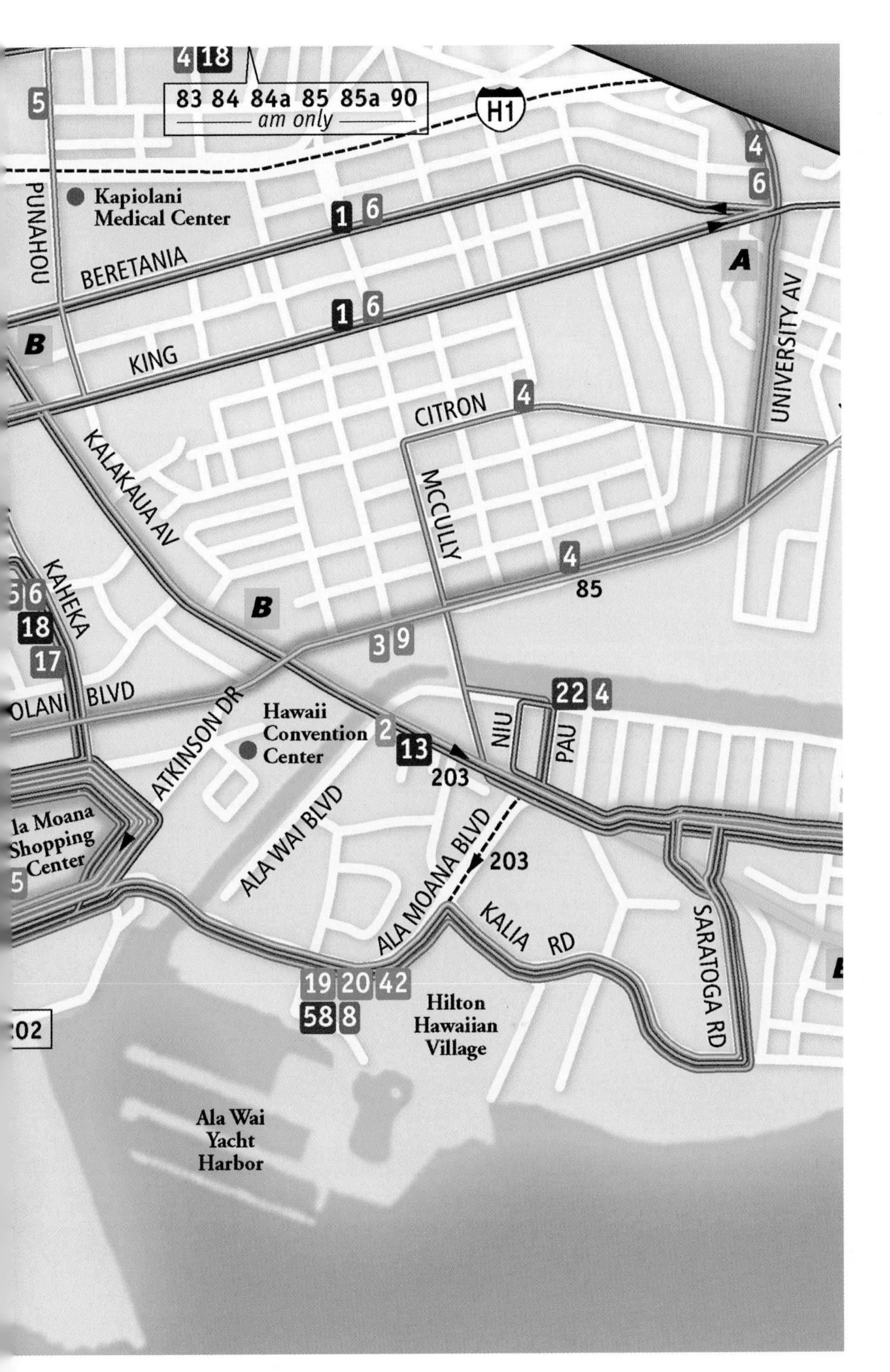

2.1 OAHU BUS MAP

A city bus map must contend with making sense of many intertwined routes. The Oahu map employs saturated hues to differentiate bus lines. Lines are tagged with identification codes in boxes with matching hues, and these labels are grouped along lines for the many bus routes that run together. Routes along the same street are displaced so each color is clearly visible, forming a wider multicolor line group. Only streets along which buses run are labeled. In contrast to the bus routes, background land colors are light and desaturated. Flat areas are only dark enough that white roads are visible against them, backed by hazy shadows that raise them slightly above the background. Other subtle color effects include transparency: for color glows around express routes, large area labels, and callout boxes holding lists of peak hour bus numbers. Each saturated line for regular routes has a lighter centerline, giving them the look of neon tubes, that adds subtly to the three-dimensional character of the map. Additional vivid color is included in water areas represented by a smooth gradient from cyan to purplish blues.

Courtesy of Oahu Transit Services and International Mapping Associates.

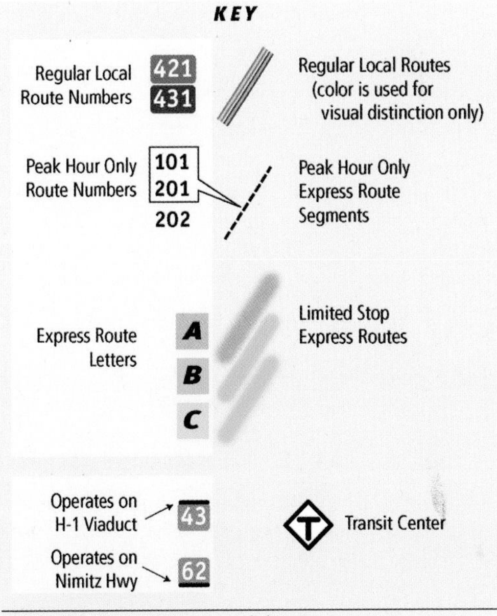

2.2 FAIRFAX BUS RIDERSHIP ANALYSIS MAP

The Fairfax, Virginia, ridership map is a more analytical project that uses good design to understand potential customer locations for connector bus routes. The hue of the bus line and stops is echoed in medium and lighter colors filling buffers of a half and quarter mile around the stops. White roads are visible against background colors, and gray casing adds emphasis to major roads. Against this light background, black dots jump forward. They each represent a point of origin of commuters with a vehicle at the Metro stop served by these bus routes. The bus routes hover over the background with a blurred shadow, bringing them forward as the main map topic and also separating them from the buffer areas lying on the surface. Bus lines that are not part of the analysis are included but represented in gray to deemphasize them.

Courtesy of Fairfax County Department of Transportation.

FINDING POTENTIAL CUSTOMERS

This map illustrates the results of a vehicle point-of-origin analysis conducted by FCDOT in November 2004 at Vienna Metro. Specifically, the map focuses on the Chantilly and Centreville areas of Fairfax County. These areas receive peak hour bus service to Vienna Metro on 13 Metrobus routes called the 12's and 20's. The black dots represent vehicle point-of-origin and are plotted on this map to show proximity to both current Metrobus routes and Metrobus stops. The light color shading shows point-of-origin of all commuters driving to Vienna Metro who reside within ¼ mile of a bus stop, the darker color shading shows commuters who reside within ½ mile of a bus stop. The results are summarized in the table below.

Distance	Metrobus 12		Metrobus 20	
	Stops	Routes	Stops	Routes
< 1/4 mi	235 (6%)	590 (14%)	63 (2%)	387 (9%)
1/4 - 1/2 mi	141 (3%)	375 (9%)	113 (3%)	377 (9%)
Totals	460 (11%)	965 (23%)	337 (8%)	764 (18%)

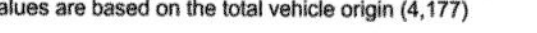

Values are based on the total vehicle origin (4,177)

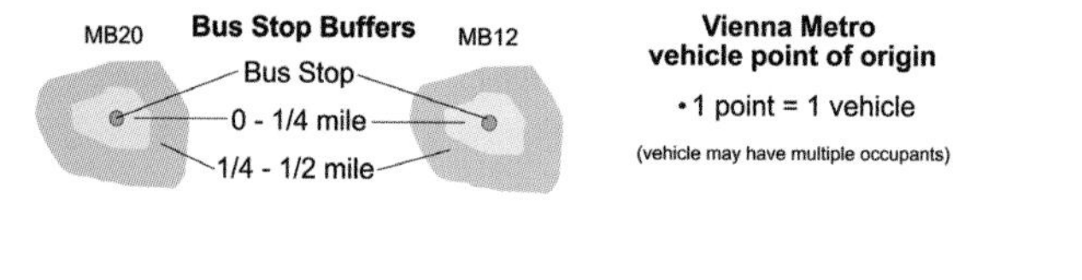

Vienna Metro vehicle point of origin

• 1 point = 1 vehicle

(vehicle may have multiple occupants)

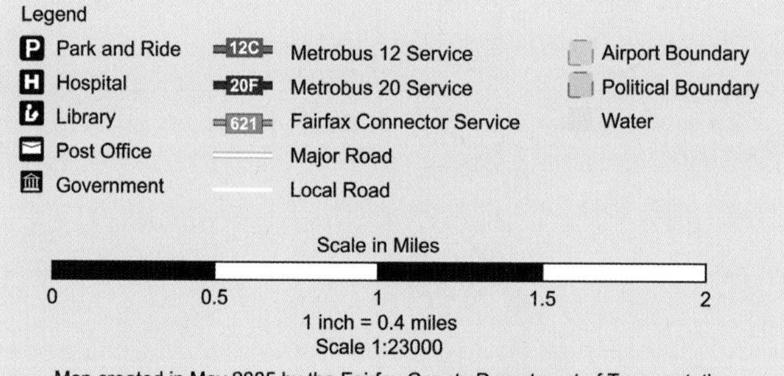

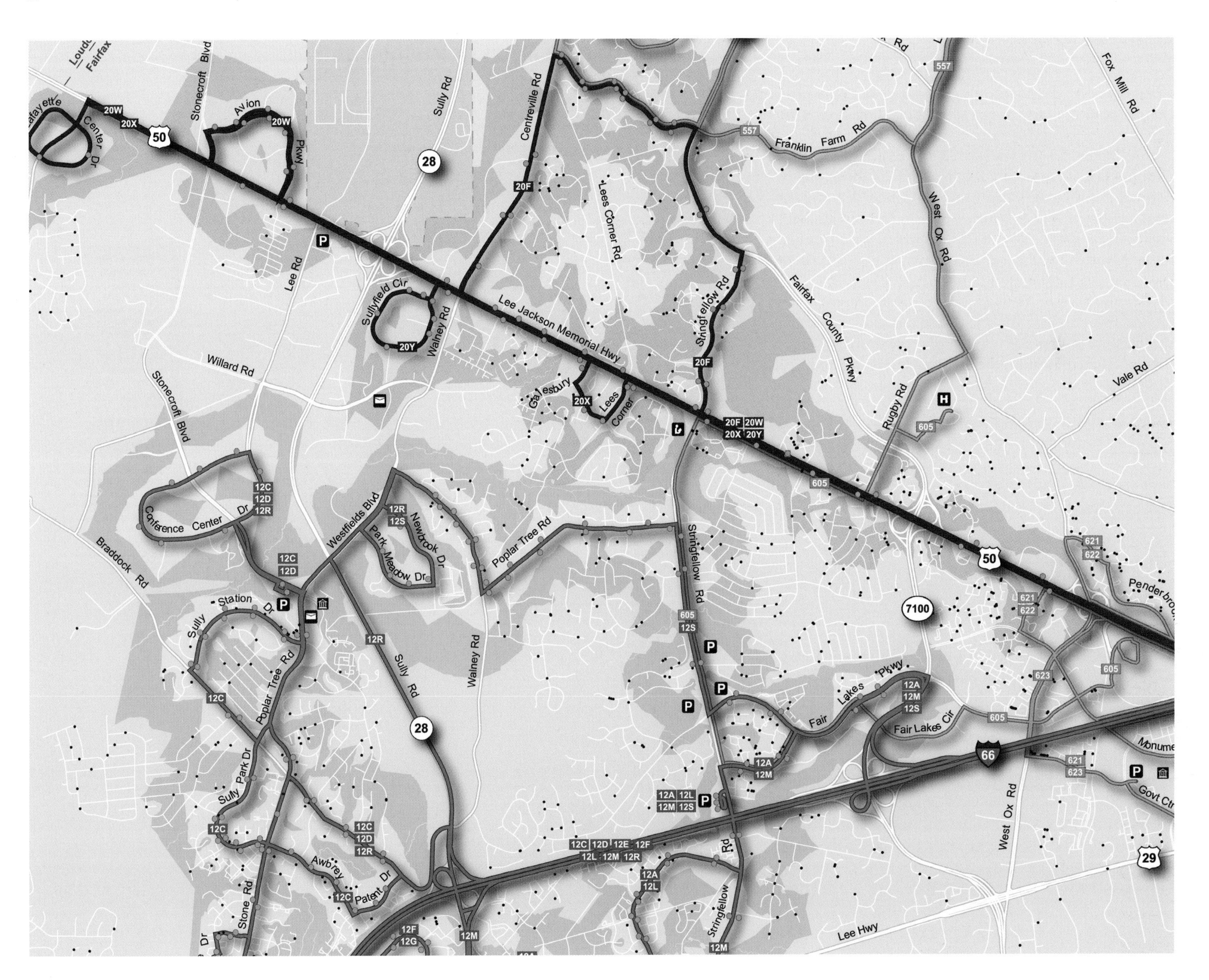
Centreville Rd
Sully Rd
Stonecroft Blvd
Avion Pkwy
Center Dr
Lee Rd
Sullyfield Cir
Walney Rd
Lee Jackson Memorial Hwy
Lees Corner Rd
Stringfellow Rd
Franklin Farm Rd
West Ox Rd
Fox Mill Rd
Fairfax County Pkwy
Rugby Rd
Vale Rd
Willard Rd
Galesbury
Lees Corner
Conference Center Dr
Braddock Rd
Westfields Blvd
Park Meadow Dr
Newbrook Dr
Poplar Tree Rd
Sully Station Dr
Sully Park Dr
Awbrey Patent Dr
Stone Rd
Fair Lakes Pkwy
Fair Lakes Cir
Monument
Govt Ctr
Lee Hwy
Penderbrook
Stringfellow

2.3 TORONTO CITY MAP

Road maps by MapArt are meticulously designed for the purpose of navigating cities. Roads are bold and labels are clear on this map of Toronto, Canada. Background colors are saturated, but the road hierarchy is clear against these colors. Roads are cased in red, with highways in orange (shown in the legend but not on this segment of map), major arteries highlighted in vivid yellow, and other roads filled white, with thinner white lines for local roads. Expressways are separated from the road network as lines cased and filled with blue or purple (see legend). The colors on the map are chosen from around the entire hue circle at consistent saturations, giving the map a coherent and attractive design. Going around the hue circle with examples shows how thoroughly organized this map is:

Reds: road casings, buildings, built-up areas

Oranges: highways, institution areas

Yellows: major roads, attraction areas

Greens: parks, golf courses, and open land areas

Blues: expressways, points-of-interest symbols, streams, water areas

Purples: toll expressways, shopping center areas, boundary lines

There are at most a few feature types in each hue, which allows clear differentiation between a lighter background use, such as blue water, and a darker foreground use, such as blue points-of-interest symbols. This care in organizing symbols lets a reader focus on a theme and search it throughout the whole map without being distracted by the other features. Black labels are almost all the same font, with italic for places, regular for roads, and large bold labels for more important roads. Road lines are wide to accommodate labels well.

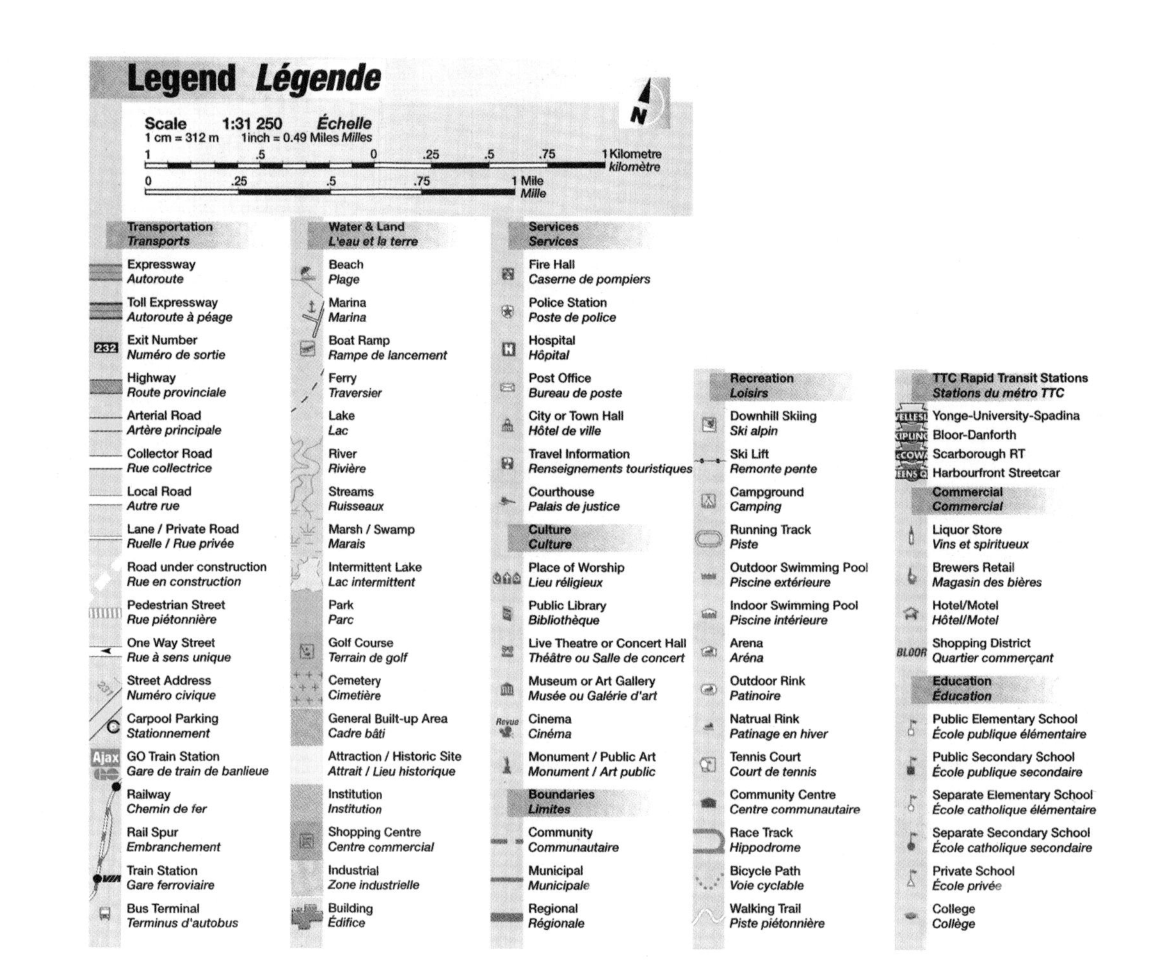

Courtesy of MapArt, copyright Mapmobility.

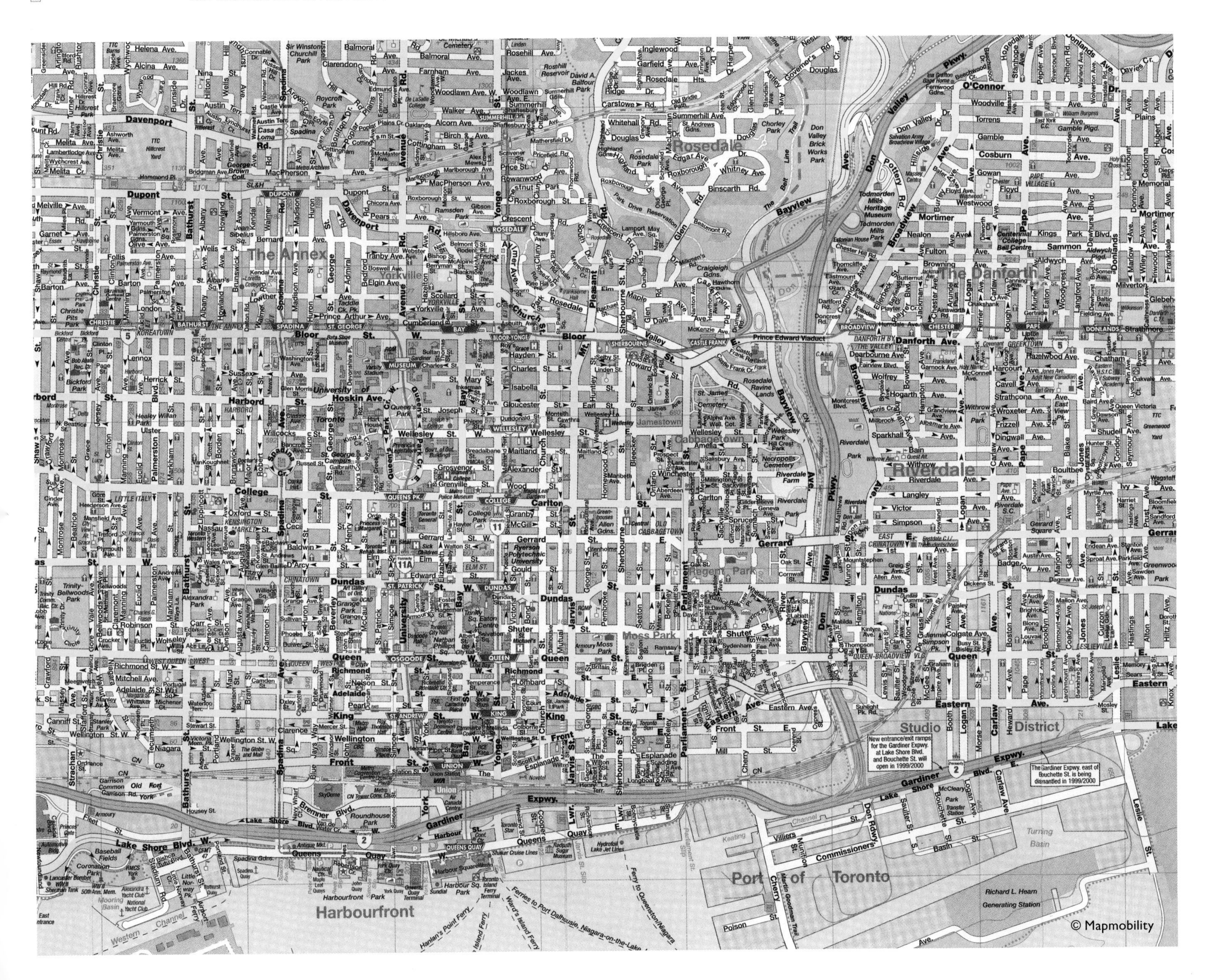
The Annex
Yorkville
Rosedale
The Danforth
Riverdale
Cabbagetown
Moss Park
Studio District
Harbourfront
Port of Toronto
New entrance/exit ramps for the Gardiner Expwy. at Lake Shore Blvd. and Bouchette St. will open in 1999/2000
The Gardiner Expwy. east of Bouchette St. is being dismantled in 1999/2000
Gardiner Expwy.
Lake Shore Blvd. W.
Queens Quay
Bloor St. W.
Danforth Ave.
Prince Edward Viaduct
Don Valley Pkwy.
Bayview Ave.
Davenport Rd.
Dupont
College
Carlton
Gerrard
Dundas
Queen
King
Front
Spadina Ave.
Yonge
Bathurst
Broadview
Richard L. Hearn Generating Station
Turning Basin
© Mapmobility

2.4 PARIS CITY MAP

The Rough Guide city map of Paris, France, is for tourists and is a delight to use because it is clearly labeled and embeds many details. Points of interest are carefully placed without interfering with road labels. They are searchable by category (green dots for restaurants, blue for entertainment, red for shops), labeled by name, and also categorized by a small letter in the dot. The reader can find the general category, catch the name right at the location, but then read deeper to focus on the type of place within the category; for example, a J in a blue circle identifies jazz clubs within the entertainment venue category. The background is subdued on this map with a subtle checkerboard of two grays to maintain the structure of the index squares when grid lines fall below roads. Area hues highlight main sights and buildings, shopping, and pedestrian areas in signature colors. Major neighborhoods are labeled in large transparent letters that get the user in the right general place but are easy to read through to other features.

- Main sight
- Notable / public building
- Shop / market
- Shopping / market street
- Pedestrian access only
- † Church
- Post office
- Police station
- Tourist information
- Hospital
- Batobus boarding point
- Cruise boat boarding point
- Taxi rank
- Parking
- Railway station
- Metro station (metro / RER map overleaf shows route numbers)
- RER station (metro / RER map overleaf shows route numbers)
- Airport bus departure point
- 70 Bus number
- 88 Bus terminus
- Airport
- Viewpoint
- 41 Street number
- « One-way street
- ★ Highly recommended
- Hotel
- Restaurant
- Bar / café
- Entertainment / nightlife venue
- Shop / market
  (Letter in dot indicates restaurant, bar, entertainment or shop category in index).

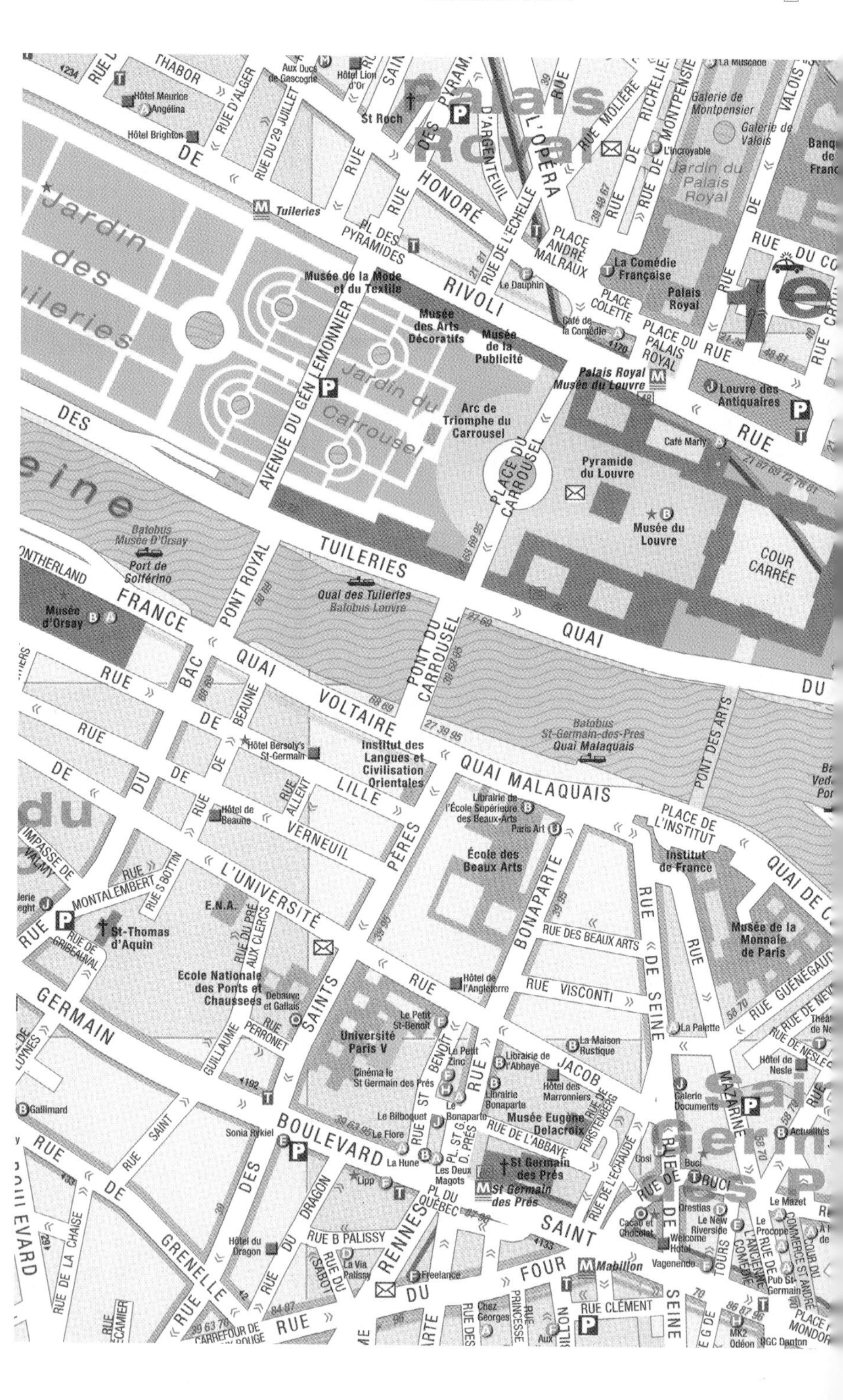

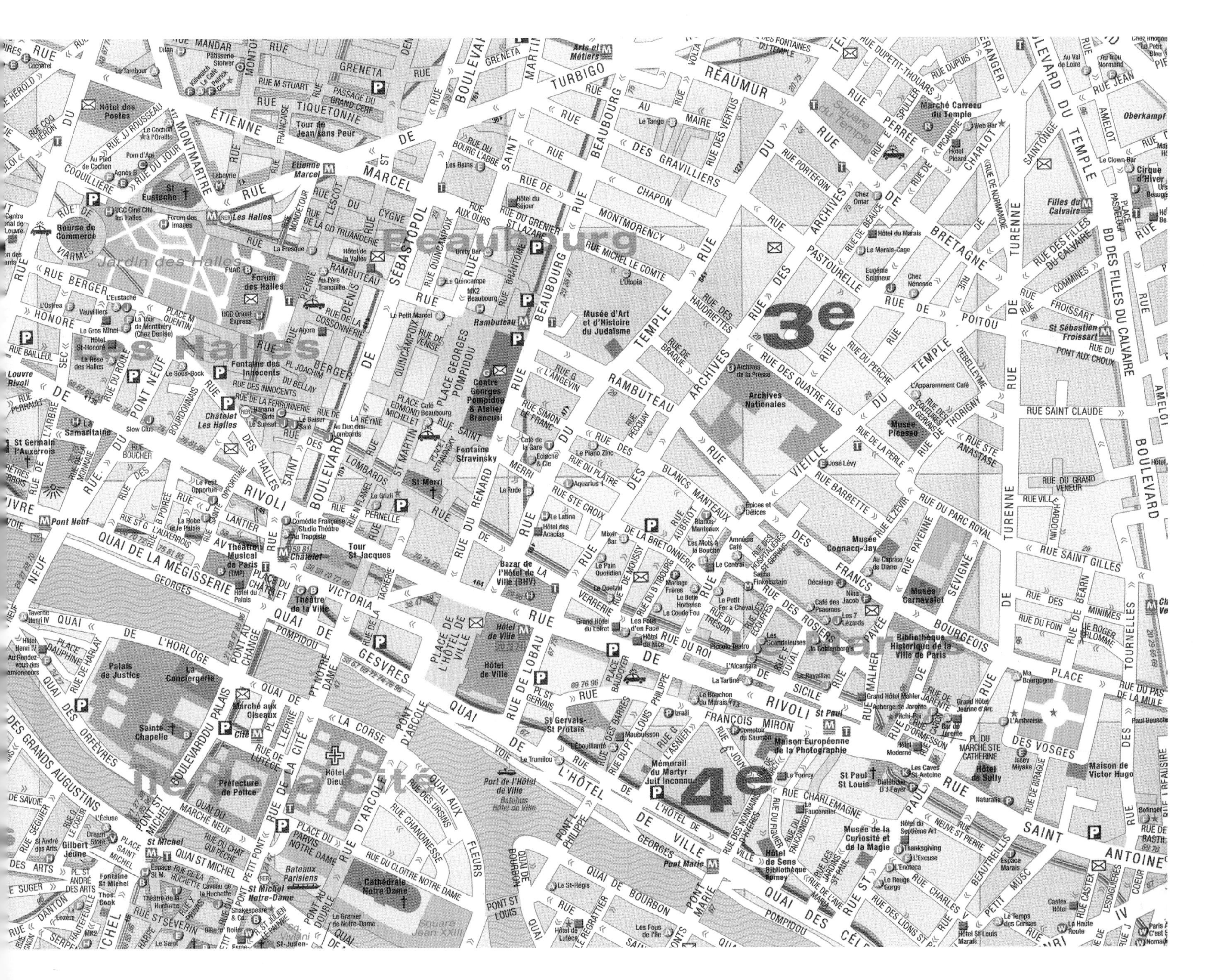
Beaubourg
Les Halles
Le Marais
Île de la Cité
3e
4e
Jardin des Halles
Forum des Halles
Centre Georges Pompidou & Atelier Brancusi
Musée d'Art et d'Histoire du Judaïsme
Archives Nationales
Musée Picasso
Musée Cognacq-Jay
Musée Carnavalet
Bibliothèque Historique de la Ville de Paris
Hôtel de Ville
Palais de Justice
La Conciergerie
Sainte Chapelle
Préfecture de Police
Hôtel Dieu
Cathédrale Notre Dame
Maison Européenne de la Photographie
Mémorail du Martyr Juif Inconnu
Hôtel de Sully
Maison de Victor Hugo
Square du Temple
Square Jean XXIII
Tour St Jacques
Fontaine Stravinsky
St Eustache
St Merri
Bourse de Commerce
RUE DE RIVOLI
RUE SAINT ANTOINE
BOULEVARD SÉBASTOPOL
RUE DU TEMPLE
RUE RÉAUMUR
RUE DE TURBIGO
RUE RAMBUTEAU
RUE DES FRANCS BOURGEOIS
PLACE DES VOSGES
QUAI DE LA MÉGISSERIE
QUAI DE L'HORLOGE
QUAI DE L'HÔTEL DE VILLE
QUAI DE BOURBON
BOULEVARD DU TEMPLE
BD DES FILLES DU CALVAIRE
RUE VIEILLE DU TEMPLE
RUE DES ARCHIVES
RUE BEAUBOURG
RUE SAINT MARTIN

2.5 SPAIN CITY MAPS

Michelin city maps are designed to ensure that the distinctive street patterns are clearly readable in a small space. Design specifics are seen in these examples for Pontevedra and Vigo, Spain. Light orange fills indicate built-up areas and also help define the street pattern by contrast with the street colors. Differences in casings and fills are used to categorize streets, such as yellow or orange fills on major streets and dashed casings for pedestrian areas. Labels do not clutter or break road lines. Within tight areas of the street network, some streets are labeled with a number and a thin black leader line that finishes with a tiny dot at the identified street. That dot distinguishes these fine black lines from dash segments, bridges, arrows, and other small symbol elements. An index next to each map indicates streets called out with numbers. Simple index grid notations (AY, for example) direct readers to the correct portion of the map to find a particular street. Locations for parking, information offices, post offices, views, and points of interest are marked with compact symbols.

Signes particuliers		Special signs
Hôtel de ville	H	Town hall
Musée	M	Museum
Université	U	University
Office de tourisme		Tourist information
Bureau de poste		Post office
Parking	P	Car park
Hôpital		Hospital
Stade		Stadium

Pontevedra

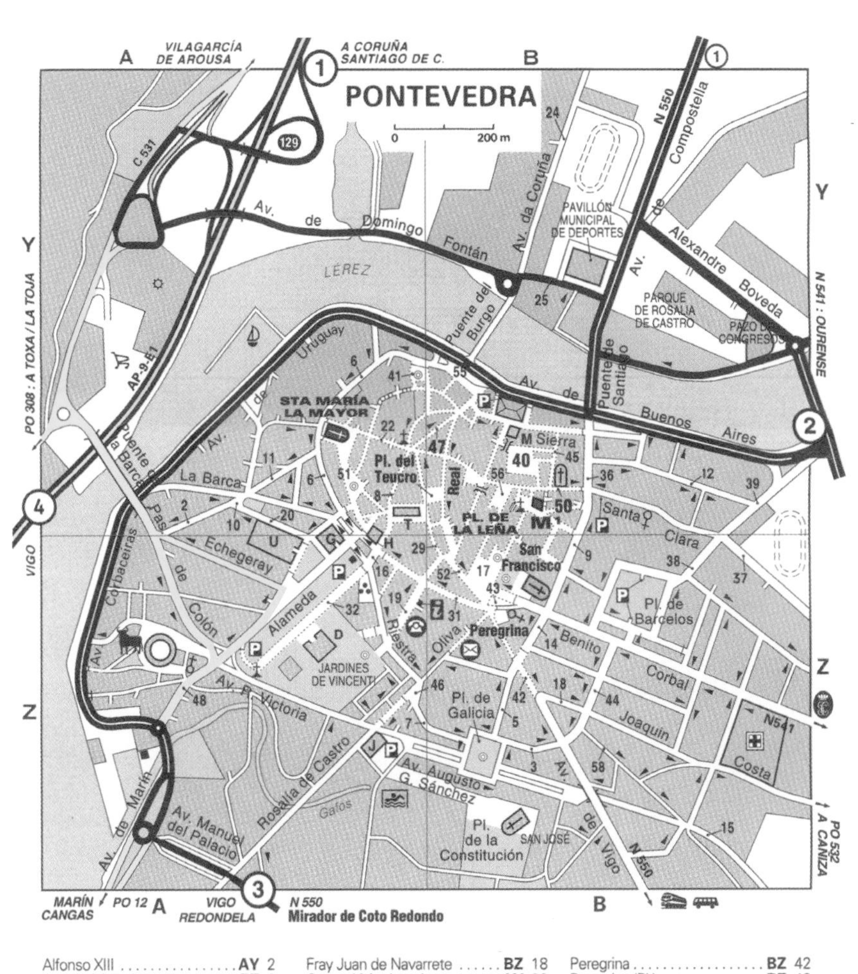

Vigo

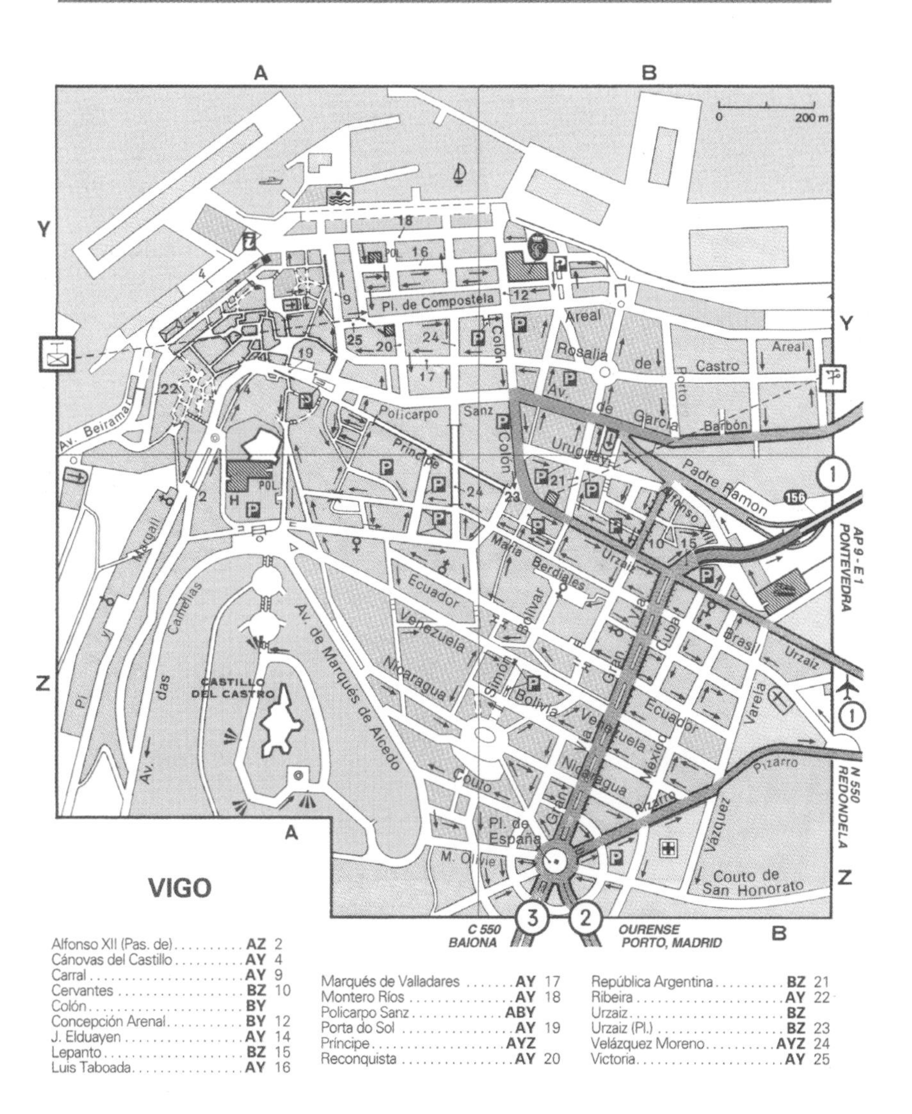

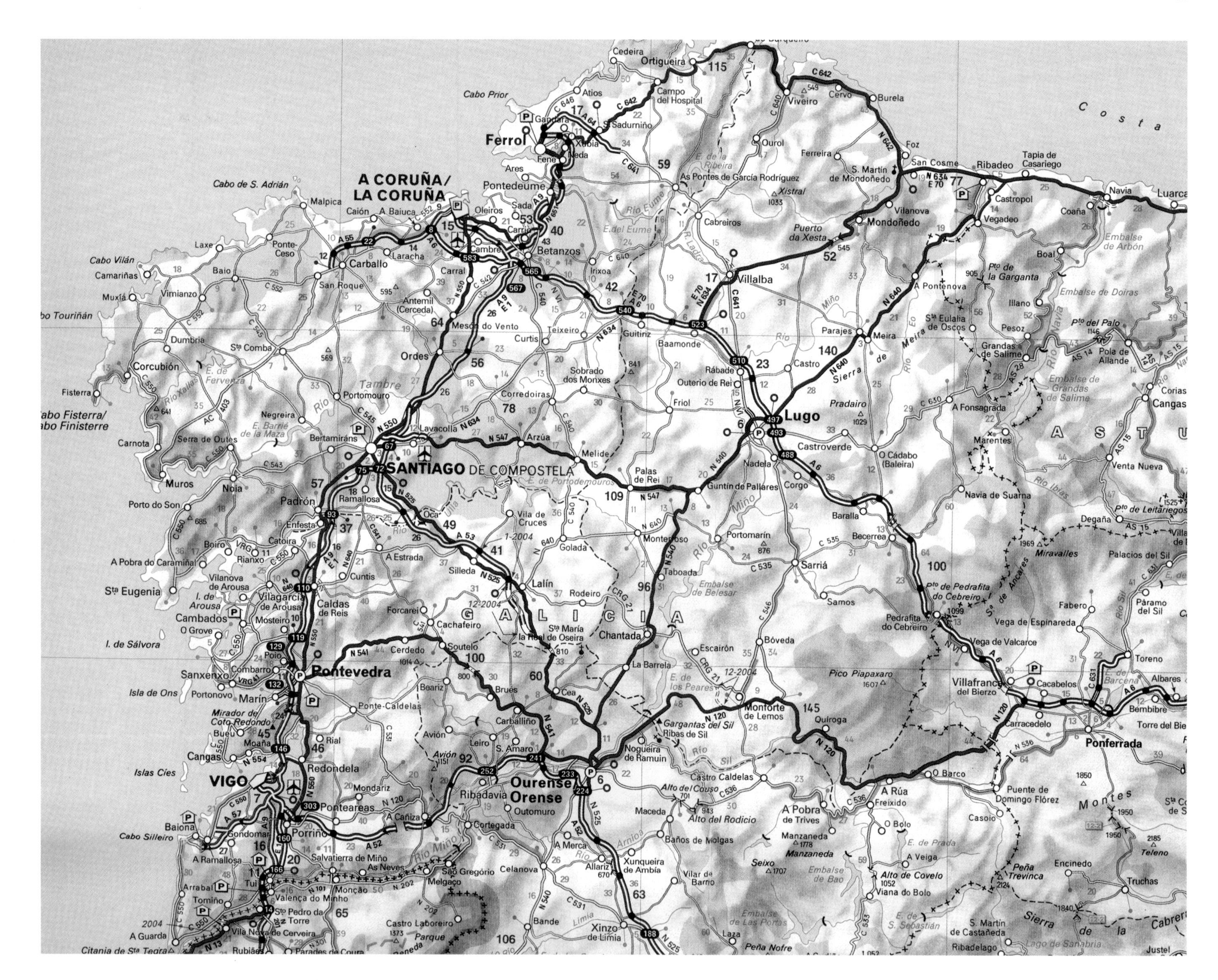
Cedeira
Ortigueira
Campo del Hospital
Cabo Prior
Atios
Ferrol
S. Sadurniño
Xubia
Fene
Neda
Ares
Pontedeume
Viveiro
Cervo
Burela
Ourol
Ferreira
S. Martín de Mondoñedo
Foz
San Cosme
Ribadeo
Tapia de Casariego
Costa
As Pontes de García Rodríguez
Xistral
E. de la Ribeira
Cabreiros
Puerto da Xesta
Mondoñedo
Vilanova
Castropol
Vegadeo
Navia
Luarca
Coaña
Embalse de Arbón
Boal
A CORUÑA/
LA CORUÑA
Cabo de S. Adrián
Malpica
Caión
A Baiuca
Oleiros
Sada
Cambre
Betanzos
Laxe
Ponte-Ceso
Carballo
Laracha
Carral
Irixoa
Cabo Vilán
Camariñas
Baio
San Roque
Antemil (Cerceda)
Muxía
Vimianzo
Villalba
Pto de la Garganta
A Pontenova
Embalse de Doiras
Illano
Tourinán
Dumbría
Sta Comba
Mesón do Vento
Ordes
Curtis
Teixeiro
Guitiriz
Baamonde
Parajes
Meira
Sta Eulalia de Oscos
Pesoz
Pto del Palo
Grandas de Salime
Pola de Allande
Corcubión
E. de Fervenza
Río Xallas
Tambre
Portomouro
Corredoiras
Sobrado dos Monxes
Rábade
Outerio de Rei
Castro
Sierra de Meira
Embalse de Grandas de Salime
Fisterra
Cabo Fisterra/
Cabo Finisterre
Negreira
E. Barrié de la Maza
Serra de Outes
Carnota
Bertamiráns
Lavacolla
Arzúa
Melide
Friol
Lugo
Pradairo
A Fonsagrada
Corias
Cangas
ASTU
SANTIAGO DE COMPOSTELA
E. de Portodemouros
Palas de Rei
Guntín de Pallares
Nadela
Castroverde
O Cádabo (Baleira)
Corgo
Marentes
Venta Nueva
Río Ibias
Muros
Noia
Porto do Son
Padrón
Ramallosa
Oca
Vila de Cruces
Enfesta
Navia de Suarna
Baralla
Pto de Leitariegos
Degaña
Boiro
Catoira
Rianxo
A Pobra do Caramiñal
A Estrada
Golada
Monterroso
Portomarín
Becerrea
Miravalles
Palacios del Sil
Silleda
Taboada
Sarriá
Cuntis
Vilanova de Arousa
Sta Eugenia
I. de Arousa
Vilagarcía de Arousa
Lalín
Rodeiro
Embalse de Belesar
Samos
Pto de Pedrafita do Cebreiro
Sa. de Ancares
Fabero
Páramo del Sil
Cambados
Caldas de Reis
Mosteiro
Forcarei
Cachafeiro
GALICIA
Sta María la Real de Oseira
Chantada
Pedrafita do Cebreiro
Vega de Espinareda
O Grove
I. de Sálvora
Cerdedo
Soutelo
Bóveda
Escairón
Vega de Valcarce
Toreno
Poio
Combarro
Pontevedra
La Barrela
Pico Piapaxaro
Sanxenxo
Isla de Ons
Portonovo
Marín
Beariz
Brués
Cea
E. de los Peares
Monforte de Lemos
Villafranca del Bierzo
Cacabelos
E. de Bárcena
Albares
Bembibre
Mirador de Coto Redondo
Bueu
Moaña
Cangas
Islas Cíes
Ponte-Caldelas
Avión
Carballiño
Gargantas del Sil
Ribas de Sil
Quiroga
Carracedelo
Torre del Bierzo
Ponferrada
Rial
Redondela
Leiro
S. Amaro
Nogueira de Ramuín
Río Sil
VIGO
Mondariz
Ribadavia
Ourense
Orense
Castro Caldelas
Alto del Couso
O Barco
Puente de Domingo Flórez
Montes
Baiona
Cabo Silleiro
Gondomar
Ponteareas
A Cañiza
Porriño
Cortegada
Outomuro
Maceda
Alto del Rodicio
A Pobra de Trives
A Rúa
Freixido
Casoio
A Ramallosa
Salvatierra de Miño
As Neves
Río Miño
São Gregório
Celanova
A Merca
Allariz
Xunqueira de Ambía
Baños de Molgas
Manzaneda
O Bolo
E. de Prada
A Veiga
Teleno
Arrabal
Tui
Monção
Melgaço
Vilar de Barrio
Seixo
Embalse de Bao
Alto de Covelo
Viana do Bolo
Peña Trevinca
Encinedo
Truchas
Tominõ
Valença do Minho
Sto Pedro da Torre
Castro Laboreiro
Bande
Xinzo de Limia
Embalse de Las Portas
E. de S. Sebastián
S. Martín de Castañeda
Sierra de la Cabrera
A Guarda
Vila Nova de Cerveira
Parque
Laza
Lago de Sanabria
Citania de Sta Tegra
Rubiães
Paredes de Coura
Peña Nofre
Ribadelago
Justel

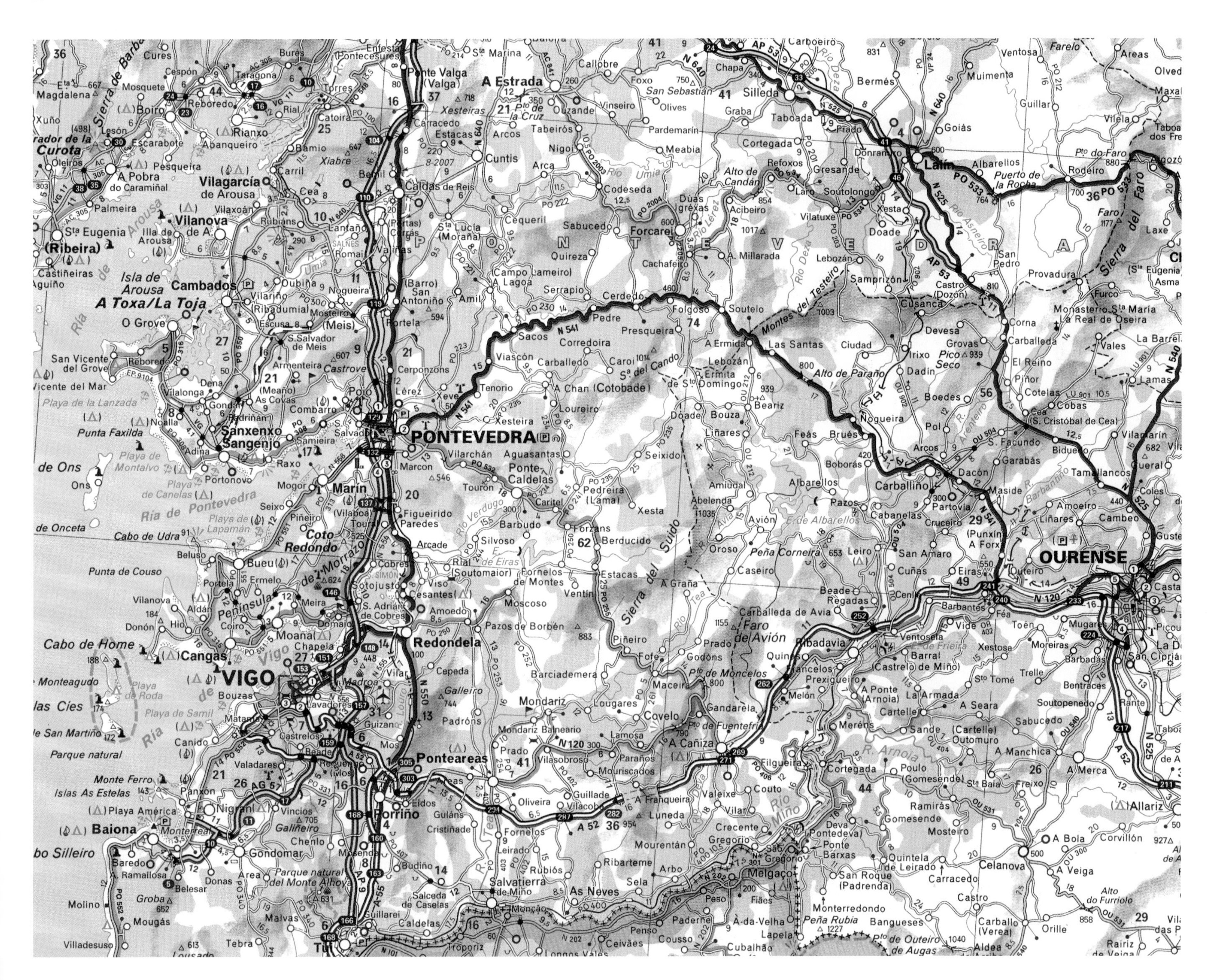
A Estrada
Silleda
Lalín
Cuntis
Caldas de Reis
Vilagarcía de Arousa
Vilanova de Arousa
Cambados
A Toxa/La Toja
O Grove
Isla de Arousa
Sta Eugenia (Ribeira)
A Pobra do Caramiñal
Boiro
Rianxo
Forcarei
Cerdedo
Pontevedra
Poio
Sanxenxo Sangenjo
Marín
Ría de Pontevedra
Bueu
Península de Morrazo
Moaña
Cangas
Vigo
Ría de Vigo
Redondela
Ponte Caldelas
Sierra del Suido
Ribadavia
Carballiño
Ourense
Mondariz
Ponteareas
Porriño
Nigrán
Baiona
Gondomar
Tui
Salvatierra de Miño
Melgaço
Celanova
Allariz
A Merca
Monte Ferro
Islas As Estelas
Cabo de Home
Islas Cíes
Parque natural
Cabo de Udra
Punta Faxilda
Coto Redondo
Faro de Avión
Sierra del Faro
Montes del Testeiro
Alto de Paraño
Río Miño
R. Arnoia
Río Umia
Río Deza
Río Lérez

2.6 SPAIN ROAD MAPS AT TWO SCALES (PREVIOUS PAGES)

Michelin road maps at two scales—1:1,000,000 for Spain (page 38; the smaller-scale map), and 1:400,000 for Galicia, Spain (page 39)—are shown here to emphasize the types of cartographic decisions made when changing scale. Many small towns and local roads are omitted from the smaller-scale map. Highway lines are simplified in shape, maintaining their most characteristic curves and also maintaining relative positions. A road is shown on the correct side of another road, river, or coast line, and these lines are often displaced to maintain those topological relationships. The roads near the town of Vigo are a good example of a simplified network where topology is maintained and smaller roads are omitted to accomplish that goal. The two major roads running north past Pontevedra are noticeably displaced at the smaller scale since the lines are shown as much farther apart on the landscape.

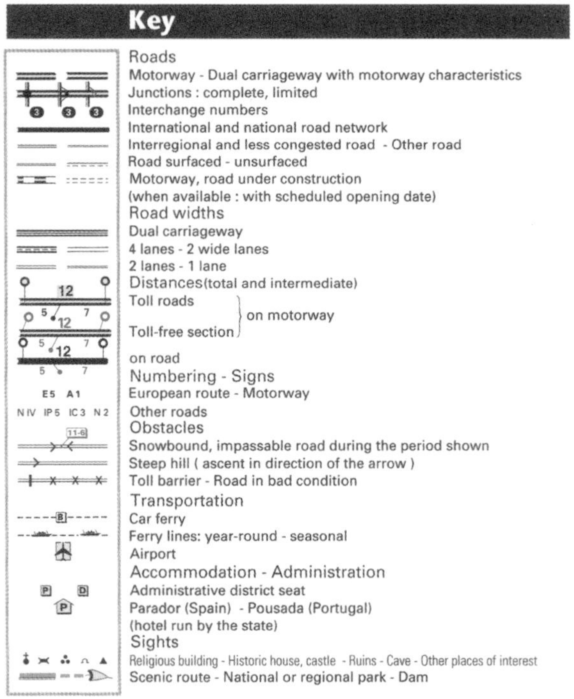

Key

Roads
Motorway - Dual carriageway with motorway characteristics
Junctions : complete, limited
Interchange numbers
International and national road network
Interregional and less congested road - Other road
Road surfaced - unsurfaced
Motorway, road under construction
(when available : with scheduled opening date)
Road widths
Dual carriageway
4 lanes - 2 wide lanes
2 lanes - 1 lane
Distances(total and intermediate)
Toll roads } on motorway
Toll-free section }
on road
Numbering - Signs
E5 A1 — European route - Motorway
N IV IP5 IC3 N2 — Other roads
Obstacles
Snowbound, impassable road during the period shown
Steep hill (ascent in direction of the arrow)
Toll barrier - Road in bad condition
Transportation
Car ferry
Ferry lines: year-round - seasonal
Airport
Accommodation - Administration
Administrative district seat
Parador (Spain) - Pousada (Portugal)
(hotel run by the state)
Sights
Religious building - Historic house, castle - Ruins - Cave - Other places of interest
Scenic route - National or regional park - Dam

See map on page 38.

Légende		Key
Routes	LA SAFOR	**Roads**
Autoroute - Aires de service		Motorway - Service areas
Double chaussée de type autoroutier		Dual carriageway with motorway characteristics
Échangeurs: complet - partiels		Interchanges : complete, limited
Numéros d'échangeurs	1 2 3	Interchange numbers
Route de liaison internationale ou nationale		International and national road network
Route de liaison interrégionale ou de dégagement		Interregional and less congested road
Route revêtue - non revêtue		Road surfaced - unsurfaced
Route en mauvais état		Road in bad condition
Chemin d'exploitation - Sentier		Rough track - Footpath
Autoroute - Route en construction		Motorway / Road under construction
(le cas échéant : date de mise en service prévue)		(when available : with scheduled opening date)
Largeur des routes		**Road widths**
Chaussées séparées		Dual carriageway
4 voies - 2 voies larges		4 lanes - 2 wide lanes
2 voies - 2 voies étroites		2 lanes - 1 lane
Distances (totalisées et partielles)		**Distances** (total and intermediate)
sur autoroute { Section à péage	14 10	Toll roads } on motorway
Section libre	14 24 10	Toll-free section }
sur route	14 24 10 24	on road
Numérotation - Signalisation		**Numbering - Signs**
Route européenne - Autoroute	E 25 A 4	European route - Motorway
Route nationale radiale - Route nationale	N IV N 301	National radial - National road
Autres routes	C 437 SE 138	Other roads
Obstacles		**Obstacles**
Forte déclivité (flèches dans le sens de la montée)	7-12% +12%	Steep hill (ascent in direction of the arrow)
Col - Altitude	915 (304)	Pass - Altitude
Parcours difficile ou dangereux		Difficult or dangerous section of road
Passages de la route: à niveau - supérieur - inférieur		Level crossing: railway passing, under road, over road
Route interdite - Route réglementée		Prohibited road - Road subject to restrictions
Barrière de péage - Route à sens unique		Toll barrier - One way road
Gué		Ford
Enneigement : période probable de fermeture	12-4	Snowbound, impassable road during the period shown
Transports		**Transportation**
Voie ferrée - Station voyageurs		Railway - Passenger station
Transport des autos:		Transportation of vehicles:
par bateau		by boat
par bac (charge maximum en tonnes)	15	by ferry (load limit in tons)
Bac pour piétons		Passenger ferry
Aéroport - Aérodrome		Airport - Airfield
Hébergement - Administration		**Accommodation-Administration**
Localité possédant un plan dans le Guide MICHELIN	1 2	Town plan featured in THE MICHELIN GUIDE
Parador (Espagne) - Pousada (Portugal) (établissement hôtelier géré par l'état)	P	Parador (Spain) - Pousada (Portugal) (hotel run by the state)
Capitale de division administrative	P D	Administrative district seat
Limites administratives		Administrative boundaries
Frontière		National boundary
Sports - Loisirs		**Sport & Recreation Facilities**
Arènes (plaza de toros) - Golf		Bullring - Golf course
efuge de montagne - Camping, caravaning		Mountain refuge hut - Caravan and camping sites
Port de plaisance - Plage		Pleasure boat harbour - Beach
Téléphérique, télésiège		Cable car, chairlift
Funiculaire - Voie à crémaillère		Funicular - Rack railway
Curiosités		**Sights**
Édifice religieux - Château - Ruines		Religious building - Historic house, castle - Ruins
Grotte - Monument mégalithique		Cave - Prehistoric monument
Autres curiosités		Other places of interest
Panorama - Point de vue		Panoramic view - Viewpoint
Parcours pittoresque		Scenic route
Signes divers		**Other signs**
Édifice religieux - Château - Ruines		Religious building - Castle - Ruins
Grotte - Monument mégalithique		Cave - Prehistoric monument
Transporteur industriel aérien		Industrial cable way
Tour ou pylône de télécommunications		Telecommunications tower or mast
Industries - Centrale électrique		Industrial activity - Power station
Raffinerie - Puits de pétrole ou de gaz		Refinery - Oil or gas well
Mine - Carrière		Mine - Quarry
Phare - Barrage		Lighthouse - Dam
Parc national - Réserve de chasse		National park - Game reserve

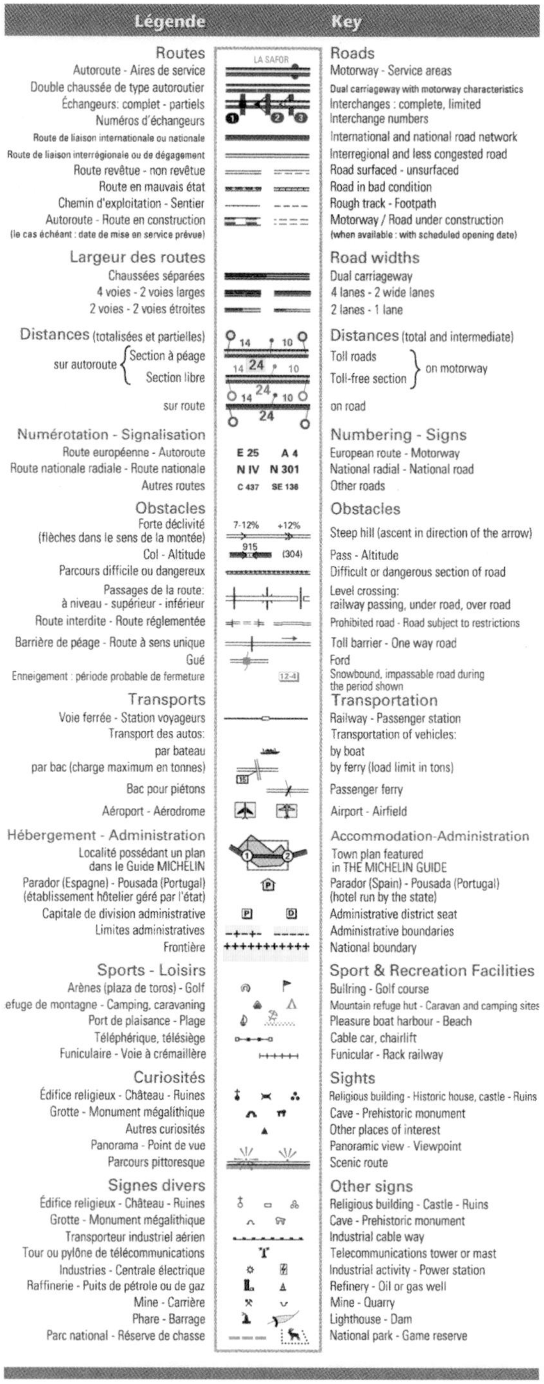

See map on page 39.

2.7 OREGON ROAD MAP (FOLLOWING PAGES)

The Benchmark Oregon map design includes expert labeling, with places categorized and ordered using differences in typeface, color hue, size, case, italics, boldness, lightness, and transparency. City and town labels are ordered by size, with larger cities also bold, and the largest cities labeled in uppercase letters. Physical features such as capes, peaks, and hydrography are labeled with italics. Points of interest, parks, recreation areas, and forests are all represented with a narrow font that varies in color hue for feature category and in type size for feature prominence. The map includes base information on landownership with transparent hues that overlay detailed but light terrain shading. Administrative area boundaries and labels are represented in brown, but differentiated by width of both lines and letters in a well-matched manner: extra-bold transparent labels designate state game management areas, which are outlined with transparent wide lines; regular type with spaced letters designates counties, which are outlined with thin dashed lines. Labels in areas dense with features arc in toward the named location rather than using leader lines. Tight halos ensure that each label is readable against a complex background.

Map image from the Oregon Recreation Map, copyright 2007 Benchmark Maps. Used by permission.

Legend

Limited Access Highway
Primary Through Highway
Secondary Through Road
Paved Road
Unpaved and 4WD Road
Pacific Crest Trail
Oregon Trail
25 Miles Between Markers
5 101 Interstate, U.S. Highway
62 N8 State, County Highway
State Boundary
County Boundary
LINN County Name
State Game Management Unit Boundary
HOOD 42 State Game Management Unit Name
State Capital, County Seat
Incorporated City, Unincorporated Town
Site/Locale
River
Stream
Intermittent Stream
Ocean, Lake
Intermittent Lake
Marsh
R Rest Area, Major Airport
Campground, RV Park
State Park/Wayside, Covered Bridge
Mountain Peak, Dam
Fishing, Boating
Ski Area, Snopark/Winter Recreation
Hiking, Rafting
Birdwatching, Zoo
Museum, Other Attraction
Lighthouse, Natural Wonder
Bureau of Land Management Land
U.S. Forest Service
Wilderness
Military Land
State Forest/Land
State Park
National Park/Monument/Recreation Area
Wildlife Area
Indian Land

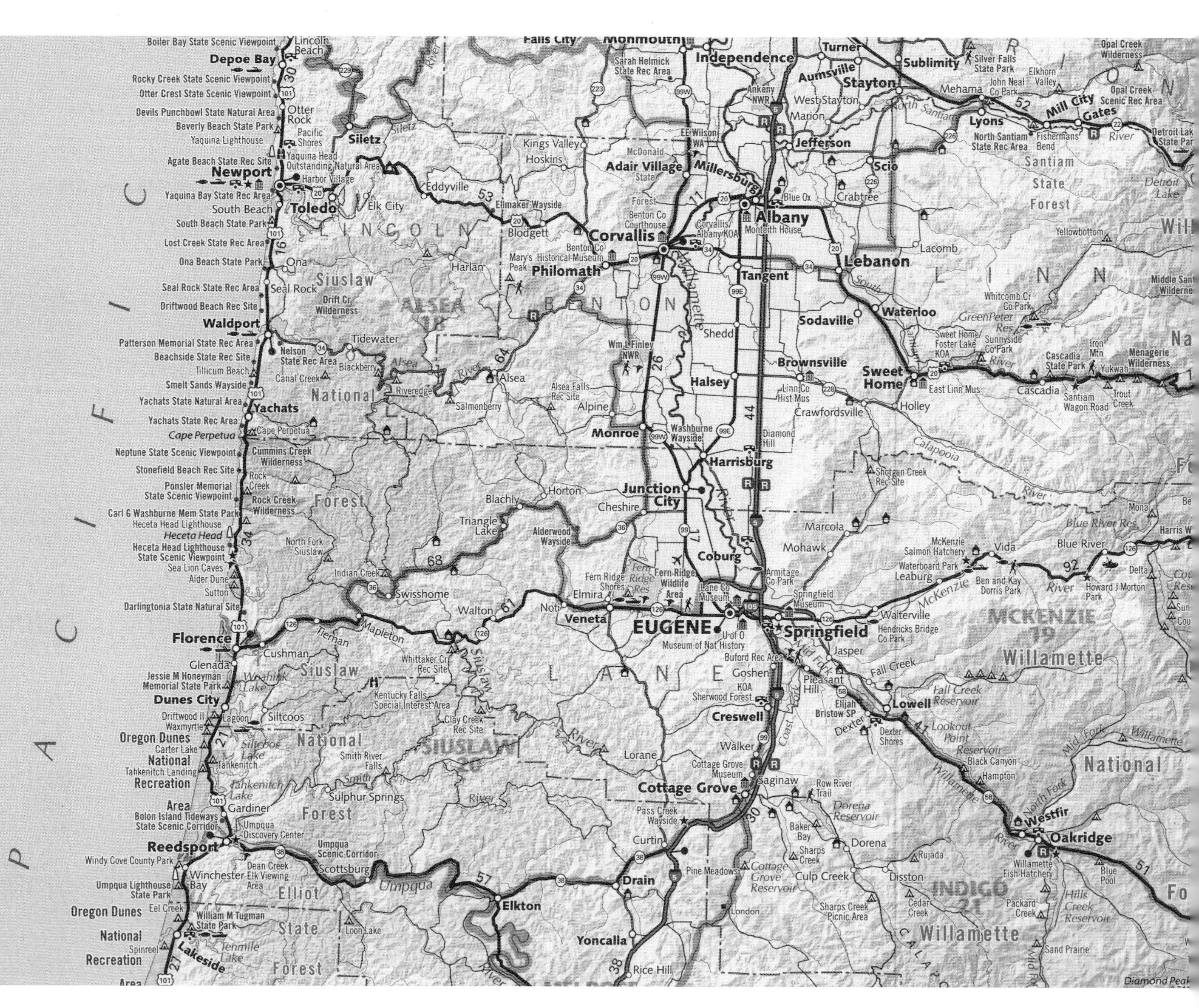
Boiler Bay State Scenic Viewpoint
Lincoln Beach
Depoe Bay
Rocky Creek State Scenic Viewpoint
Otter Crest State Scenic Viewpoint
Devils Punchbowl State Natural Area
Otter Rock
Beverly Beach State Park
Yaquina Lighthouse
Pacific Shores
Agate Beach State Rec Site
Yaquina Head Outstanding Natural Area
Newport
Harbor Village
Yaquina Bay State Rec Area
South Beach
South Beach State Park
Toledo
Lost Creek State Rec Area
Ona Beach State Park
Ona
Seal Rock State Rec Area
Seal Rock
Driftwood Beach Rec Site
Waldport
Patterson Memorial State Rec Area
Beachside State Rec Site
Tillicum Beach
Smelt Sands Wayside
Yachats State Natural Area
Yachats
Yachats State Rec Area
Cape Perpetua
Neptune State Scenic Viewpoint
Cummins Creek Wilderness
Stonefield Beach Rec Site
Rock Creek
Ponsler Memorial State Scenic Viewpoint
Rock Creek Wilderness
Carl G Washburne Mem State Park
Heceta Head Lighthouse
Heceta Head
Heceta Head Lighthouse State Scenic Viewpoint
Sea Lion Caves
Alder Dune
Sutton
Darlingtonia State Natural Site
Florence
Glenada
Jessie M Honeyman Memorial State Park
Dunes City
Driftwood II
Waxmyrtle
Lagoon
Siltcoos
Oregon Dunes National Recreation Area
Carter Lake
Tahkenitch
Tahkenitch Landing
Gardiner
Bolon Island Tideways State Scenic Corridor
Umpqua Discovery Center
Reedsport
Windy Cove County Park
Winchester Bay
Dean Creek Elk Viewing Area
Umpqua Lighthouse State Park
Eel Creek
William M Tugman State Park
Spinreel
Lakeside
P A C I F I C
Siletz
Eddyville
Elk City
Ellmaker Wayside
Blodgett
LINCOLN
Harlan
Siuslaw
Drift Cr Wilderness
ALSEA 18
Tidewater
Nelson State Rec Area
Blackberry
Canal Creek
Alsea
Riveredge
Salmonberry
National
Forest
North Fork Siuslaw
Indian Creek
Swisshome
Triangle Lake
Blachly
Horton
Alderwood Wayside
Walton
Noti
Elmira
Veneta
Cushman
Mapleton
Whittaker Cr Rec Site
Kentucky Falls Special Interest Area
Clay Creek Rec Site
SIUSLAW 20
Smith River Falls
Sulphur Springs
Umpqua Scenic Corridor
Scottsburg
Elliot State Forest
Loon Lake
Elkton
Drain
Yoncalla
Rice Hill
Falls City
Monmouth
Sarah Helmick State Rec Area
Independence
Kings Valley
Hoskins
McDonald
Adair Village
Forest
Benton Co Courthouse
Corvallis
Benton Co Historical Museum
Mary's Peak
Philomath
BENTON
Wm L Finley NWR
Alsea Falls Rec Site
Alpine
Monroe
Washburne Wayside
Harrisburg
Junction City
Cheshire
Coburg
Fern Ridge Shores
Fern Ridge Res
Fern Ridge Wildlife Area
Lane Co Museum
EUGENE
U of O Museum of Nat History
Springfield
Springfield Museum
Armitage Co Park
Buford Rec Area
Goshen
Sherwood Forest KOA
Creswell
Pleasant Hill
Lorane
LANE
Walker
Cottage Grove Museum
Cottage Grove
Saginaw
Pass Creek Wayside
Curtin
Pine Meadows
Cottage Grove Reservoir
London
Turner
Aumsville
Stayton
Ankeny NWR
West Stayton
Marion
E E Wilson WA
Jefferson
Millersburg
Scio
Blue Ox
Albany
Monteith House
Corvallis/Albany KOA
Crabtree
Tangent
Lebanon
Lacomb
Shedd
Sodaville
Waterloo
Brownsville
Halsey
Linn Co Hist Mus
Crawfordsville
Diamond Hill
Sweet Home
East Linn Mus
Holley
Sweet Home/Foster Lake KOA
Sunnyside Co Park
GreenPeter Res
Whitcomb Cr Co Park
Shotgun Creek Rec Site
Calapooia River
Marcola
Mohawk
Waterboard Park
Leaburg
McKenzie Salmon Hatchery
Vida
Ben and Kay Dorris Park
Walterville
Hendricks Bridge Co Park
Jasper
Fall Creek
Fall Creek Reservoir
Lowell
Elijah Bristow SP
Dexter
Dexter Shores
Lookout Point Reservoir
Black Canyon
Hampton
Row River Trail
Dorena Reservoir
Baker Bay
Sharps Creek
Dorena
Culp Creek
Disston
Sharps Creek Picnic Area
Cedar Creek
Sublimity
Silver Falls State Park
John Neal Co Park
Mehama
Elkhorn Valley
Lyons
Mill City
Gates
North Santiam State Rec Area
Fishermans Bend
Opal Creek Wilderness
Opal Creek Scenic Rec Area
Detroit Lake State Park
Detroit Lake
Santiam State Forest
Yellowbottom
LINN
Middle Santiam Wilderness
Iron Mtn
Menagerie Wilderness
Cascadia State Park
Yukwah
Cascadia
Santiam Wagon Road
Trout Creek
Mona
Blue River Res
Blue River
Delta
Howard J Morton Park
McKenzie River
MCKENZIE 19
Willamette National Forest
Westfir
Oakridge
Rujada
Willamette Fish Hatchery
Blue Pool
Packard Creek
Hills Creek Reservoir
INDIGO 21
Sand Prairie
Diamond Peak
Harris W

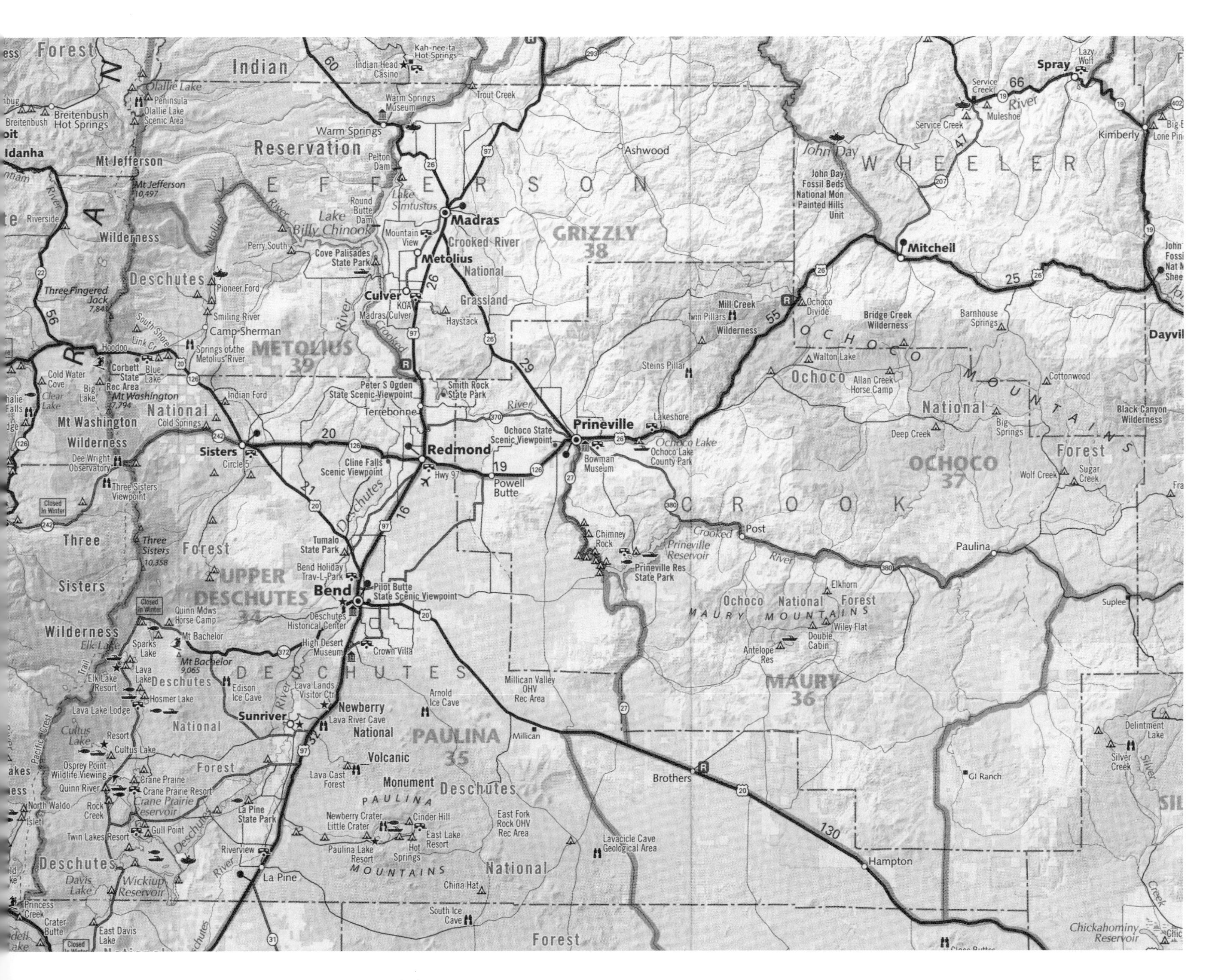
Indian
Reservation
Kah-nee-ta Hot Springs
Indian Head Casino
Warm Springs Museum
Trout Creek
Warm Springs
Pelton Dam
Olallie Lake
Peninsula
Olallie Lake Scenic Area
Breitenbush
Breitenbush Hot Springs
Idanha
Mt Jefferson
Mt Jefferson 10,497
JEFFERSON
Lake Simtustus
Round Butte Dam
Lake Billy Chinook
Madras
Mountain View
Metolius
Crooked River National Grassland
Cove Palisades State Park
Perry South
Culver
KOA
Madras/Culver
Haystack
Riverside
Wilderness
Deschutes
Pioneer Ford
Three Fingered Jack 7,841
Smiling River
Camp Sherman
Springs of the Metolius River
METOLIUS 39
Hoodoo
Corbett State Rec Area
Blue Lake
Cold Water Cove
Big Lake
Clear Lake
Mt Washington 7,794
Mt Washington Wilderness
National
Cold Springs
Indian Ford
Sisters
Circle 5
Peter S Ogden State Scenic Viewpoint
Smith Rock State Park
Terrebonne
Redmond
Cline Falls Scenic Viewpoint
Hwy 97
Powell Butte
Ochoco State Scenic Viewpoint
Prineville
Bowman Museum
Lakeshore
Ochoco Lake
Ochoco Lake County Park
Dee Wright Observatory
Three Sisters Viewpoint
Closed In Winter
Three
Sisters
Three Sisters 10,358
Forest
Tumalo State Park
Bend Holiday Trav-L-Park
Pilot Butte State Scenic Viewpoint
Bend
UPPER DESCHUTES 34
Quinn Mdws Horse Camp
Mt Bachelor
Mt Bachelor 9,065
Deschutes Historical Center
High Desert Museum
Crown Villa
Elk Lake
Sparks Lake
Lava Lake
Elk Lake Resort
Hosmer Lake
Edison Ice Cave
Lava Lands Visitor Ctr
Newberry
Lava River Cave
National
Volcanic
Monument
Sunriver
Lava Lake Lodge
Cultus Lake
Resort
Cultus Lake
Osprey Point Wildlife Viewing
Quinn River
Crane Prairie
Crane Prairie Resort
Crane Prairie Reservoir
North Waldo
Rock Creek
Islet
Twin Lakes Resort
Gull Point
La Pine State Park
Riverview
La Pine
Davis Lake
Wickiup Reservoir
Princess Creek
Crater Butte
East Davis Lake
Arnold Ice Cave
PAULINA 35
Lava Cast Forest
PAULINA MOUNTAINS
Newberry Crater
Little Crater
Cinder Hill
East Lake Resort
Paulina Lake Resort
Hot Springs
China Hat
South Ice Cave
National
Forest
Millican Valley OHV Rec Area
Millican
East Fork Rock OHV Rec Area
Lavacicle Cave Geological Area
Brothers
Hampton
Ashwood
GRIZZLY 38
Mill Creek
Twin Pillars
Wilderness
Steins Pillar
Ochoco Divide
Chimney Rock
Prineville Reservoir
Prineville Res State Park
Crooked River
Post
CROOK
Paulina
Elkhorn
Ochoco National Forest
MAURY MOUNTAINS
Wiley Flat
Double Cabin
Antelope Res
MAURY 36
John Day Fossil Beds National Mon Painted Hills Unit
John Day River
WHEELER
Service Creek
Mulesheo
Spray
Lazy Wolf
Kimberly
Lone Pine
Mitchell
Bridge Creek Wilderness
Barnhouse Springs
Walton Lake
OCHOCO MOUNTAINS
Ochoco
Allan Creek Horse Camp
Cottonwood
National
Deep Creek
Big Springs
Black Canyon Wilderness
Forest
OCHOCO 37
Wolf Creek
Sugar Creek
Dayville
Suplee
Delintment Lake
Silver Creek
GI Ranch
Chickahominy Reservoir
Pacific Crest Trail

2.8 WASHINGTON, D.C., VICINITY MAP

Smaller scale than many AAA maps, the D.C. vicinity map was produced to demonstrate versatility for multiscale production from AAA's database. The map is designed to emphasize town names adjacent to D.C., which are labeled with large, bold type. A network of major streets is shown outside the inset area that directs the reader to a larger scale map. Labels are categorized by using italics for points of interest and magenta italics for institutions such as universities and hospitals.

The map design facilitates the printing process. Most maps are printed using four process ink colors: cyan, magenta, yellow, and black (CMYK). By using these core colors and few combinations, this map minimizes the overprinting of inks needed to produce a wide range of hues. Hospital symbols and water features are in solid or screened cyan ink only. Place-names and major road lines are in magenta ink only. Gray minor roads and black place-names only require black ink. Green wooded areas and exit numbers are a combination of screened yellow and cyan, but yellow is such a light ink that errors in color registration do not compromise reading these symbols. This simple design would also allow the printer to remove a feature at the last minute at the printing plant and affect only one of the four separations used for CMYK printing.

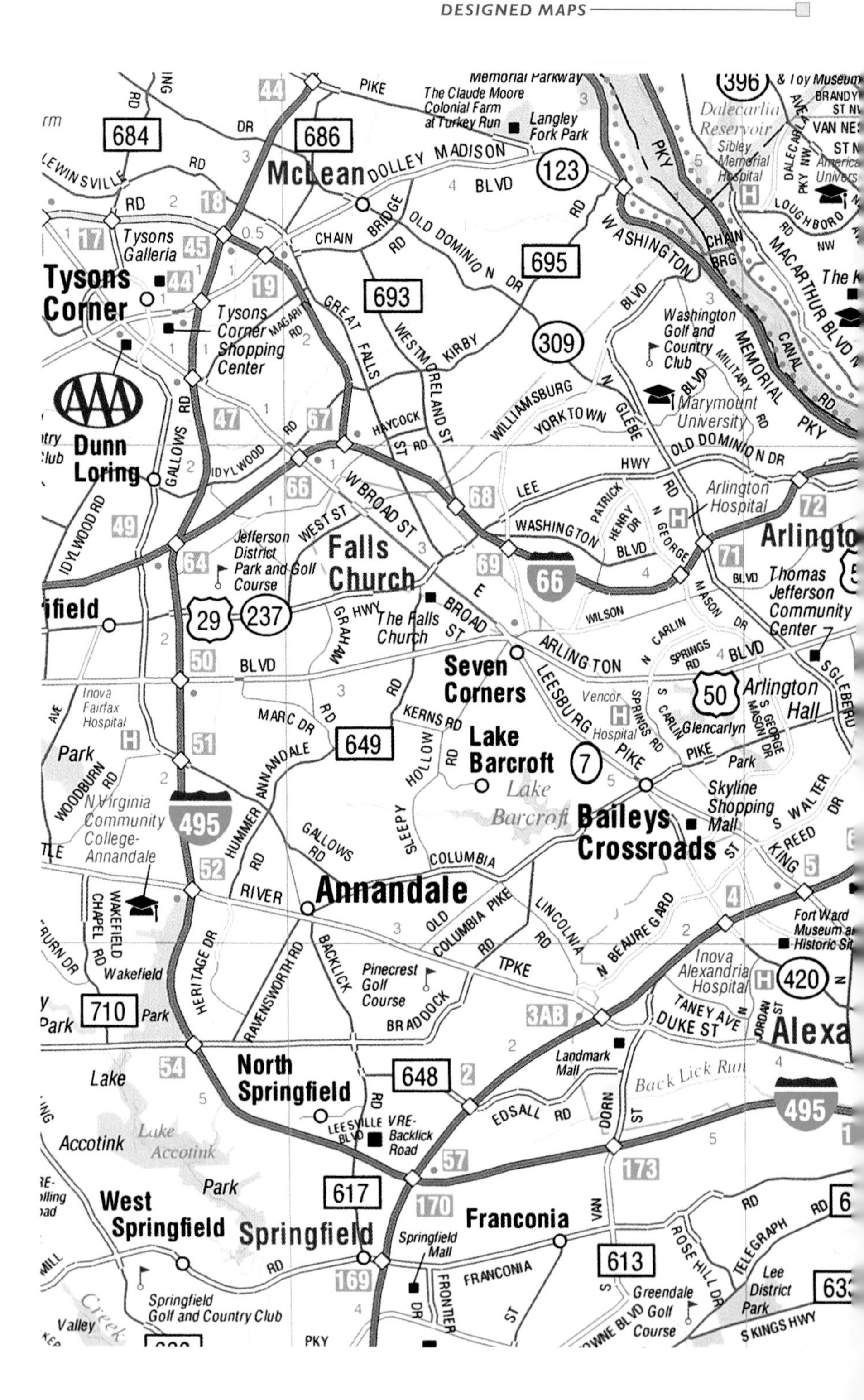

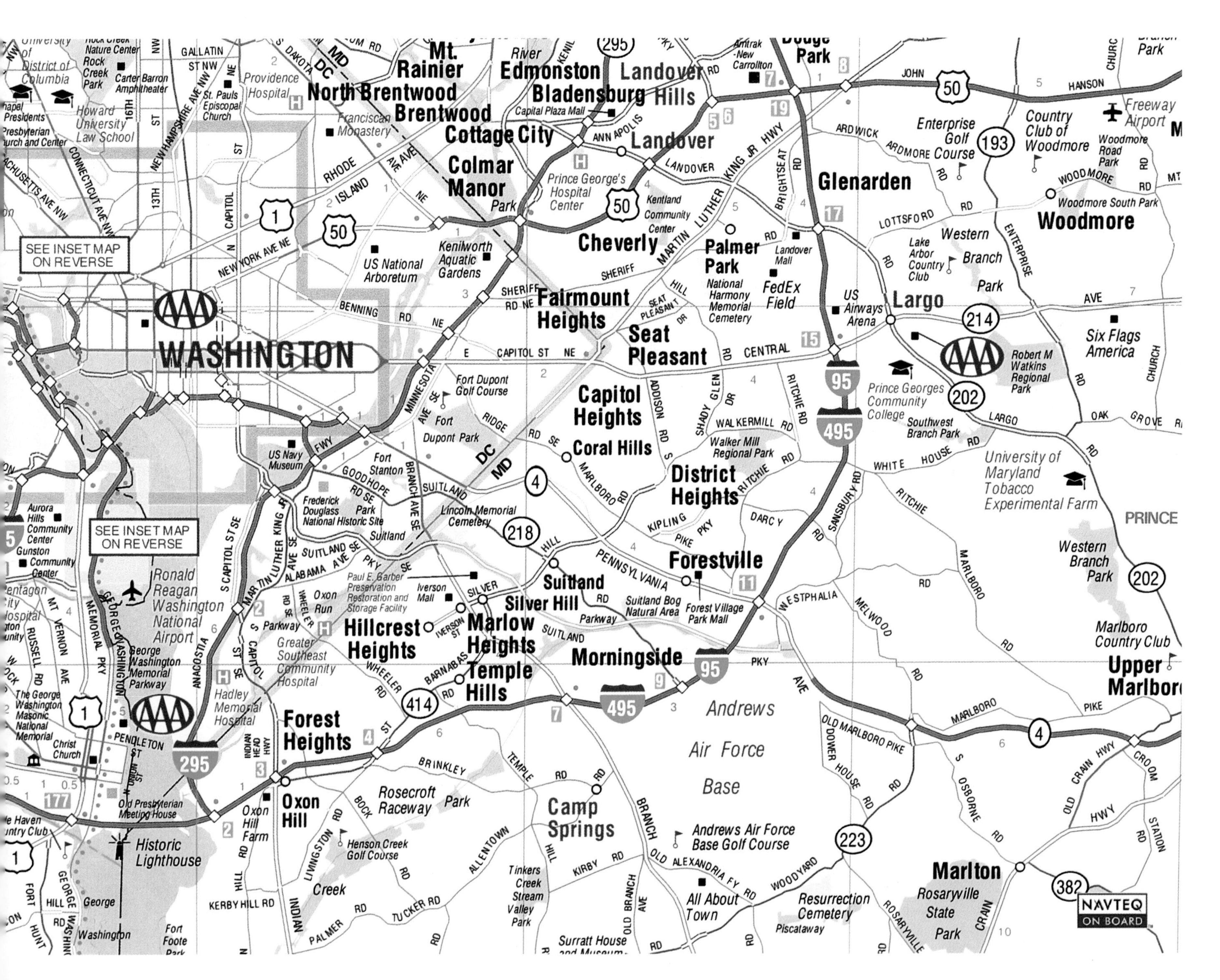
Rock Creek Nature Center
Carter Barron Amphitheater
Howard University
Howard University Law School
Providence Hospital
St. Pauls Episcopal Church
Franciscan Monastery
Mt. Rainier
North Brentwood
Brentwood
Cottage City
Colmar Manor
River
Edmonston
Bladensburg
Landover Hills
Landover
Capital Plaza Mall
Amtrak -New Carrollton
Dodge Park
Prince George's Hospital Center
Kentland Community Center
Cheverly
Palmer Park
Landover Mall
FedEx Field
National Harmony Memorial Cemetery
Glenarden
Enterprise Golf Course
Country Club of Woodmore
Freeway Airport
Woodmore Road Park
Woodmore South Park
Woodmore
Lake Arbor Country Club
Western Branch Park
Largo
US Airways Arena
Six Flags America
Robert M Watkins Regional Park
Prince Georges Community College
Southwest Branch Park
University of Maryland Tobacco Experimental Farm
PRINCE
Western Branch Park
Marlboro Country Club
Upper Marlboro
SEE INSET MAP ON REVERSE
US National Arboretum
Kenilworth Aquatic Gardens
WASHINGTON
Fairmount Heights
Seat Pleasant
Capitol Heights
Fort Dupont Golf Course
Fort Dupont Park
Coral Hills
District Heights
Walker Mill Regional Park
US Navy Museum
Fort Stanton
Frederick Douglass National Historic Site
Lincoln Memorial Cemetery
Suitland
Forestville
Suitland Bog Natural Area
Forest Village Park Mall
Paul E. Garber Preservation Restoration and Storage Facility
Iverson Mall
Oxon Run
Silver Hill
Hillcrest Heights
Marlow Heights
Temple Hills
Morningside
Greater Southeast Community Hospital
Hadley Memorial Hospital
Ronald Reagan Washington National Airport
George Washington Memorial Parkway
The George Washington Masonic National Memorial
Christ Church
Aurora Hills Community Center
Gunston Community Center
Forest Heights
Oxon Hill
Oxon Hill Farm
Old Presbyterian Meeting House
Historic Lighthouse
Rosecroft Raceway Park
Henson Creek Golf Course
Creek
Camp Springs
Andrews Air Force Base
Andrews Air Force Base Golf Course
Tinkers Creek Stream Valley Park
All About Town
Resurrection Cemetery
Piscataway
Marlton
Rosaryville State Park
Surratt House and Museum
George Washington
Fort Foote Park
NAVTEQ ON BOARD

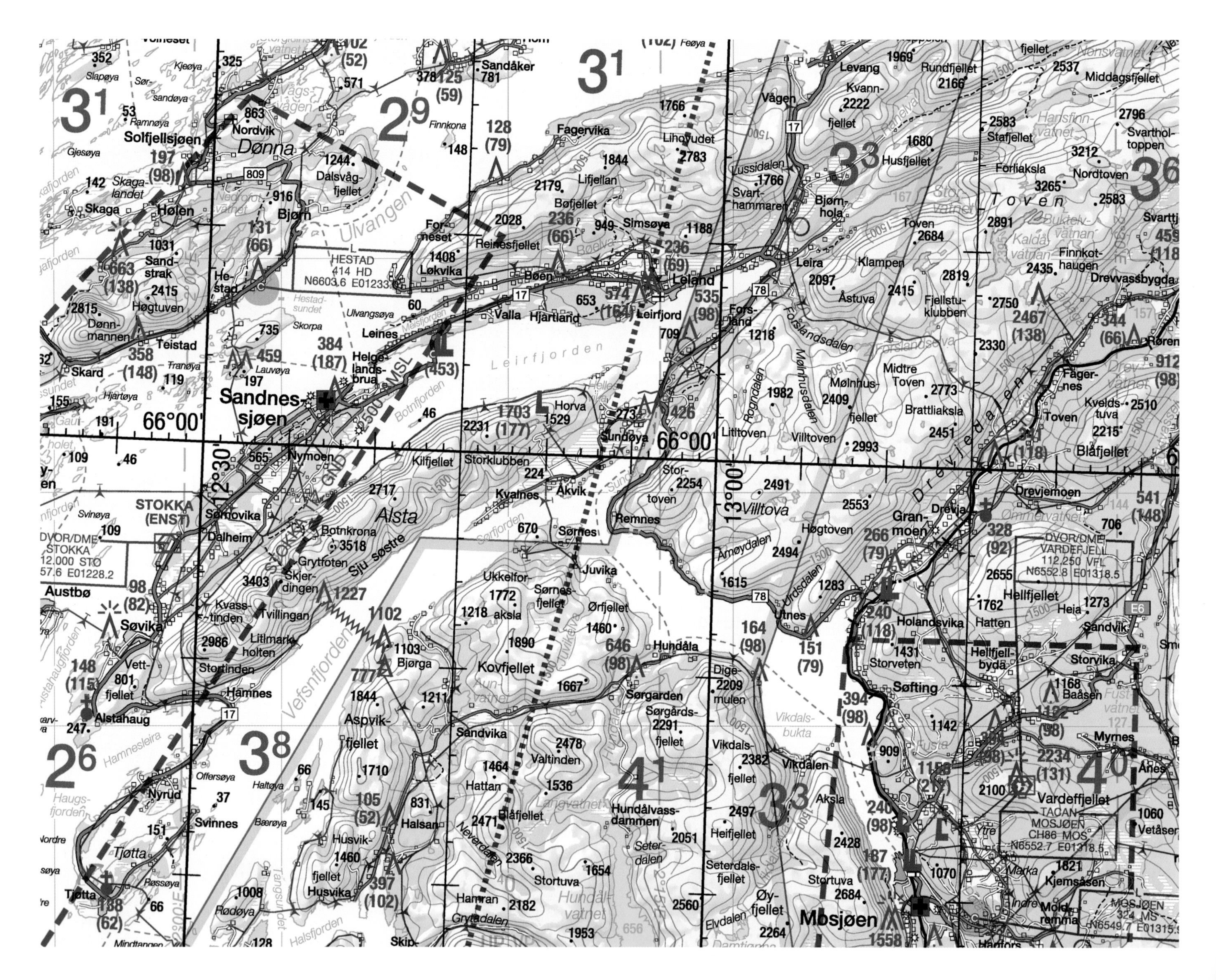

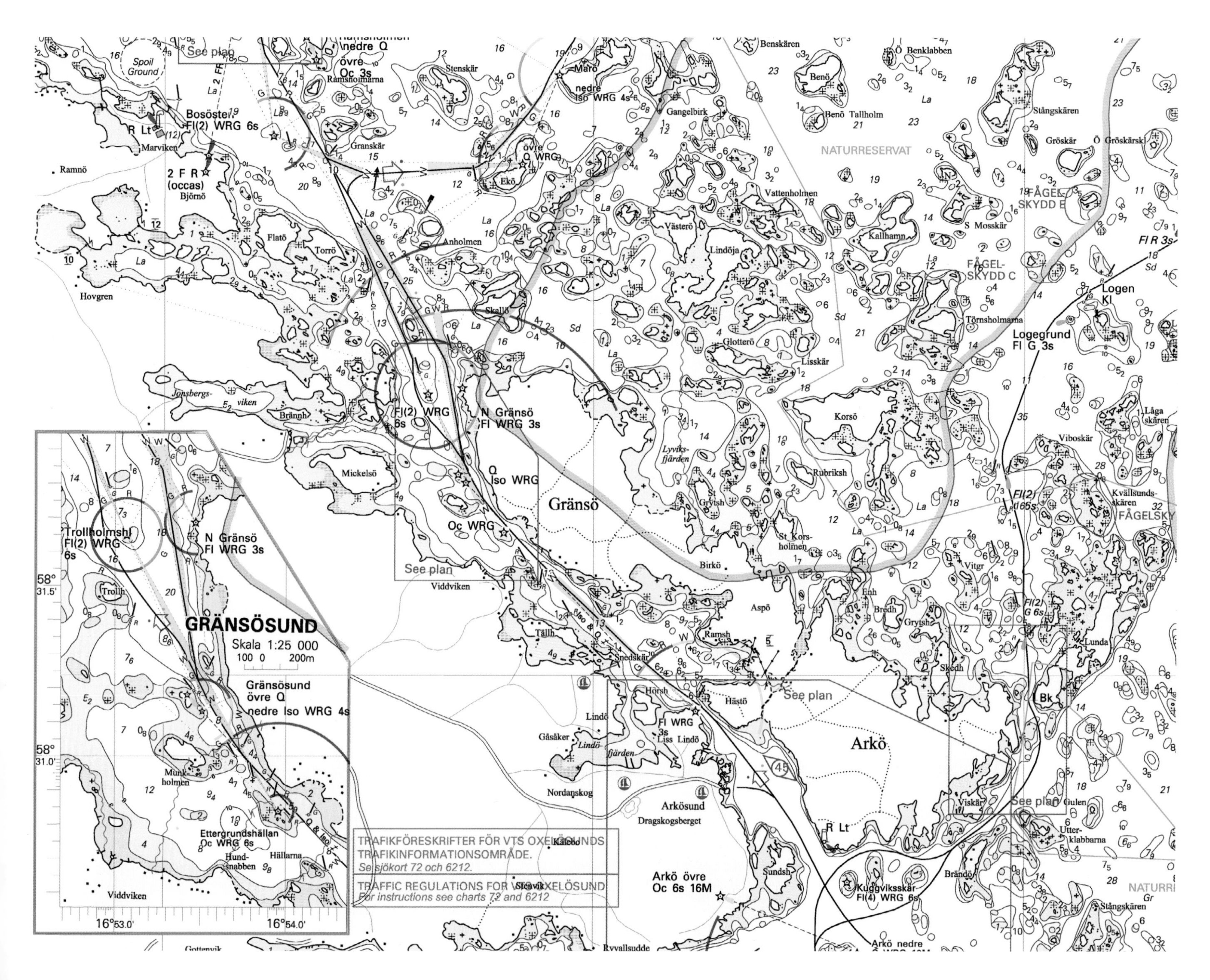

GRÄNSÖSUND
Skala 1:25 000
Gränsö
Arkö
TRAFIKFÖRESKRIFTER FÖR VTS OXELÖSUNDS TRAFIKINFORMATIONSOMRÅDE.
Se sjökort 72 och 6212.
TRAFFIC REGULATIONS FOR VTS OXELÖSUND
For instructions see charts 72 and 6212
See plan
NATURRESERVAT
FÅGELSKYDD C
Arkö övre Oc 6s 16M
N Gränsö Fl WRG 3s
Logegrund Fl G 3s
Arkösund

2.9 AND 2.10 NORWEGIAN AERONAUTICAL AND SWEDISH NAUTICAL CHARTS (PREVIOUS PAGES)

These maps are examples of special-purpose navigation charts. The pair show different selections of base information suited to the map purposes. The Norwegian aeronautical chart on page 46 emphasizes topographic and settlement detail that a pilot would see flying over an area. The 1:50,000-scale map is intended for low-flying aircraft, such as a helicopter on a rescue mission. The Swedish nautical chart on page 47 emphasizes bathymetric detail used by boat pilots with almost no detail in yellow land areas except generalized road and settlement information. Both maps show a bold overlay of navigation detail that assist in taking readings from navigation points, avoiding areas with navigation restrictions, avoiding barriers such as power lines or submerged rocks, and making precise evaluations of location. The aeronautical chart includes a detailed latitude–longitude grid and the nautical chart includes that detail at its edges (see the edge of the inset map).

Charts courtesy of T-Kartor Sweden AB.

3

SPECIAL-PURPOSE MAPS

VISITOR AND RECREATION

Visitor and recreation maps are akin to reference maps, but they emphasize subsets of reference information that support their special purposes. These maps are designed to guide map users with specific goals such as hiking or mountain biking on trails, sightseeing in a park, visiting a university, touring a city, or biking city streets. Visitor and recreation maps incorporate a strong visual hierarchy that downplays supporting information and emphasizes specific routes and amenities the visitor will most likely seek, rather than offering a wide-ranging mix of features better suited to the varied uses of general reference maps.

Some special-purpose maps emphasize particular routes and sites. For example, bike maps emphasize routes that are safest for cyclists on and off roads. The basemap is important for seeing how the cycling routes thread through the city streets, but the city streets are not presented in a way that best supports general navigation. Maps with routes through natural areas are supported by terrain information, and the maps display only the features appropriate to the typical user's needs. For instance, a scenic driving loop in red will stand out from other roads in gray, which may only be included because they provide needed access to the loop. City maps for visitors have more emphasis on built-up areas, building architecture, or building uses, and this detail may dominate and even obscure the base network of roads.

Special-purpose maps meet the challenge of displaying the overlap and mixture of showcased features and base information. Wide lines may be positioned underneath the regular network of roads to characterize particular roads as scenic or as suitable cycling routes. Different patterns set within a route line may show key information, such as hiking routes that switch from trail to road. Light residential building footprints will provide a context for the colorful buildings of interest for an urban area so they do not look like isolated features.

Clear color categorization and label styles assist the user in interpreting visitor and recreation maps and searching for features. When all hotels are yellow, that makes it easier to find them among the many buildings on a city map. The bold type of tourist site labels facilitates a user's search, and nearby regular labels help the map user know what else is in the vicinity of a destination.

The special-purpose maps gathered in this chapter mix general reference and thematic characteristics. They are not designed to show the general distribution of features in the category of interest, whether that is parking, camping, and ranger stations along a trail, or shopping and tourist destinations in a city. Rather, they are designed to emphasize these destinations within the local context so a map user can find them quickly and get to them easily.

3.0 PENN STATE CAMPUS MAP

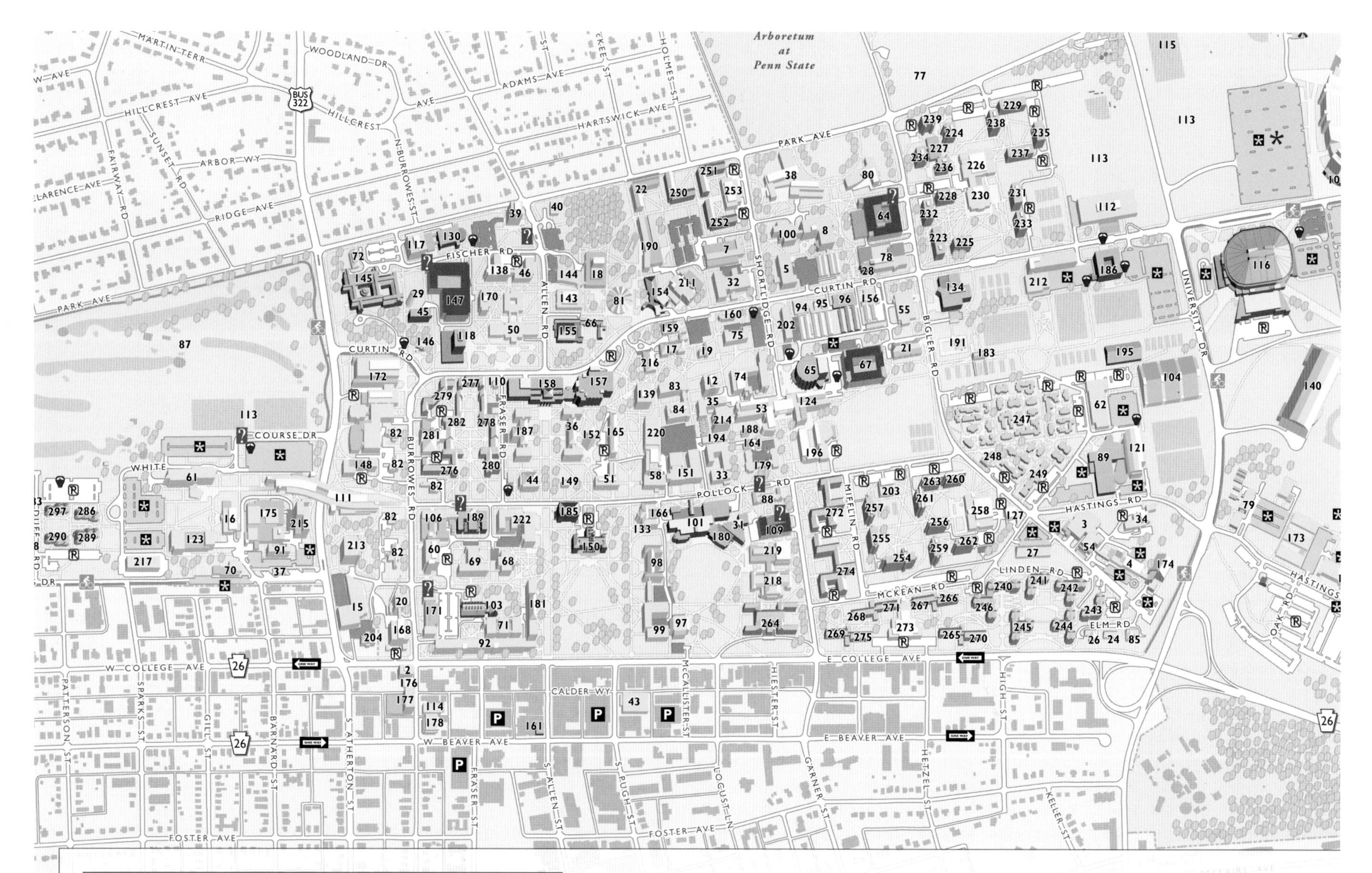

BUILDING INDEX

1 103 Building, The *N1*
2 300 Building, The *C6*
3 Academic Activities *F5*
4 Academic Projects *G5*
5 Agricultural Administration *E4*
6 Agricultural Arena *I2*
7 Agricultural Engineering *D4*

43 Calder Square II *D6*
44 Carnegie *C5*
45 Carpenter *C4*
46 CEDAR *C4*
47 Centralized Biological Lab *H6*
48 Central Milk Testing Lab *I1*
49 Centre County—Penn State

84 Frear South *D5*
85 Gardner House *G6*
86 General Stores *H2*
87 Golf Course (White) *A4*
88 Grange *E5*
89 Greenberg (Ice Pavilion) *G5*
90 Gypsy Moth Research Lab *I1*

126 Lubert *N1*
127 Maintenance I (Pollock) *F5*
128 Mascaro/Steiniger Turfgrass Equip. Museum *J1*
129 Mastitis Research Barn *J1*
130 Mateer *C4*
131 Materials Research Institute *N1*

165 Pond Lab *D5*
166 Post Office *D5*
167 Poultry Research and Education Center *F1*
168 Power Plant *C6*
169 Procurement Services *F2*
170 Rackley *C4*

206 Univ
207 USD
208 USD
209 Valer
210 Veter
211 Visua
212 Wag

3.0 PENN STATE CAMPUS MAP

VISITOR PARKING INFORMATION

With a visitor parking permit (see reverse for details):

Valid lot, 7:00 a.m. to 5:00 p.m., when using an appropri[...] visitor permit

Without a visitor parking permit (see reverse for de[...]

Free visitor parking after 5:00 p.m. and on weekends; RESERVED Staff or Student parking (7:00 a.m. to 5:00 [...]

Free visitor parking after 9:00 p.m., visitor permits valic[...] 4:00 p.m.; RESERVED Staff parking (7:00 a.m. to 9:00 p.m[...]

Paid visitor parking garage; Parking by permit only 2:00 a.m. to 4:00 a.m.

Parking meters are enforced from 7:00 a.m. to 9:00 p.m., M-F

No parking 2:00 a.m. to 4:00 a.m.

No parking at any time; LOT RESERVED 24 HOURS/DAY

State College Municipal Parking Garages

Points of Interest (see reverse for details)

Residence Halls

General Use Buildings

Information Booths

The Pennsylvania State University campus map combines small, generalized, three-dimensional buildings with detailed information on roads, parking, walkways, open spaces, and the surrounding town. Roads are represented with polygons that vary in width and shape throughout the map. The four map designs on the following pages primarily change the emphasis on the buildings and the detail selected for each version.

Courtesy of Erin Greb, Gould Center, Department of Geography, The Pennsylvania State University. Initial design by David DiBiase, Deasy GeoGraphics Lab.

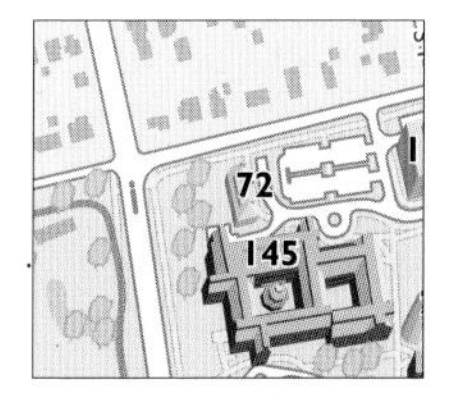

ORIGINAL DESIGN (Pages 50 and 52)

This version of the Penn State campus map, produced by the Department of Geography, uses three color hues to categorize buildings as points of interest (purple), residence halls (blue), and general-use buildings (teal). A second color sequence from dark to light orange-brown categorizes parking areas. Other campus features are symbolized with light browns, and generalized features of the area surrounding campus are rendered in light grays. Black capitalized labels stand out well against this detailed base.

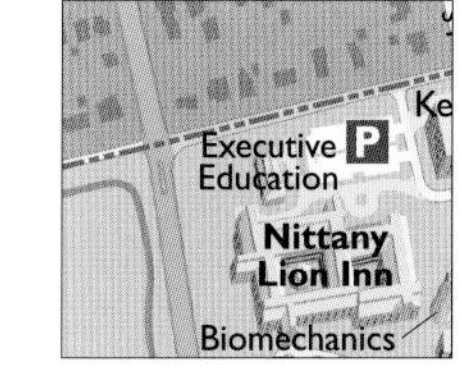

EMPHASIS ON NAMING (Page 53)

Going back and forth between numbers on buildings and the map index can be laborious for a map reader. This version of the map includes names for all buildings on the map. Placement is a challenge and labels are small. The map index is still needed to specify the cell in the index grid along the map edges (not shown) that map users rely on to find their destination buildings. Campus architecture has a coherent red-brick theme, and this map picks up that hue in its color choices. Parking has an opposite contrast than on the other map versions; parking lots stand out in white rather than dark colors.

EMPHASIS ON ONE COLLEGE (Page 52)

An advantage of a GIS-based campus map is the ease with which it can be tailored to the needs of an event or particular unit on the campus. This version of the map highlights buildings in which departments and institutes of the College of Earth and Mineral Sciences (EMS) are located. These buildings are named on the map and boldly colored. Other buildings are included for context but are pushed to the background using a series of light grays for shading. Visitor parking near college buildings is also emphasized.

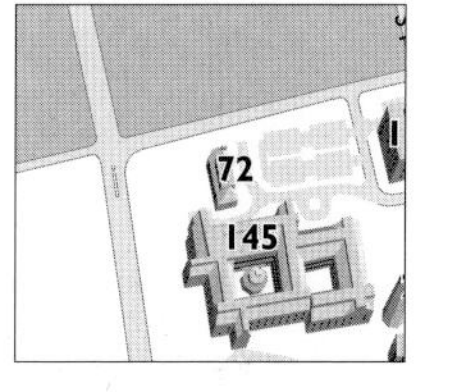

BLACK-AND-WHITE DESIGN (Page 53)

The black-and-white version of the map removes campus walkways and building footprints outside campus. The simple shading model, using light to dark gray for the buildings, brings them forward from the flat roads, parking, and town area. Fine halos on building numbers keep them readable against the buildings. The distinction between campus and town areas is bold, and road grays contrast with both of these light and dark areas.

REDESIGNS

3A ORIGINAL DESIGN

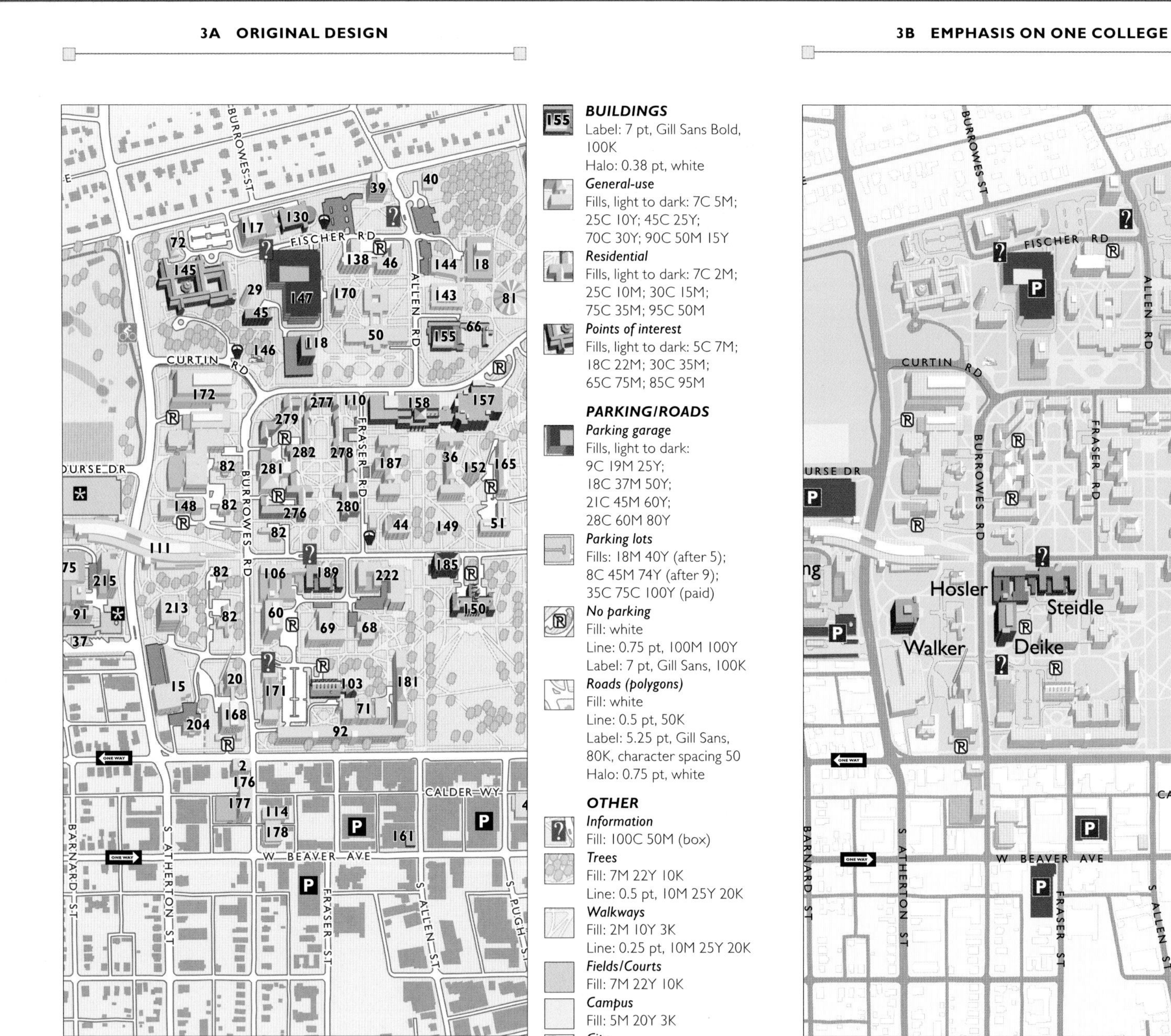

BUILDINGS
Label: 7 pt, Gill Sans Bold, 100K
Halo: 0.38 pt, white

General-use
Fills, light to dark: 7C 5M; 25C 10Y; 45C 25Y; 70C 30Y; 90C 50M 15Y

Residential
Fills, light to dark: 7C 2M; 25C 10M; 30C 15M; 75C 35M; 95C 50M

Points of interest
Fills, light to dark: 5C 7M; 18C 22M; 30C 35M; 65C 75M; 85C 95M

PARKING/ROADS

Parking garage
Fills, light to dark: 9C 19M 25Y; 18C 37M 50Y; 21C 45M 60Y; 28C 60M 80Y

Parking lots
Fills: 18M 40Y (after 5); 8C 45M 74Y (after 9); 35C 75C 100Y (paid)

No parking
Fill: white
Line: 0.75 pt, 100M 100Y
Label: 7 pt, Gill Sans, 100K

Roads (polygons)
Fill: white
Line: 0.5 pt, 50K
Label: 5.25 pt, Gill Sans, 80K, character spacing 50
Halo: 0.75 pt, white

OTHER

Information
Fill: 100C 50M (box)

Trees
Fill: 7M 22Y 10K
Line: 0.5 pt, 10M 25Y 20K

Walkways
Fill: 2M 10Y 3K
Line: 0.25 pt, 10M 25Y 20K

Fields/Courts
Fill: 7M 22Y 10K

Campus
Fill: 5M 20Y 3K

City
Fills: 28K (buildings); 2M 10Y 3K (blocks)

ArcMap Tips (see pages 161 to 167)

16 Random fill	30 Custom symbols
31 Embed font	

3B EMPHASIS ON ONE COLLEGE

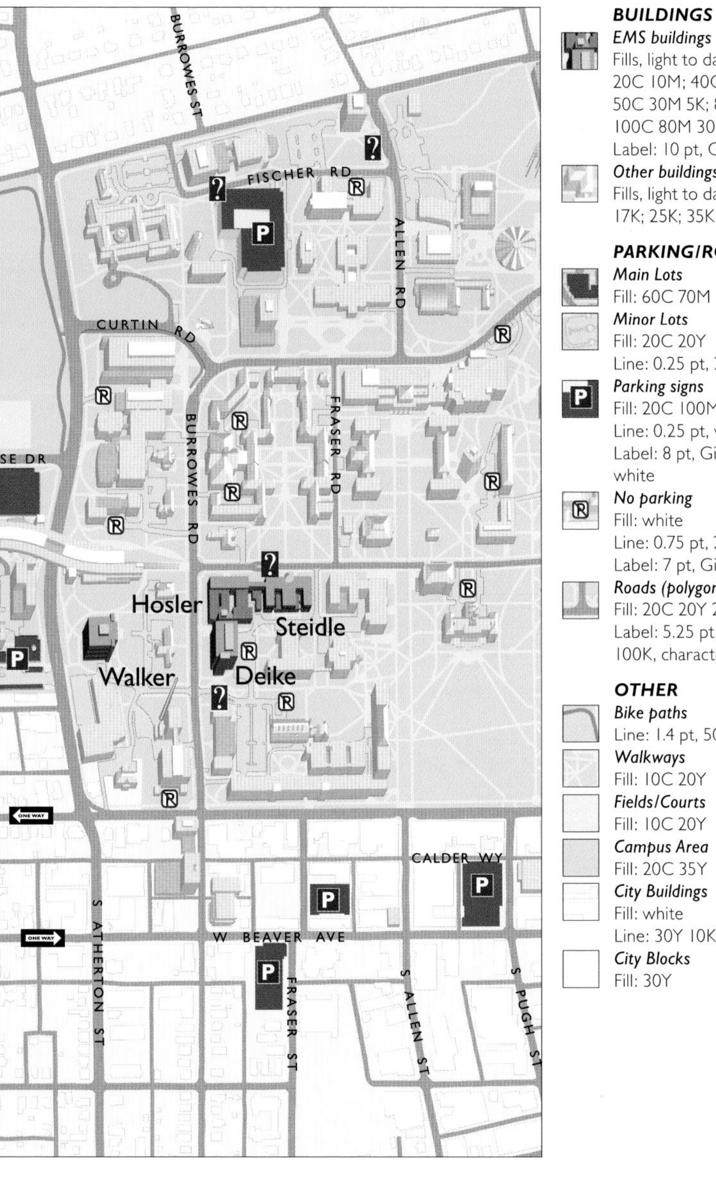

BUILDINGS

EMS buildings
Fills, light to dark: 20C 10M; 40C 20M; 50C 30M 5K; 80C 60M 15K; 100C 80M 30K
Label: 10 pt, Gill Sans, 100K

Other buildings
Fills, light to dark: 1K; 7K; 17K; 25K; 35K

PARKING/ROADS

Main Lots
Fill: 60C 70M

Minor Lots
Fill: 20C 20Y
Line: 0.25 pt, 20C 20Y 20K

Parking signs
Fill: 20C 100M
Line: 0.25 pt, white
Label: 8 pt, Gill Sans Bold, white

No parking
Fill: white
Line: 0.75 pt, 20C 100M
Label: 7 pt, Gill Sans, 100K

Roads (polygons)
Fill: 20C 20Y 20K
Label: 5.25 pt, Gill Sans, 100K, character spacing 50

OTHER

Bike paths
Line: 1.4 pt, 50C 20M 50Y

Walkways
Fill: 10C 20Y

Fields/Courts
Fill: 10C 20Y

Campus Area
Fill: 20C 35Y

City Buildings
Fill: white
Line: 30Y 10K

City Blocks
Fill: 30Y

ArcMap Tips

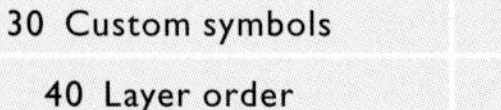

30 Custom symbols	35 Annotation
40 Layer order	

3C EMPHASIS ON NAMING

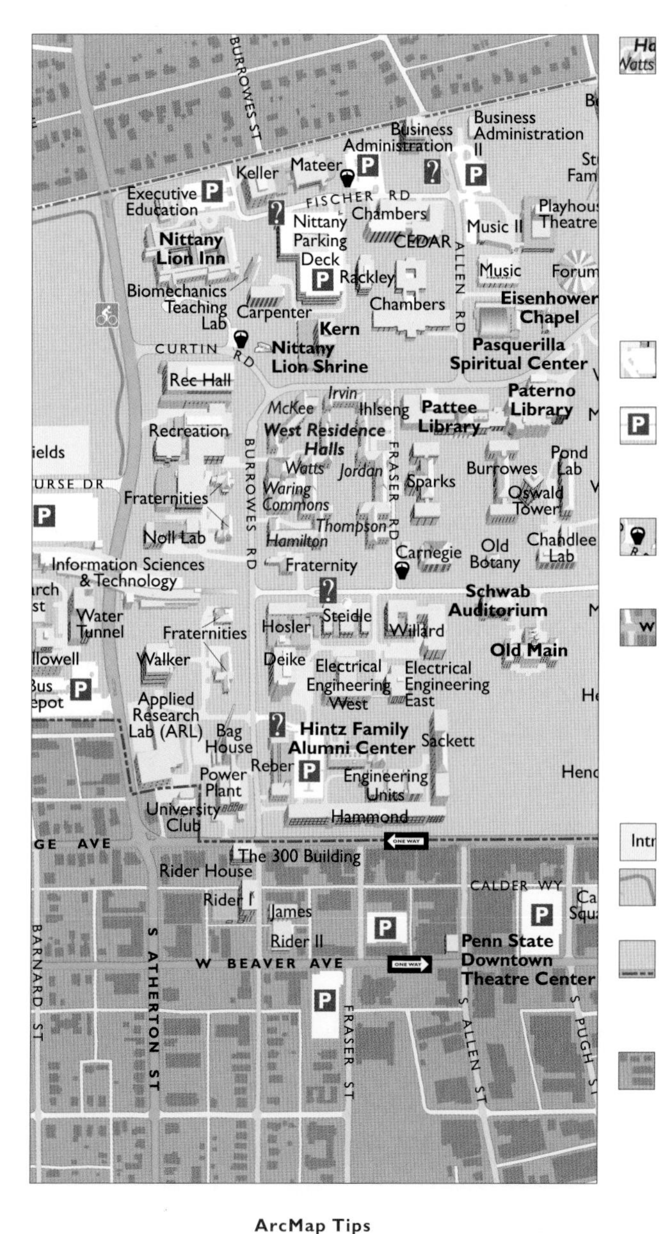

BUILDINGS
Fills, light to dark: 3M 3Y; 10M 10Y; 20M 20Y 5K; 20M 30Y 10K; 40M 40Y 20K
Line: 0.1 pt, 100M 100Y (fenestration)
Labels: 7 pt, Gill Sans Bold, 100K (emphasis); Gill Sans Italic and Bold Italic, 100C 70M 30K (residence); Gill Sans, 100K (other)
Leader line: 0.25 pt, 100K

PARKING/ROADS
Parking lots
Fills: white (recommended); 10C 20Y (other)
Parking signs
Fill: 80C 60M 10K
Line: 0.25 pt, white
Label: 8.5 pt, Gill Sans Bold, white
Parking meters
Fill: 100C 100M
Line: 0.35 pt, 100C 100M 30K
Roads (polygons)
Fills: 25M 100Y (main); 10M 50Y (secondary); 30Y (local)
Line: 0.25 pt, 40C 20M
Labels: 5.25 pt, Gill Sans and Gill Sans Bold, 100K, character spacing 50

OTHER
Fields/Courts
Fill: 10C 20Y
Bike paths
Line: 1.4 pt, 50C 10M 50Y 10K
Campus
Fill: 10C 20Y 10K
Line: 1 pt, 80C 60M 10K, dash pattern 6/1/2/1/2/1 *(dash/gap in pts)*
City
Fills: 40M 40Y 20K (buildings); 15M 15Y 20K (blocks)

ArcMap Tips

33 Leader lines	36 Convert to annotation
40 Layer order	

3D BLACK-AND-WHITE DESIGN

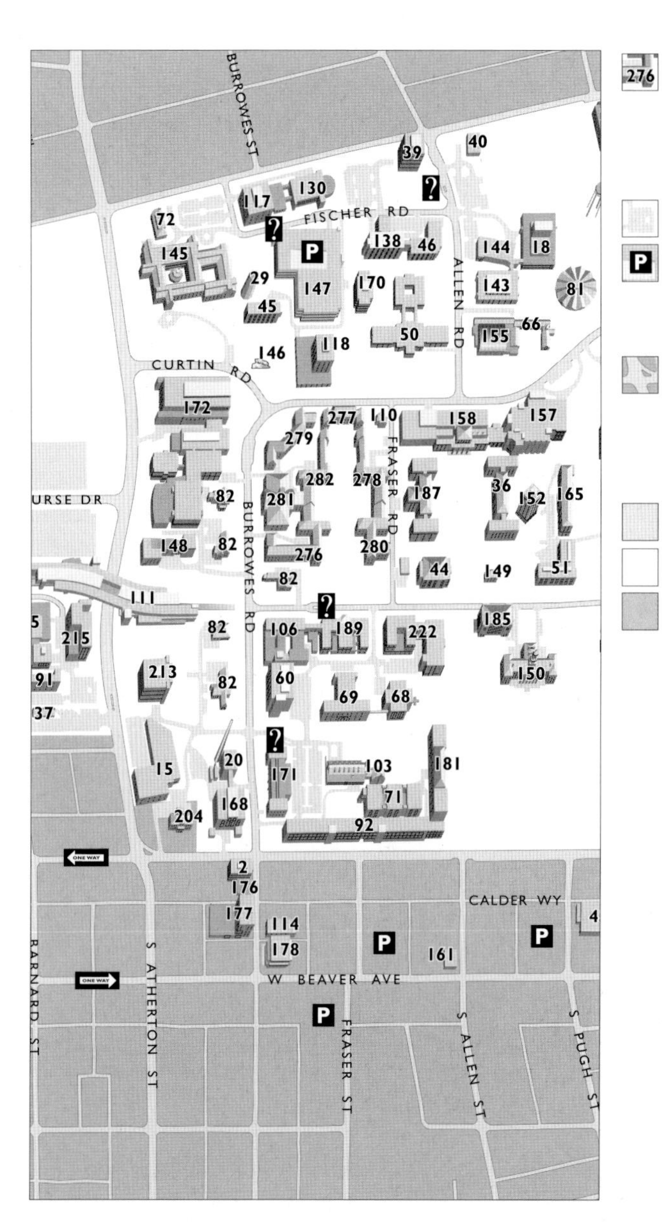

BUILDINGS
Fills, light to dark: 5K; 15K; 30K; 50K; 70K;
Label: 7 pt, Gill Sans, 100K
Halo: 0.38 pt, white

PARKING/ROADS
Parking lots
Fill: 10K
Parking signs
Fill: 100K
Line: 0.25 pt, white
Label: 8 pt, Gill Sans Bold, white
Roads (polygons)
Fill: 10K
Line: 0.25 pt, 50K
Label: 5.25 pt, Gill Sans, 100K, character spacing 50

OTHER
Fields/Courts
Fill: 10K
Campus
Fill: white
City
Fill: 25K

ArcMap Tips

30 Custom symbols	42 3D buildings

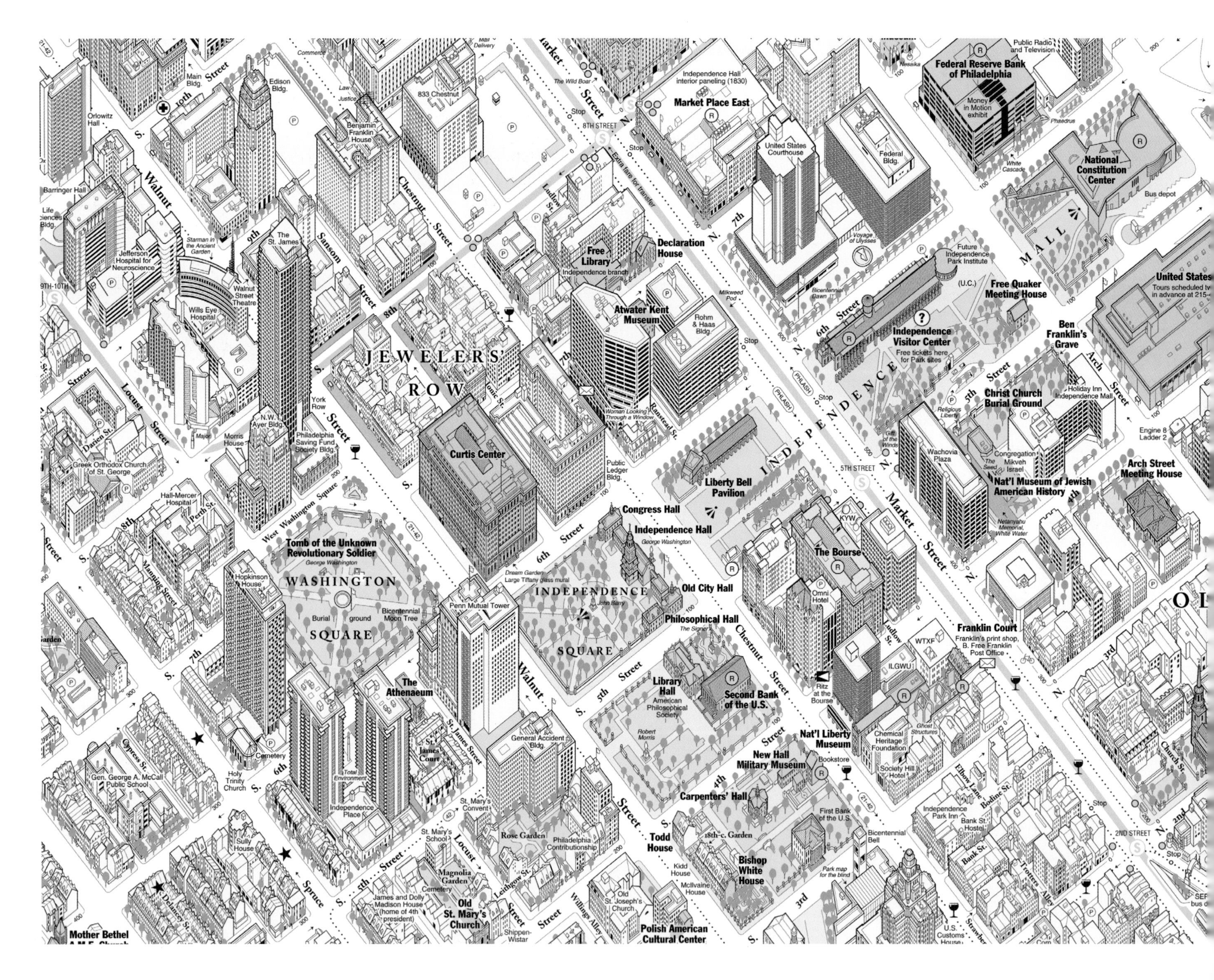
Federal Reserve Bank of Philadelphia
Money in Motion exhibit
Market Place East
Independence Hall interior paneling (1830)
United States Courthouse
Federal Bldg.
National Constitution Center
Bus depot
MALL
Public Radio and Television
833 Chestnut
Benjamin Franklin House
Edison Bldg.
Main Bldg.
Orlowitz Hall
Barringer Hall
Jefferson Hospital for Neuroscience
Wills Eye Hospital
Walnut Street Theatre
The St. James
Free Library
Independence branch
Declaration House
Atwater Kent Museum
Rohm & Haas Bldg.
Future Independence Park Institute
Free Quaker Meeting House
Independence Visitor Center
Free tickets here for Park sites
Ben Franklin's Grave
Christ Church Burial Ground
Holiday Inn Independence Mall
United States
JEWELERS' ROW
Curtis Center
York Row
N.W. Ayer Bldg.
Philadelphia Saving Fund Society Bldg.
Morris House
Greek Orthodox Church of St. George
Hall-Mercer Hospital
Public Ledger Bldg.
Liberty Bell Pavilion
INDEPENDENCE
Wachovia Plaza
Congregation Mikveh Israel
Nat'l Museum of Jewish American History
Arch Street Meeting House
Engine 8 Ladder 2
Congress Hall
Independence Hall
Tomb of the Unknown Revolutionary Soldier
WASHINGTON SQUARE
Burial ground
Bicentennial Moon Tree
Penn Mutual Tower
INDEPENDENCE SQUARE
Old City Hall
Philosophical Hall
The Bourse
Omni Hotel
Franklin Court
Franklin's print shop, B. Free Franklin Post Office
Hopkinson House
The Athenaeum
Library Hall
American Philosophical Society
Second Bank of the U.S.
Nat'l Liberty Museum
New Hall Military Museum
Carpenters' Hall
Chemical Heritage Foundation
Society Hill Hotel
General Accident Bldg.
St. James Court
Independence Place
St. Mary's Convent
Rose Garden
Philadelphia Contributionship
Todd House
18th-c. Garden
Bishop White House
First Bank of the U.S.
Bicentennial Bell
Independence Park Inn
Bank St. Hostel
Gen. George A. McCall Public School
Holy Trinity Church
Sully House
Magnolia Garden
James and Dolly Madison House (home of 4th president)
Old St. Mary's Church
Old St. Joseph's Church
Polish American Cultural Center
Kidd House
McIlvaine House
Mother Bethel
Walnut Street
Chestnut Street
Market Street
Locust Street
Spruce Street
Sansom Street
Arch Street
U.S. Customs House

3.1 3D PHILADELPHIA MAP

For this distinctive map the detailed drawings of city buildings in Philadephia, Pennsylvania, are composed using an equal-area axonometric projection that combines a planimetric view of the streets with a three-dimensional view of each building. Looking past the building detail, the grid of streets maintains square blocks and right angles but is tilted about 40 degrees from a horizontal/vertical grid alignment (the Philadelphia basemap has north angled to the right). Vertical building corners are aligned vertically on the page, rising out of the angled grid of blocks, and rooftops retain planimetrically correct shapes as well. This is not a perspective view with vanishing points that distort the basemap. The detail in fenestration and rooftops yields a view of the city that entices visitors to explore and provides them with many landmarks by which to navigate. Places open to the public are highlighted with desaturated colors and annotated with bold labels to draw map users' attention. Symbols for restaurant areas, viewpoints, and transit station entrances are also useful to tourists.

Courtesy of Bill Marsh, Marsh Maps.

3.2 HARVARD UNIVERSITY MAP

The large-scale map of the Harvard University campus and its Cambridge, Massachusetts, environs is rich in detail on building types, both on campus and in the business district. The scale is large enough that road shapes widen for more lanes and echo curved curb lines. The road hierarchy is dramatic, from black for main roads to gray for secondary streets. White labels are used where lines are wide enough to contain labels, which is most places. Halos on labels are applied very selectively, only when a label passes over a similar color, such as a black label over a black road or the dark red outline for the inset map area. Color categories for buildings are carefully chosen to be of similar lightness, with more saturated colors for places of particular interest to the public, such as theaters and museums on campus. In contrast, residential homes are pushed to the background with simple gray outlines that barely contrast with the yellow base. Labels for groups of buildings take curvilinear paths across their spaces, which separate them from straight building labels and provide a graceful indication of the extent of the areas. Label categories and hierarchies are fully structured by careful assignment of type size, color, font, style (bold and italic), and character spacing.

Academic Buildings
Museums, Arts, Etc.
Medical Buildings
Academic Housing
Parking Lot
Parking Structure
Public Parking P
Pedestrian Areas
Harvard Land
Other Campuses
Park Land
MUNICIPAL BOUNDARY
Hotels
Shops, Restaurants, Bars
Other Offices and Industry
Churches, Schools, Etc.
Residential
Historic Houses/Sites
Arterial Street
Secondary Street
One Way Street →
Limited Access Road
Path
T Subway (MBTA)* Entrance
Z Zipcar Parking**
Scale: 10.5 inches = about 1 mile (1:6000)
0
1/8 mile
1/4 mile
0
500 feet
1,000 feet
*Massachusetts Bay Transportation Authority **See note on back under "Getting Around Boston"

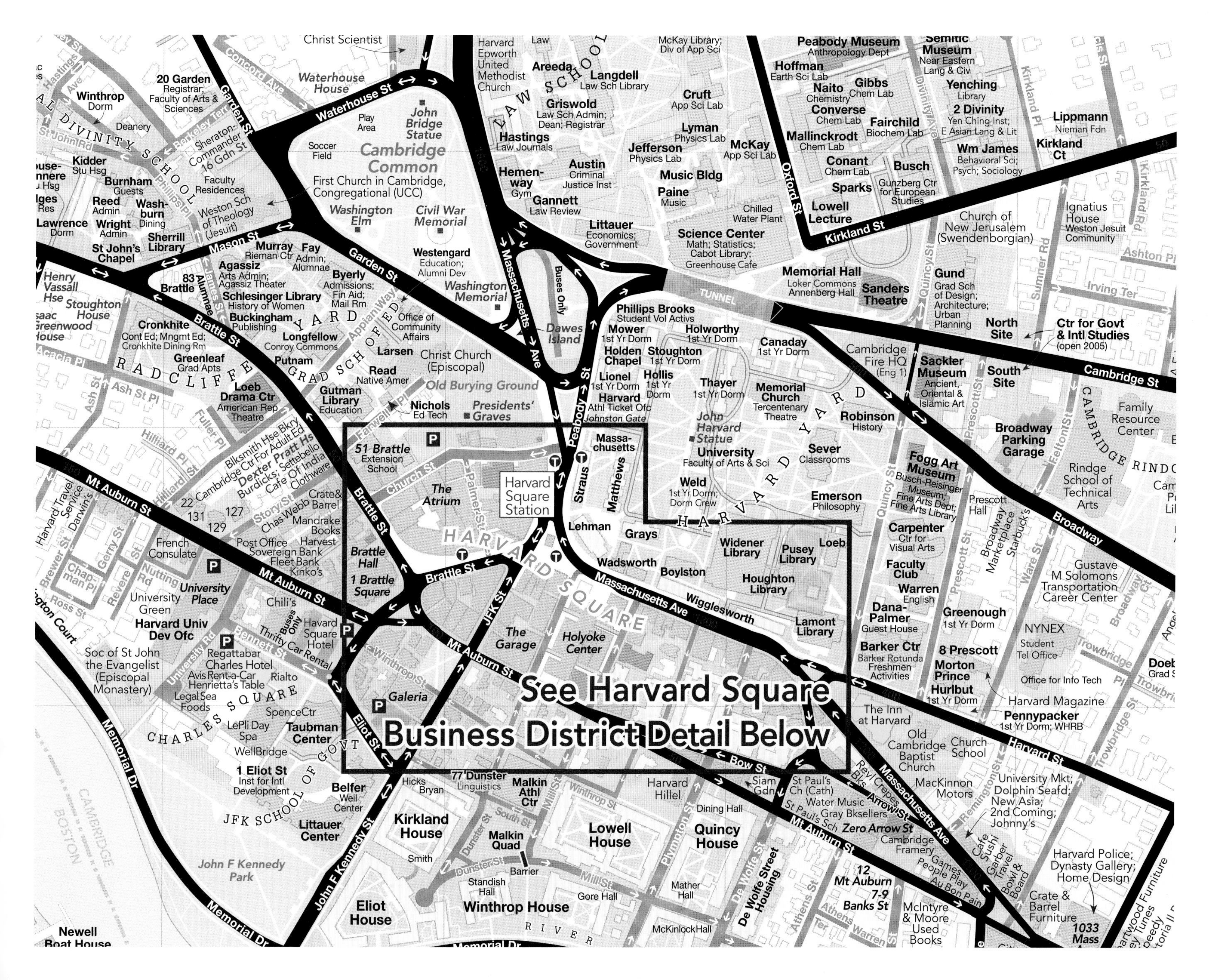
Christ Scientist
Waterhouse House
Waterhouse St
20 Garden
Registrar; Faculty of Arts & Sciences
Winthrop Dorm
Deanery
Kidder
Stu Hsg
Burnham
Guests
Reed
Admin
Wash-burn
Dining
Lawrence
Dorm
Wright
Admin
St John's Chapel
Sherrill Library
Sheraton-Commander
16 Gdn St
Faculty Residences
Weston Sch of Theology (Jesuit)
Concord Ave
Garden St
Berkeley Ter
Phillips Pl
Mason St
Play Area
Soccer Field
John Bridge Statue
Cambridge Common
First Church in Cambridge, Congregational (UCC)
Washington Elm
Civil War Memorial
Westengard
Education; Alumni Dev
Washington Memorial
Murray
Rieman Ctr
Fay
Admin; Alumnae
Agassiz
Arts Admin; Agassiz Theater
83 Brattle
Byerly
Admissions; Fin Aid; Mail Rm
Schlesinger Library
History of Women
Buckingham
Publishing
Longfellow
Conroy Commons
Office of Community Affairs
Larsen
Christ Church (Episcopal)
Putnam
Read
Native Amer
Old Burying Ground
Gutman Library
Education
Nichols
Ed Tech
Presidents' Graves
Henry Vassall Hse
Stoughton House
Cronkhite
Cont Ed; Mngmt Ed; Cronkhite Dining Rm
Greenleaf
Grad Apts
Loeb Drama Ctr
American Rep Theatre
Brattle St
James St
Alumnae
Appian Way
Farwell Pl
Ash St Pl
Ash St
Fuller Pl
Hilliard Pl
Hilliard St
Acacia Pl
RADCLIFFE YARD
GRAD SCH OF ED
Harvard Epworth United Methodist Church
Law
Areeda
Langdell
Law Sch Library
Griswold
Law Sch Admin; Dean; Registrar
Hastings
Law Journals
Hemen-way
Gym
Austin
Criminal Justice Inst
Gannett
Law Review
Littauer
Economics; Government
LAW SCHOOL
McKay Library; Div of App Sci
Cruft
App Sci Lab
Lyman
Physics Lab
Jefferson
Physics Lab
McKay
App Sci Lab
Music Bldg
Paine
Music
Chilled Water Plant
Science Center
Math; Statistics; Cabot Library; Greenhouse Cafe
Massachusetts Ave
Buses Only
Dawes Island
Peabody St
Oxford St
Peabody Museum
Anthropology Dept
Hoffman
Earth Sci Lab
Naito
Chemistry
Gibbs
Chem Lab
Converse
Chem Lab
Mallinckrodt
Chem Lab
Fairchild
Biochem Lab
Conant
Chem Lab
Busch
Sparks
Gunzberg Ctr for European Studies
Lowell Lecture
Kirkland St
Semitic Museum
Near Eastern Lang & Civ
Yenching
Library
2 Divinity
Yen Ching Inst; E Asian Lang & Lit
Wm James
Behavioral Sci; Psych; Sociology
Divinity Ave
Kirkland Pl
Lippmann
Nieman Fdn
Kirkland Ct
Ignatius House
Weston Jesuit Community
Church of New Jerusalem (Swendenborgian)
Memorial Hall
Loker Commons Annenberg Hall
Sanders Theatre
TUNNEL
Gund
Grad Sch of Design; Architecture; Urban Planning
North Site
South Site
Ctr for Govt & Intl Studies
(open 2005)
Cambridge St
Irving Ter
Ashton Pl
Sumner Rd
Quincy St
Phillips Brooks
Student Vol Activs
Mower
1st Yr Dorm
Holworthy
1st Yr Dorm
Holden Chapel
Stoughton
1st Yr Dorm
Canaday
1st Yr Dorm
Lionel
1st Yr Dorm
Hollis
1st Yr Dorm
Harvard
Athl Ticket Ofc
Thayer
1st Yr Dorm
Memorial Church
Tercentenary Theatre
Johnston Gate
John Harvard Statue
University
Faculty of Arts & Sci
Robinson
History
Sever
Classrooms
Weld
1st Yr Dorm; Dorm Crew
Emerson
Philosophy
Massa-chusetts
Matthews
Straus
Lehman
Grays
Widener Library
Pusey Library
Loeb
Wadsworth
Boylston
Houghton Library
Wigglesworth
Lamont Library
HARVARD YARD
Cambridge Fire HQ
(Eng 1)
Sackler Museum
Ancient, Oriental & Islamic Art
Fogg Art Museum
Busch-Reisinger Museum; Fine Arts Dept; Fine Arts Library
Carpenter
Ctr for Visual Arts
Faculty Club
Warren
English
Dana-Palmer
Guest House
Barker Ctr
Barker Rotunda Freshmen Activities
Prescott Hall
Broadway Marketplace
Starbuck's
Broadway Parking Garage
Family Resource Center
Rindge School of Technical Arts
CAMBRIDGE RINDGE
Broadway
Gustave M Solomons Transportation Career Center
Greenough
1st Yr Dorm
8 Prescott
Morton Prince
Hurlbut
1st Yr Dorm
NYNEX
Student Tel Office
Office for Info Tech
Harvard Magazine
Pennypacker
1st Yr Dorm; WHRB
Prescott St
Felton St
Ware St
Trowbridge St
Trowbridge Pl
Doeb Grad
51 Brattle
Extension School
Church St
The Atrium
Palmer St
Harvard Square Station
Brattle Hall
1 Brattle Square
Brattle St
JFK St
HARVARD SQUARE
The Garage
Holyoke Center
Galeria
Winthrop St
Mt Auburn St
Eliot St
See Harvard Square Business District Detail Below
Bksmith Hse Bkry
Cambridge Ctr For Adult Ed
Dexter Pratt Hs
Burdicks; Settebello
Cafe Of India
Clothware
Crate& Barrel
Story St
Chas Webb
Mandrake Books
Post Office
Harvest
Sovereign Bank
Fleet Bank
Kinko's
22
131
127
129
Harvard Travel Service
Darwin's
French Consulate
University Place
Chili's
Buses Only
Harvard Square Hotel
Thrifty Car Rental
University Green
Harvard Univ Dev Ofc
Regattabar
Charles Hotel
Avis Rent-a-Car
Rialto
Henrietta's Table
Legal Sea Foods
Spence Ctr
LePli Day Spa
Taubman Center
WellBridge
CHARLES SQUARE
Soc of St John the Evangelist (Episcopal Monastery)
Bennett St
University Rd
Mt Auburn St
Brewer St
Gerry St
Chapman Pl
Revere St
Nutting Rd
Ross St
Lexington Court
Memorial Dr
1 Eliot St
Inst for Intl Development
Belfer
Weil Center
Littauer Center
JFK SCHOOL OF GOVT
John F Kennedy Park
John F Kennedy St
CAMBRIDGE
BOSTON
Newell Boat House
Hicks
Bryan
77 Dunster
Linguistics
Malkin Athl Ctr
Kirkland House
Smith
Malkin Quad
Standish Hall
Barrier
Winthrop House
Eliot House
Lowell House
Gore Hall
Quincy House
Harvard Hillel
Dining Hall
Mather Hall
McKinlock Hall
De Wolfe Street Housing
Winthrop St
South St
Dunster St
Mill St
Plympton St
De Wolfe St
Athens St
Athens Ter
Bow St
Siam Gdn
St Paul's Ch (Cath)
Water Music
Gray Bksellers
Rev/Crepes
Arrow St
St Paul's Sch
Zero Arrow St
Cambridge Framery
Games People Play
Au Bon Pain
12 Mt Auburn
7-9 Banks St
McIntyre & Moore Used Books
Warren
The Inn at Harvard
Old Cambridge Baptist Church
Church School
MacKinnon Motors
University Mkt; Dolphin Seafd; New Asia; 2nd Coming; Johnny's
Cafe Sushi
Garber Travel
Bowl & Board
Harvard Police; Dynasty Gallery; Home Design
Crate & Barrel Furniture
1033 Mass
Harvard St
Remington St
Massachusetts Ave
RIVER
Memorial Dr

3.3 CHICAGO BIKE MAP

The mapmaker uses contrast and saturation to emphasize city cycling routes in this map of Chicago, Illinois. Details of off-street trails, bike lanes, and recommended bike routes are presented with a combination of orange, blue, and purple lines over the city road network. For example, the side of the street on which a bike lane can be found is indicated by purple lines running along the orange core of a recommended route. Rail transit lines are also prominent in the design. Roads not recommended for biking are detailed, but set as background information—the lines are light and have low contrast with the background color. Road labels are organized into stacked groups. Consistent orientation and alignment of labels in these groups is maintained through as many adjacent roads as practical until shifting to another group in another open area of the design. This predictable rhythm eases a map user's search for road names. Footprints for major buildings and area fills for parks, schools, water bodies, and other points of interest assist with location.

Courtesy of Dennis McClendon, Chicago CartoGraphics.

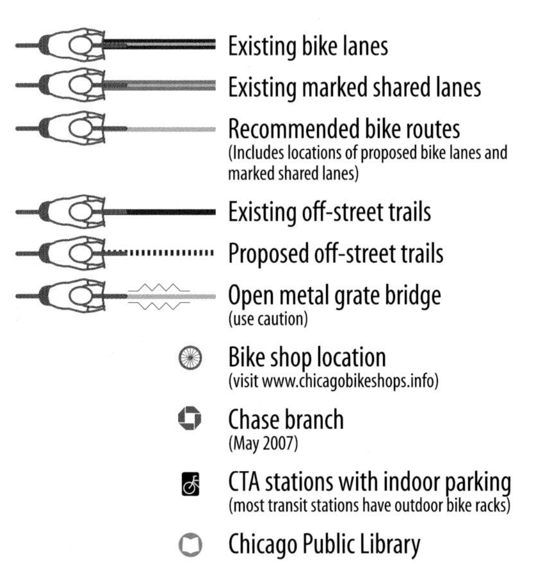

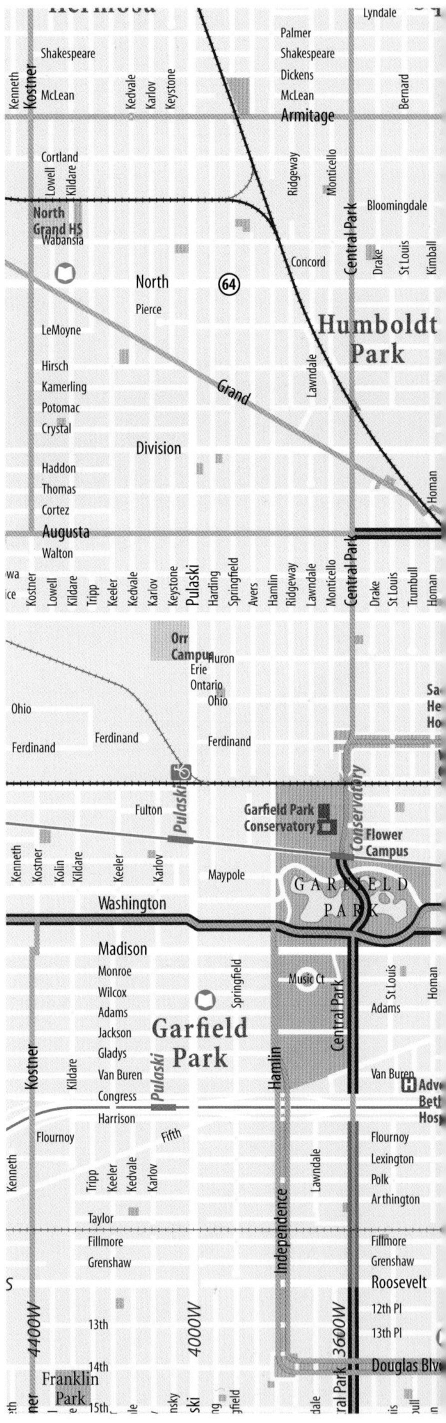

Bucktown
DePaul
Lincoln Park
Old Town
Goose Island
Wicker Park
Ukrainian Village
Near West Side
Cabrini-Green
River North
Streeterville
Loop
South Loop
Little Italy
Tri-Taylor
Humboldt Park
Douglas Park
Grant Park
Lakefront access via Passerelle bridge and ramps
North Avenue Beach
North Avenue lakefront access via tunnel and ramps
Division St lakefront access via tunnel and stairs
Oak St lakefront access via underpass
Oak Street Beach
Chicago Ave lakefront access via tunnel and stairs
Ohio St lakefront access via tunnel and ramps
Olive Park
Navy Pier
Monroe Harbor
See Loop Inset on other side
Museum Campus lakefront access via underpass
Field Museum
Shedd Aquarium
Adler Planetarium
Soldier Field
Burnham Harbor
Lakefront Trail
Lake Shore Drive
Chicago History Museum
John Hancock Center
Museum of Contemporary Art
Northwestern Memorial Hospital
Merchandise Mart
Harold Washington College
Millennium Station
Daley Center
City Hall
Civic Opera House
Ogilvie Center
Union Station
Sears Tower
Chicago Board of Trade
DePaul Univ
Chicago Cultural Center
McDonald's Cycle Center
Pritzker Music Pavilion
Art Institute of Chicago
Petrillo Bandshell
Buckingham Fountain
Roosevelt Univ
Columbia College
University of Illinois at Chicago
Rush University Medical Center
Cook County Hospital
UIC Hospital
VA Hospital
Illinois Medical District
United Center
Union Park
Eisenhower Expwy
Kennedy Expwy
CTA Green Line to Harlem/Lake
CTA Blue Line to O'Hare
Blue Line to Forest Park
Adams Park
Smith Park
Eckhart Park
Seward Park
Stanton Park
Moody Bible Inst
Loyola Univ
Northwestern Univ
Lincoln Park Zoo
Lincoln Park Hospital
St Elizabeth Hospital
St Mary of Nazareth Hospital
Norwegian American Hospital
Mt Sinai Hospital
Clemente HS
Crane HS
Wells HS
Payton HS
Jones HS
Young HS
Handlebar Restaurant
Best Practice HS
Malcolm X College
Collins HS
Division
Chicago
Grand
Madison
Roosevelt
Halsted
Ashland
Western
Milwaukee
Ogden
Clybourn
Elston
2000N
1600N
1200N
800N
400N
200N
1N/1S
300S
600S
1200S
800W
1200W
1600W
2800W

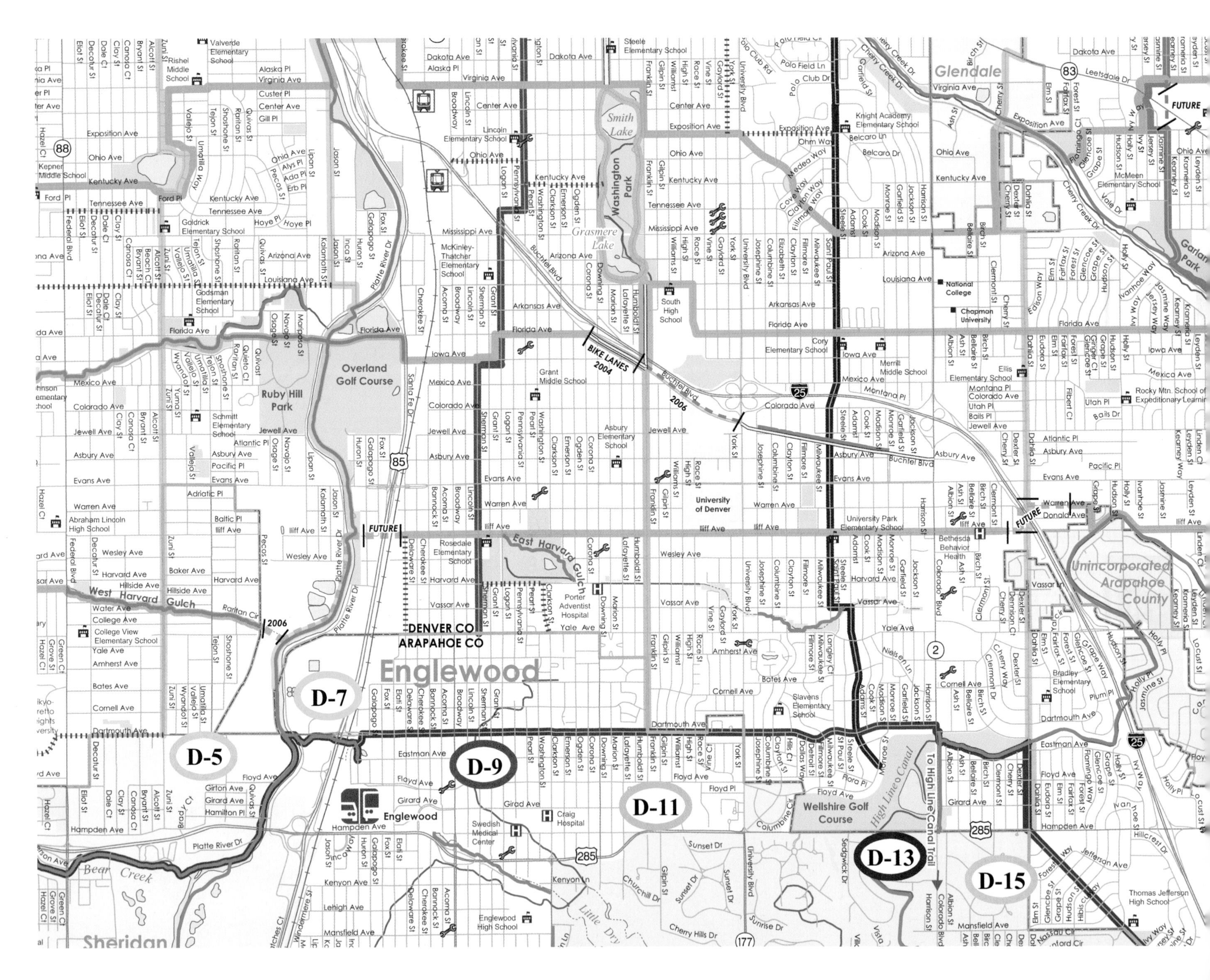
Glendale
Englewood
Sheridan
Unincorporated Arapahoe County
DENVER CO
ARAPAHOE CO
D-5
D-7
D-9
D-11
D-13
D-15
Overland Golf Course
Ruby Hill Park
Washington Park
Smith Lake
Grasmere Lake
Garland Park
Wellshire Golf Course
University of Denver
Chapman University
National College
West Harvard Gulch
East Harvard Gulch
Bear Creek
High Line Canal
To High Line Canal Trail
BIKE LANES 2004
2006
FUTURE
Buchtel Blvd
Platte River Dr
Santa Fe Dr
University Blvd
Colorado Blvd
Federal Blvd
Hampden Ave
Dartmouth Ave
Yale Ave
Evans Ave
Iliff Ave
Florida Ave
Mississippi Ave
Exposition Ave
Alameda Ave
Leetsdale Dr
Cherry Creek Dr
Broadway
Downing St
South High School
Abraham Lincoln High School
Englewood High School
Thomas Jefferson High School
Grant Middle School
Merrill Middle School
Kepner Middle School
Rishel Middle School
Porter Adventist Hospital
Swedish Medical Center
Craig Hospital
Bethesda Behavior Health
University Park Elementary School
Rocky Mtn. School of Expeditionary Learning
85
88
83
285
25
2
177

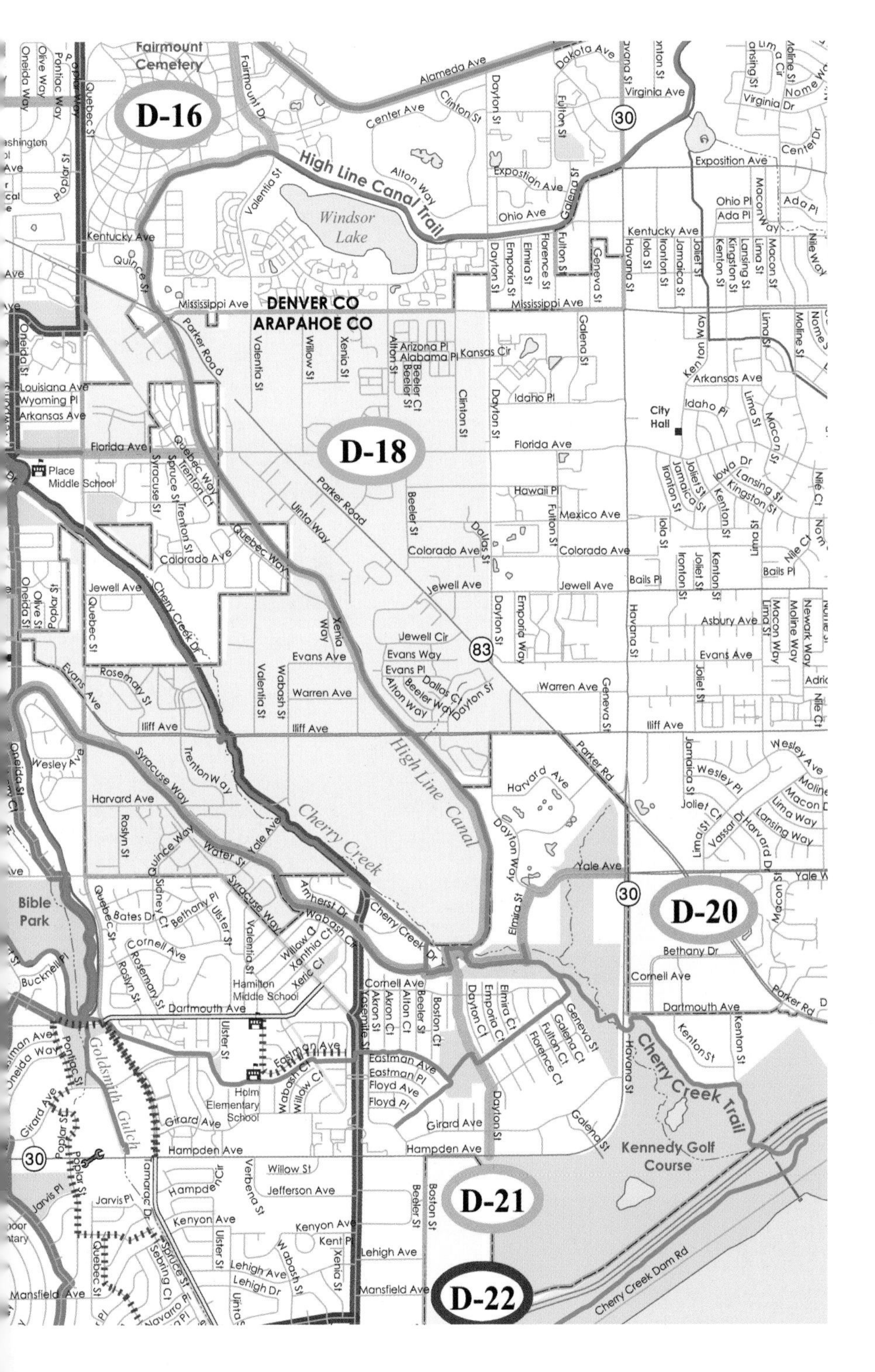

3.4 DENVER BIKE MAP

Denver, Colorado, bicycling routes are divided into a grid of numbered routes distinguished by hue. The routes are boldly labeled at the edges of the area of bike routes so it is easy to identify particular routes and follow the color across the map. The labels are a helpful courtesy; expecting users to identify numerous line colors in a legend and then find those colors on the map would create ambiguities. This approach also simplifies the map legend, since the routes do not need to be coded there. These wide, colored route lines sit underneath the road network so cyclists see they are coincident. In addition to the numbered bike routes, roads with bike lanes are symbolized with a yellow fill and off-street bike paths are in green. To keep the visual emphasis on bike routes, there is minimal hierarchy in the road network with only slightly darker grays for more major roads and no type size differences. Bicycle shops are shown with a custom wrench symbol.

 Denver thanks the Cities of Calgary, Alberta, Canada, and Chicago, Illinois, Bill Romano, and the Chicagoland Bicycle Federation for their contributions to the map.

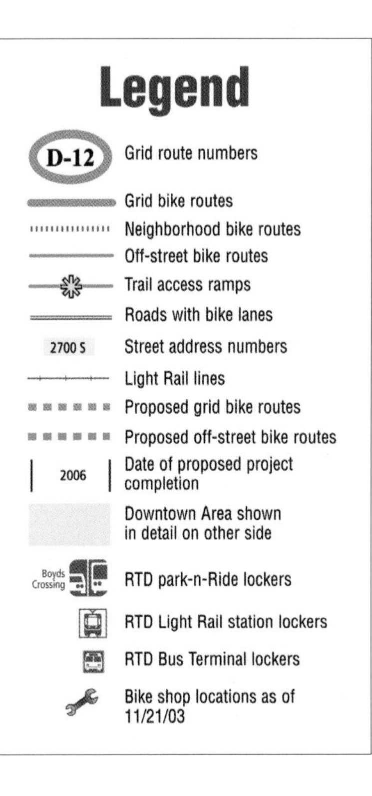

3.5 3D PRAGUE MAP

The three-dimensional building data for Prague, Czech Republic, was compiled from 1:7,000 aerial photos and simplified to balance detail and overall housing block structure. For example, roof elements such as dormers and chimneys less than two meters wide and one meter square are not represented. Building facets are shaded with a simple lighting model. Adjacent buildings do not cast shadows on each other; however, there are shadows in the aerial photo, and the direction of illumination for the buildings matches the directions of these shadows. The lightest roof and wall surfaces face generally up and right in the view, and the darkest building surfaces face down and left, with intermediate lightness for planes between. This choice deviates from the standard for terrain illumination but is essential for coherence between the photo and 3D features. Schematic trees are also included for selected areas, such as parklands, and are rendered at a similar level of simplification as the buildings. Buildings and trees are accurately registered with the photo basemap. The photo provides information on roads, parking, and open areas, which are enhanced with a subtle overlay of transparent green.

Courtesy of IMIP PRAHA.

3.6 ADIRONDACK PARK TRAIL MAP (FOLLOWING PAGES)

The Trails Illustrated series of National Geographic Society maps emphasize recreation detail. The trails on the Adirondack Park (New York) map are clearly marked with bold black dashed lines. Trail segments are numbered in bold green ovals that link them to names and description lists. Red distance measurements are listed for short segments between red pointers. Route difficulty may be interpreted from the detailed contour information supported by spot heights and hillshading. Types of property ownership are highlighted with color fills and tint bands, signaling potential regulation changes that hikers need to follow as they cross boundaries. Roads are symbolized with black casings and a red to orange to white sequence of fills. Private roads have brown casings that push them to the background. The UTM coordinate system grid in cyan and latitude–longitude black ticks at the map edge provide coordinate detail that hikers may use to locate their positions on the map with GPS equipment.

Maps used with permission of the National Geographic Society.

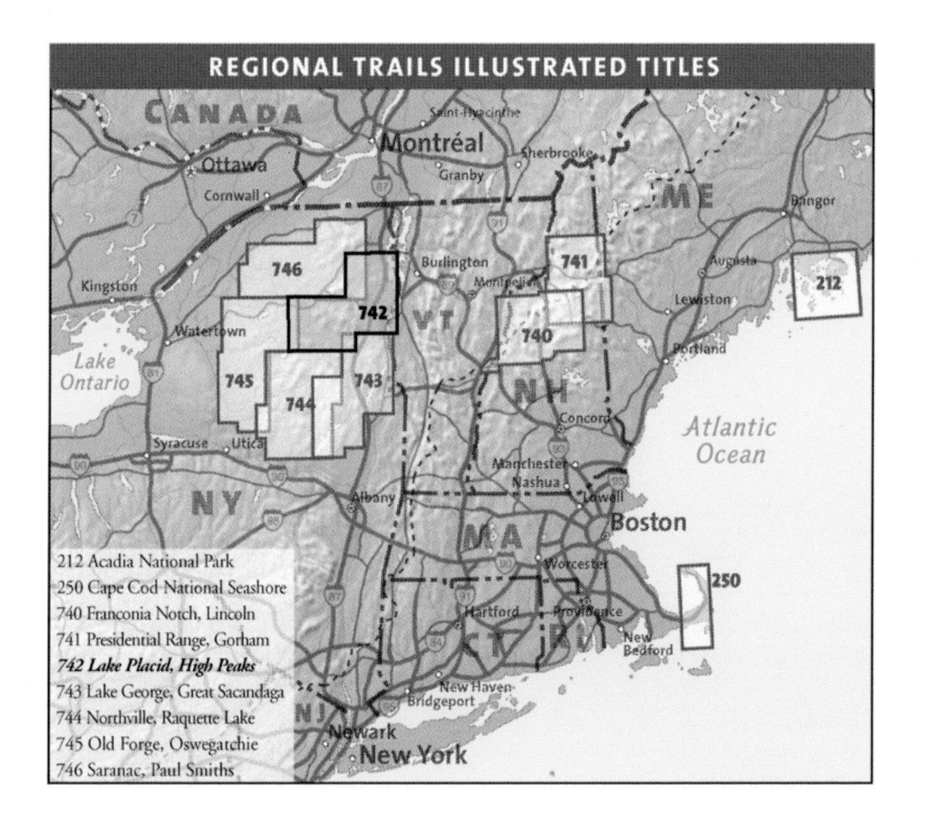

SCALE

Approximate Scale 1:75,000	Contour Interval = 50 feet
0 1/2 1 2 3 Miles	
0 1/2 1 2 3 Kilometers	
1 inch = 1.18 miles	1 centimeter = 0.75 kilometer
Projection: UTM, Zone 18 T (Central Meridian 75° W)	Datum: NAD 1927 Ellipsoid: Clarke 1866

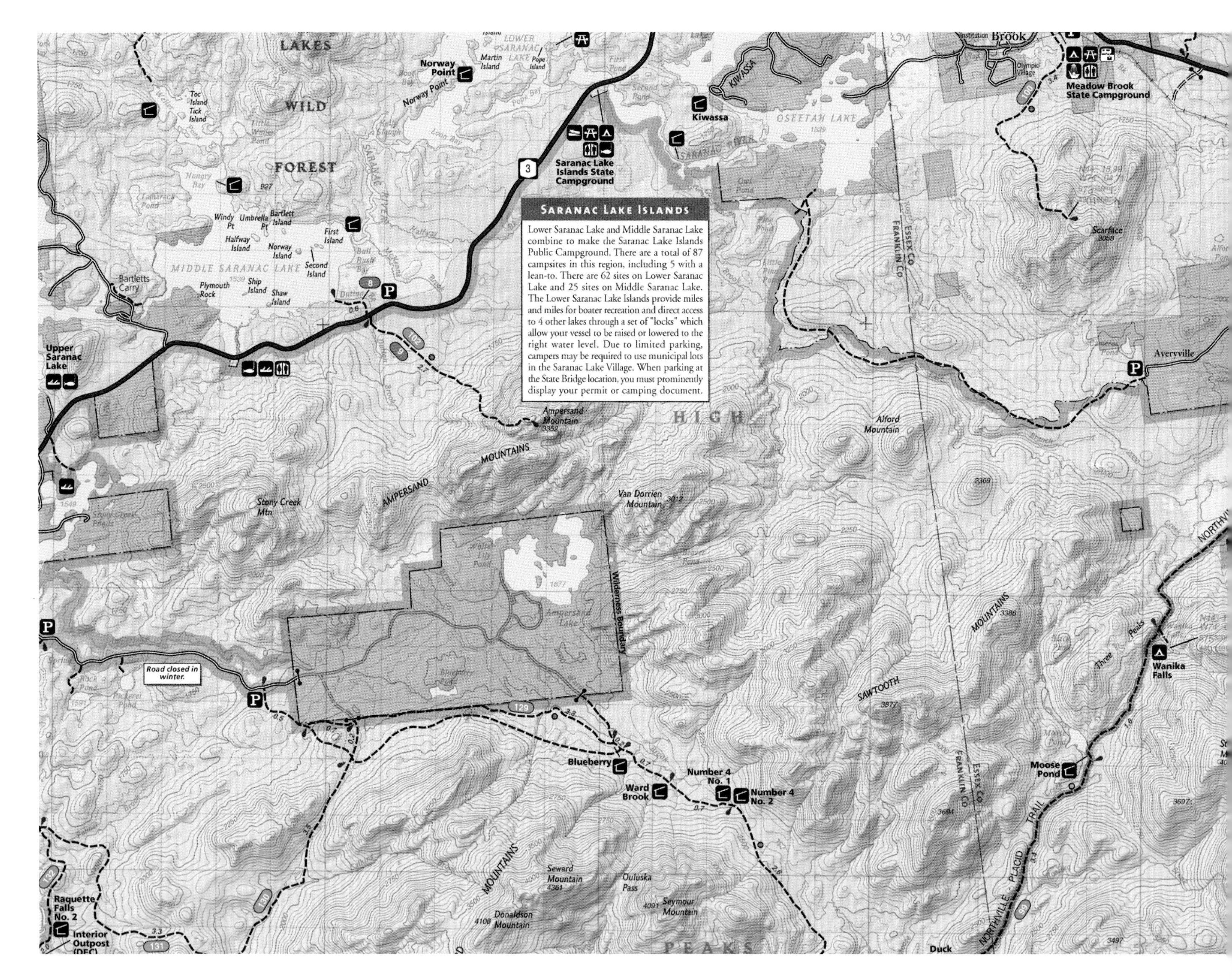
Saranac Lake Islands
Lower Saranac Lake and Middle Saranac Lake combine to make the Saranac Lake Islands Public Campground. There are a total of 87 campsites in this region, including 5 with a lean-to. There are 62 sites on Lower Saranac Lake and 25 sites on Middle Saranac Lake. The Lower Saranac Lake Islands provide miles and miles for boater recreation and direct access to 4 other lakes through a set of "locks" which allow your vessel to be raised or lowered to the right water level. Due to limited parking, campers may be required to use municipal lots in the Saranac Lake Village. When parking at the State Bridge location, you must prominently display your permit or camping document.
LAKES
WILD
FOREST
HIGH
PEAKS
Norway Point
Saranac Lake Islands State Campground
Kiwassa
OSEETAH LAKE
Meadow Brook State Campground
MIDDLE SARANAC LAKE
Upper Saranac Lake
Bartletts Carry
Ampersand Mountain
AMPERSAND
MOUNTAINS
Stony Creek Mtn
Van Dorrien Mountain
Alford Mountain
Scarface
Averyville
Wilderness Boundary
Ampersand Lake
Road closed in winter.
Blueberry
Ward Brook
Number 4 No. 1
Number 4 No. 2
SAWTOOTH
Moose Pond
Wanika Falls
Seward Mountain
Donaldson Mountain
Seymour Mountain
Ouluska Pass
Raquette Falls No. 2
Interior Outpost (DEC)
ESSEX CO
FRANKLIN CO
NORTHVILLE - PLACID TRAIL
Duck

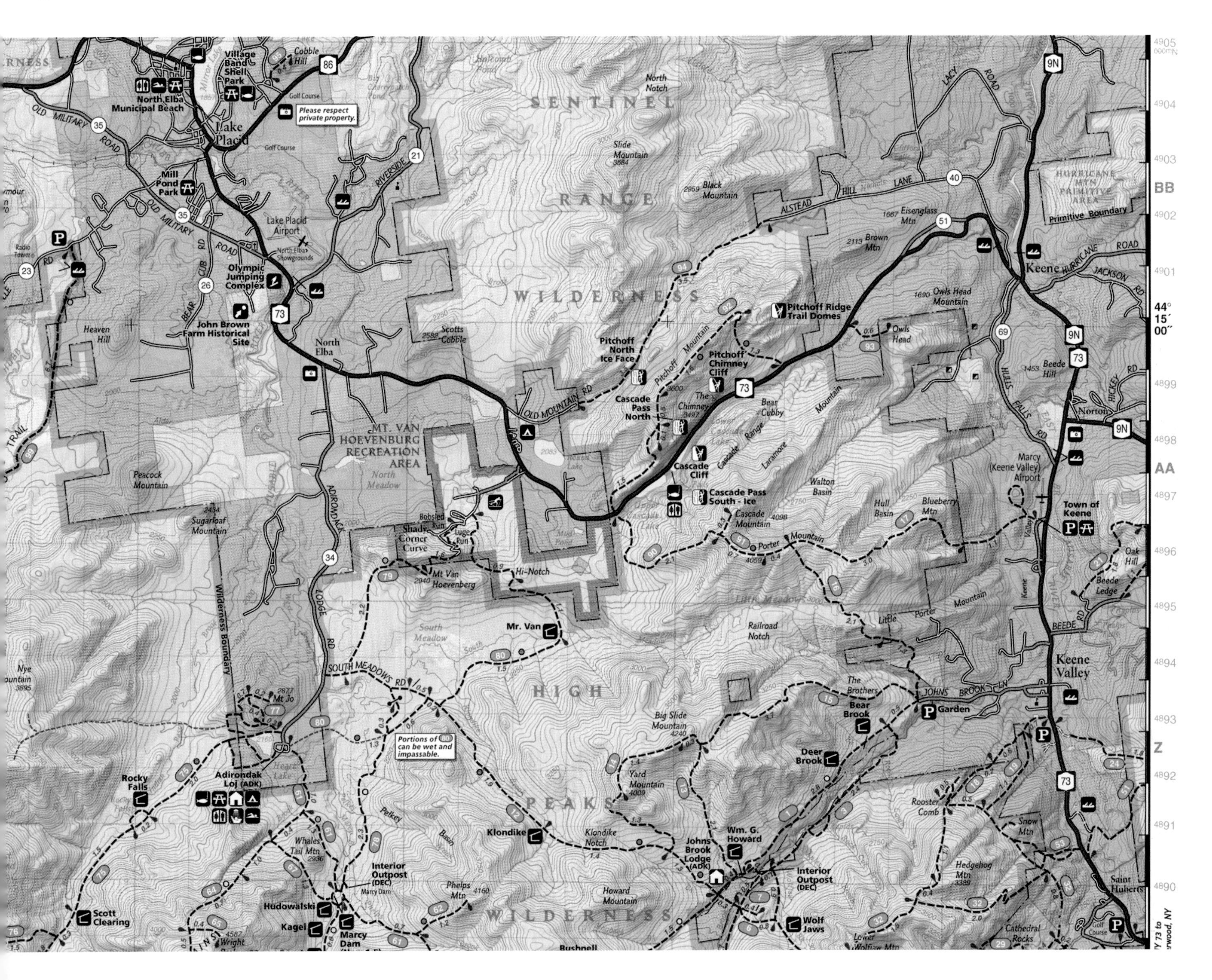

SENTINEL RANGE WILDERNESS
HIGH PEAKS WILDERNESS
Lake Placid
North Elba Municipal Beach
Village Band Shell Park
Mill Pond Park
Lake Placid Airport
Olympic Jumping Complex
John Brown Farm Historical Site
North Elba
MT. VAN HOEVENBURG RECREATION AREA
Please respect private property.
Portions of 80 can be wet and impassable.
Pitchoff Ridge Trail Domes
Pitchoff North Ice Face
Pitchoff Chimney Cliff
Cascade Pass North
Cascade Cliff
Cascade Pass South - Ice
Keene
Town of Keene
Keene Valley
Marcy (Keene Valley) Airport
HURRICANE MTN PRIMITIVE AREA
Primitive Boundary
Adirondak Loj (ADK)
Rocky Falls
Scott Clearing
Hudowalski
Kagel
Marcy Dam
Interior Outpost (DEC)
Klondike
Mr. Van
Johns Brook Lodge (ADK)
Wm. G. Howard
Bear Brook
Deer Brook
Wolf Jaws
Garden
Saint Huberts
Wilderness Boundary
44° 15′ 00″

3.7 ACADIA NATIONAL PARK MAP

The National Park Service (NPS) produces visitor maps for their parks, large and small, across the country. The town of Bar Harbor, Maine, is surrounded by the portion of Acadia National Park on Mount Desert Island. Linear features on the map are visually organized with light colors for trails and dark colors for roads. Line casing further distinguishes yellow carriage roads, for biking and hiking, from more rugged trails signified by white dashed lines. The scenic Park Loop and Cadillac Mountain roads are emphasized in dark red, and labels indicating important information associated with the routes (such as one-way traffic and bridge clearance measurements) are in the same color. Differences in contrast for road and trail lines against the green park color are maintained in the legend by placing a green rectangle behind each line symbol. Park entrances are highlighted with bold callouts that cover underlying features. Amenities, such as picnic areas, restrooms, and parking, are symbolized with high-contrast pictograms that aid users searching the map, and also help non-English-speaking visitors with map interpretation. The map is enlivened with a slightly darker blue vignette that emphasizes the rugged coastline and a light hillshade.

Courtesy of National Parks Service, U.S. Department of the Interior.

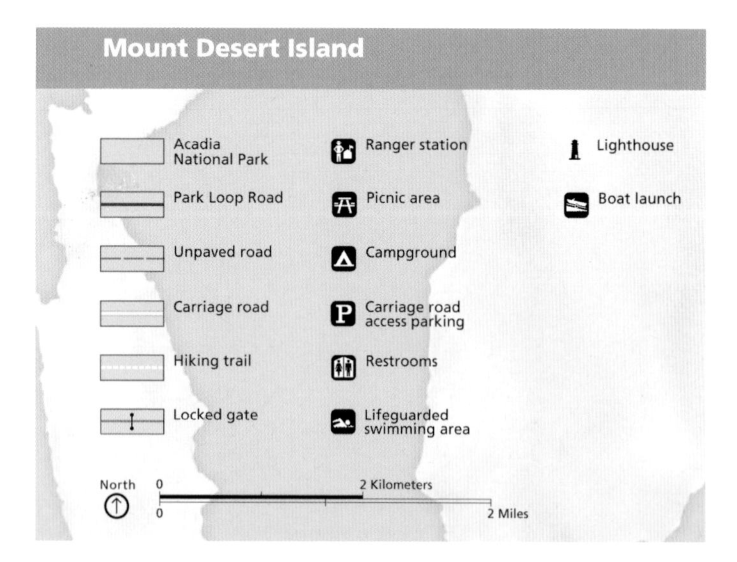

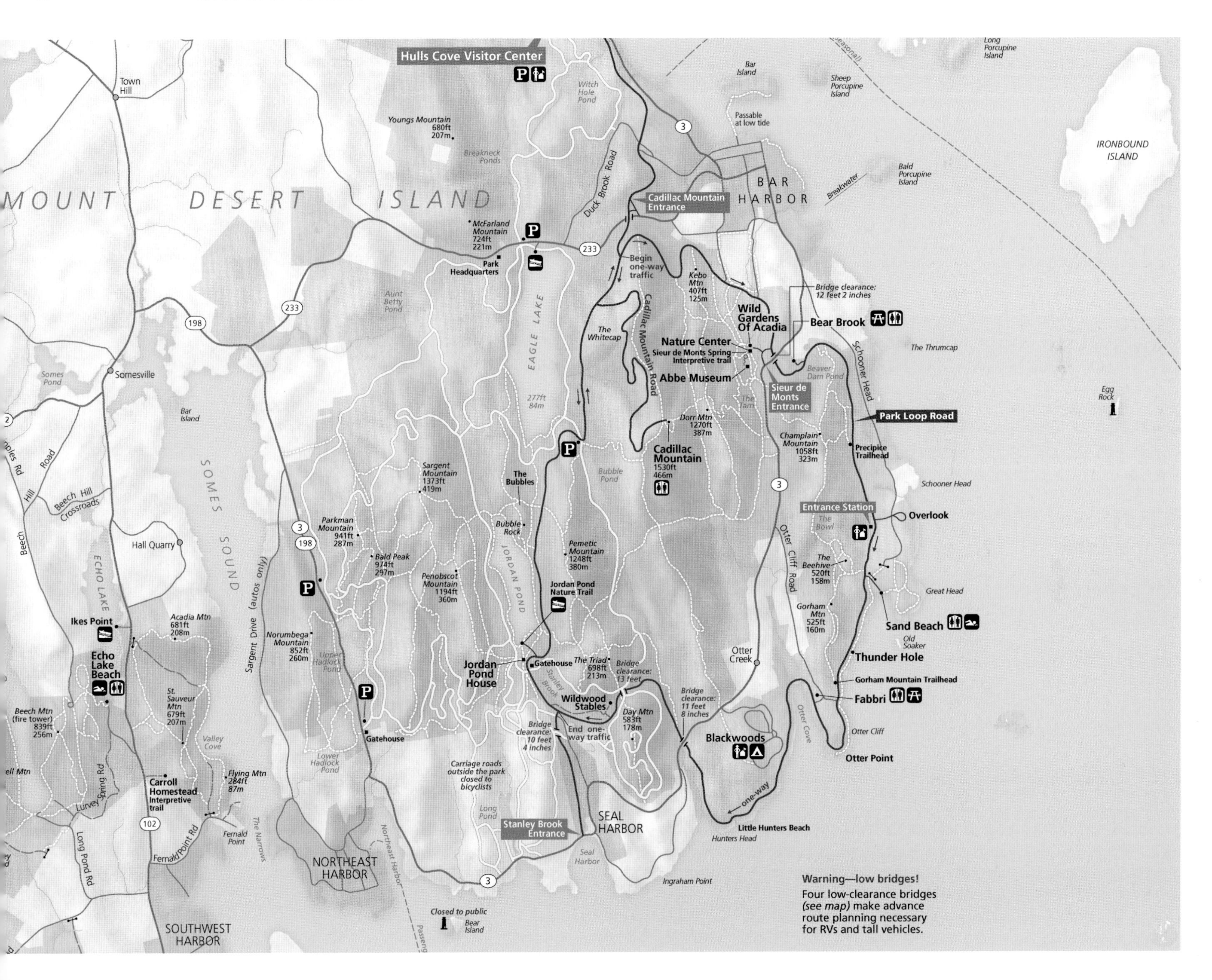
Hulls Cove Visitor Center
MOUNT DESERT ISLAND
Cadillac Mountain Entrance
BAR HARBOR
Park Loop Road
Sieur de Monts Entrance
Entrance Station
Stanley Brook Entrance
Wild Gardens Of Acadia
Nature Center
Sieur de Monts Spring Interpretive trail
Abbe Museum
Bear Brook
Cadillac Mountain 1530ft 466m
Jordan Pond House
Jordan Pond Nature Trail
Wildwood Stables
Blackwoods
Sand Beach
Thunder Hole
Gorham Mountain Trailhead
Fabbri
Otter Point
Overlook
Echo Lake Beach
Ikes Point
Carroll Homestead Interpretive trail
SEAL HARBOR
NORTHEAST HARBOR
SOUTHWEST HARBOR
SOMES SOUND
Somesville
Town Hill
Hall Quarry
IRONBOUND ISLAND
Bridge clearance: 12 feet 2 inches
Bridge clearance: 13 feet
Bridge clearance: 11 feet 8 inches
Bridge clearance: 10 feet 4 inches
Begin one-way traffic
End one-way traffic
Carriage roads outside the park closed to bicyclists
Closed to public
Warning—low bridges!
Four low-clearance bridges *(see map)* make advance route planning necessary for RVs and tall vehicles.

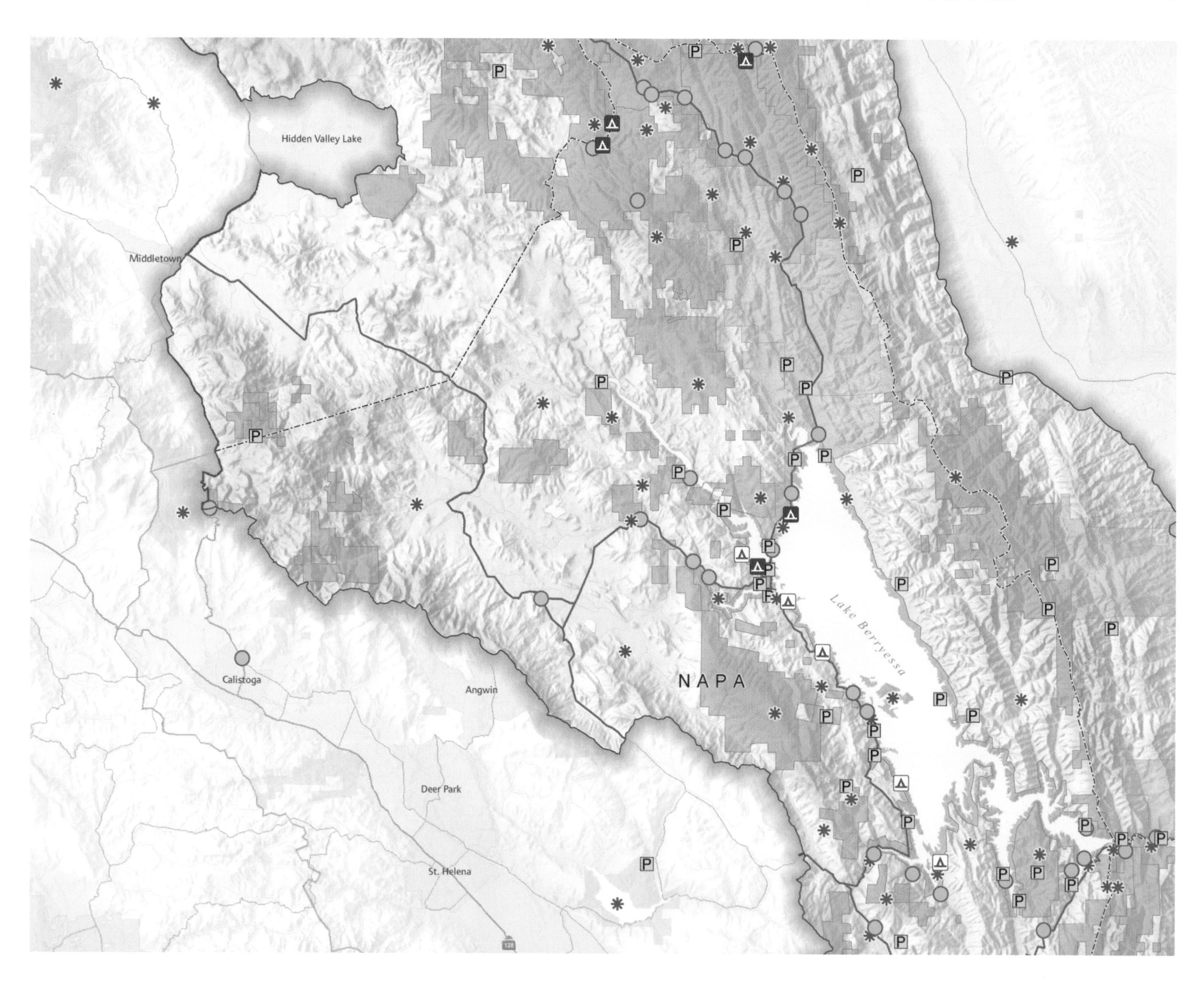
Hidden Valley Lake
Middletown
Calistoga
Angwin
NAPA
Lake Berryessa
Deer Park
St. Helena

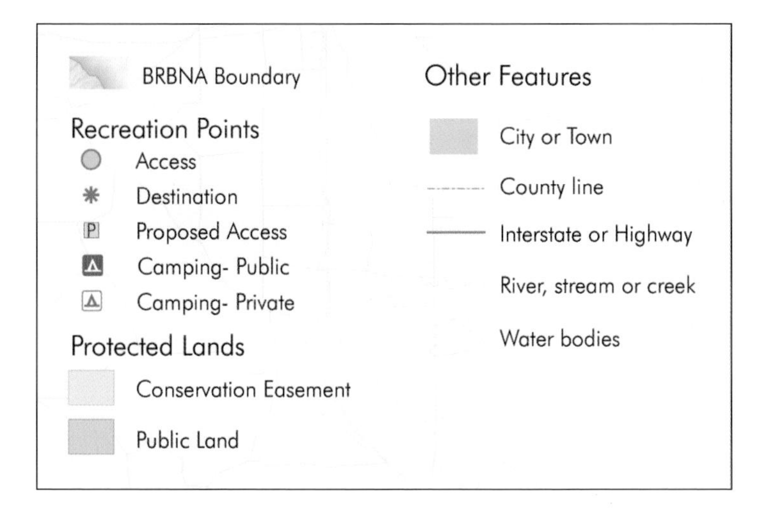

3.8 BLUE RIDGE BERRYESSA NATURAL AREA MAP

The map of Blue Ridge Berryessa Natural Area (BRBNA) in California is characterized by multiple overlaid layers that are partly transparent. For example, a green public land area outside the natural area is a stack of at least four layers: a transparent white over transparent green over transparent elevation tints over terrain (or transparent terrain over elevation tints—either order works). The map uses a muted series of the typical elevation tints from green lowlands to yellow to reddish brown highlands. This series relies on the visual effect (called chromatic aberration) that causes reds to advance and blues (and cool greens) to recede. The effect is subtle, but gives the mountainous areas a bit of a push toward the foreground. The public-land greens and conservation-easement oranges are set off from similar colors in the elevation tints by a thin darker line around each area. The transparency setting in these hues mostly masks the elevation tints but allows a hint of landform to show through with the hillshading. A narrow purple gradient feathers out from the natural area boundary and into a white transparent mask over non-BRBNA areas, establishing BRBNA as the main focus of the map while still providing the lightly rendered context of surrounding features. Pictograms for recreation points, situated above all the transparent layers for readability, are distinguished by hue as well as shape.

Courtesy of GreenInfo Network, copyright 2006.

3.9 BACKBONE TRAIL MAP

In this map for the Santa Monica Mountains National Recreation Area (California), the Backbone trail is represented by a wide white line with a black casing. This line contrasts with the variety of colors beneath it, and it is coincident with line patterns of dotted trails and dashed roads. Proposed sites that may be added along the trail, such as equestrian camps, are shown with light brown pictograms. The use of pictograms on this map varies from other National Park Service recreation maps because a single rectangle may include multiple symbols for site features, as opposed to the typical arrangement of multiple single-symbol rectangles. Types of public landownership are represented in hues with moderate saturation, since they are important for the draft map's purposes of supporting trail management planning and public review. Background features include roads, built-up areas, intermittent streams, and terrain form.

Courtesy of National Parks Service, U.S. Department of the Interior.

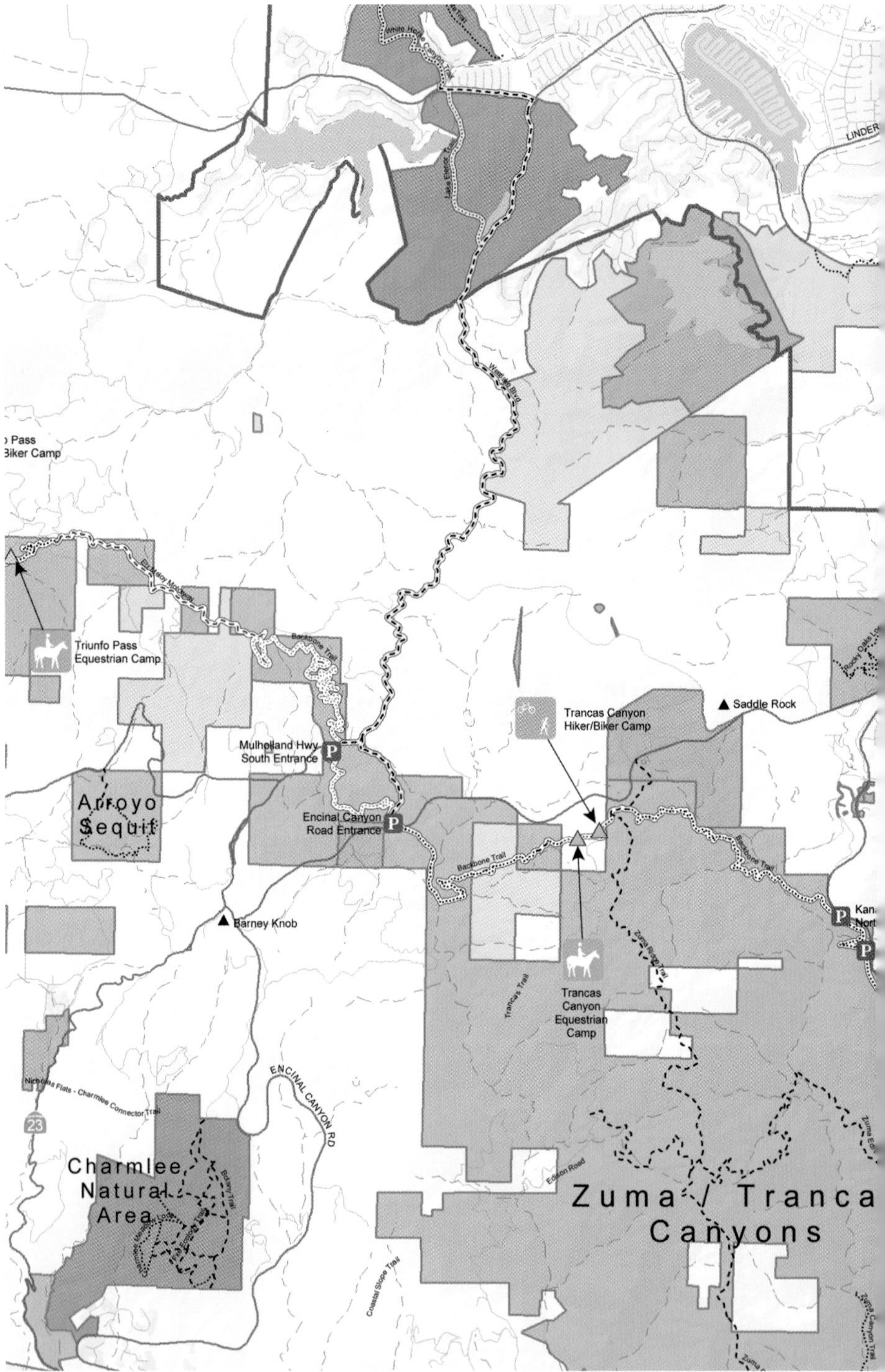

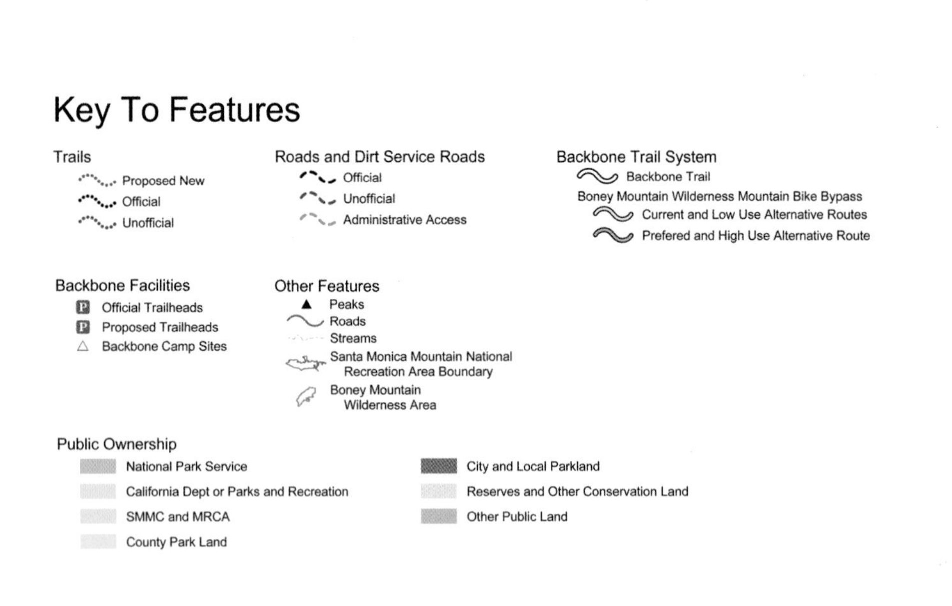

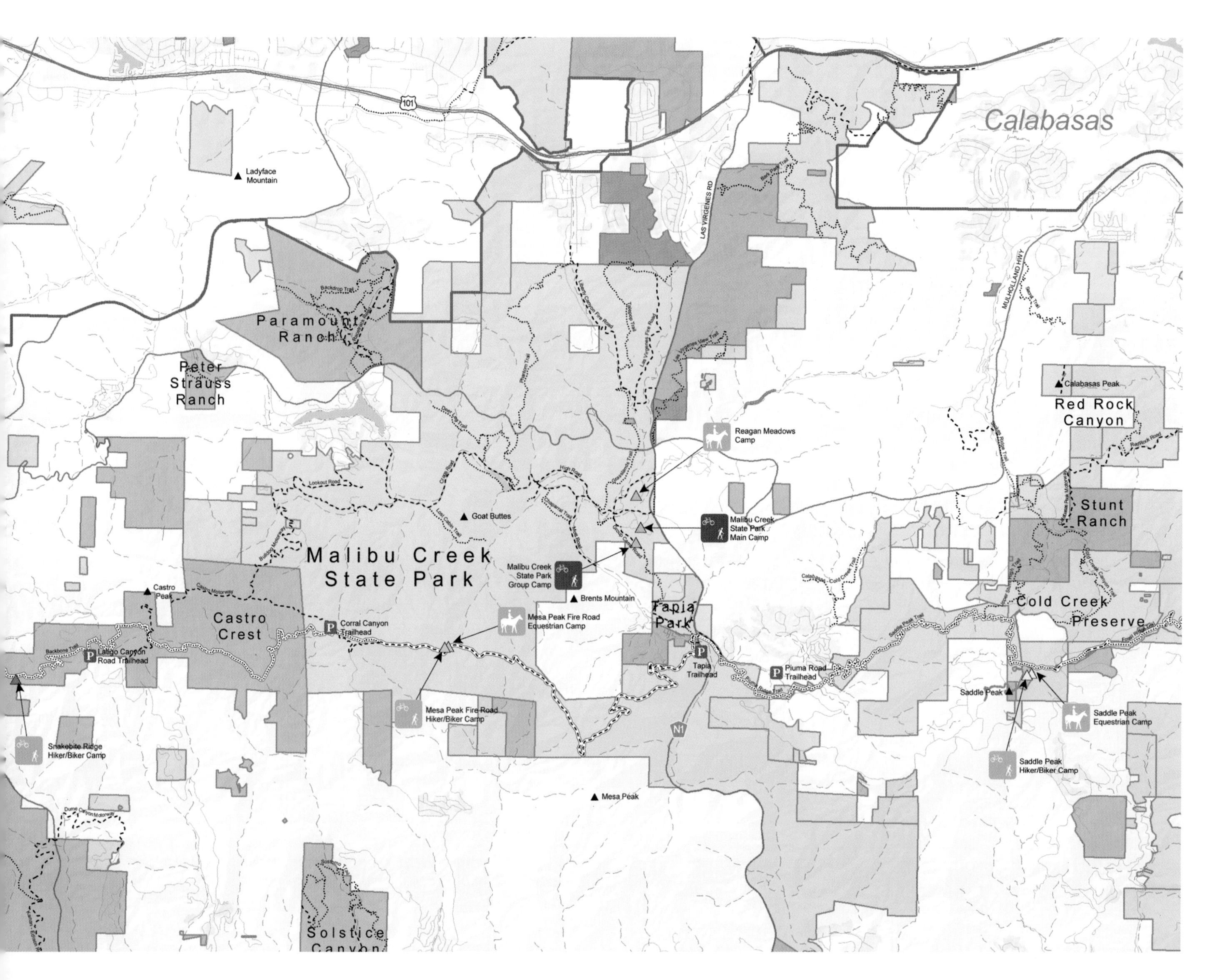

Calabasas
101
Ladyface Mountain
LAS VIRGENES RD
MULHOLLAND HWY
Paramount Ranch
Peter Strauss Ranch
Calabasas Peak
Red Rock Canyon
Reagan Meadows Camp
Goat Buttes
Malibu Creek State Park Main Camp
Stunt Ranch
Malibu Creek State Park
Malibu Creek State Park Group Camp
Brents Mountain
Castro Peak
Castro Crest
Corral Canyon Trailhead
Mesa Peak Fire Road Equestrian Camp
Tapia Park
Cold Creek Preserve
Latigo Canyon Road Trailhead
Tapia Trailhead
Piuma Road Trailhead
Saddle Peak
Saddle Peak Equestrian Camp
Mesa Peak Fire Road Hiker/Biker Camp
Snakebite Ridge Hiker/Biker Camp
Saddle Peak Hiker/Biker Camp
N1
Mesa Peak
Solstice Canyon

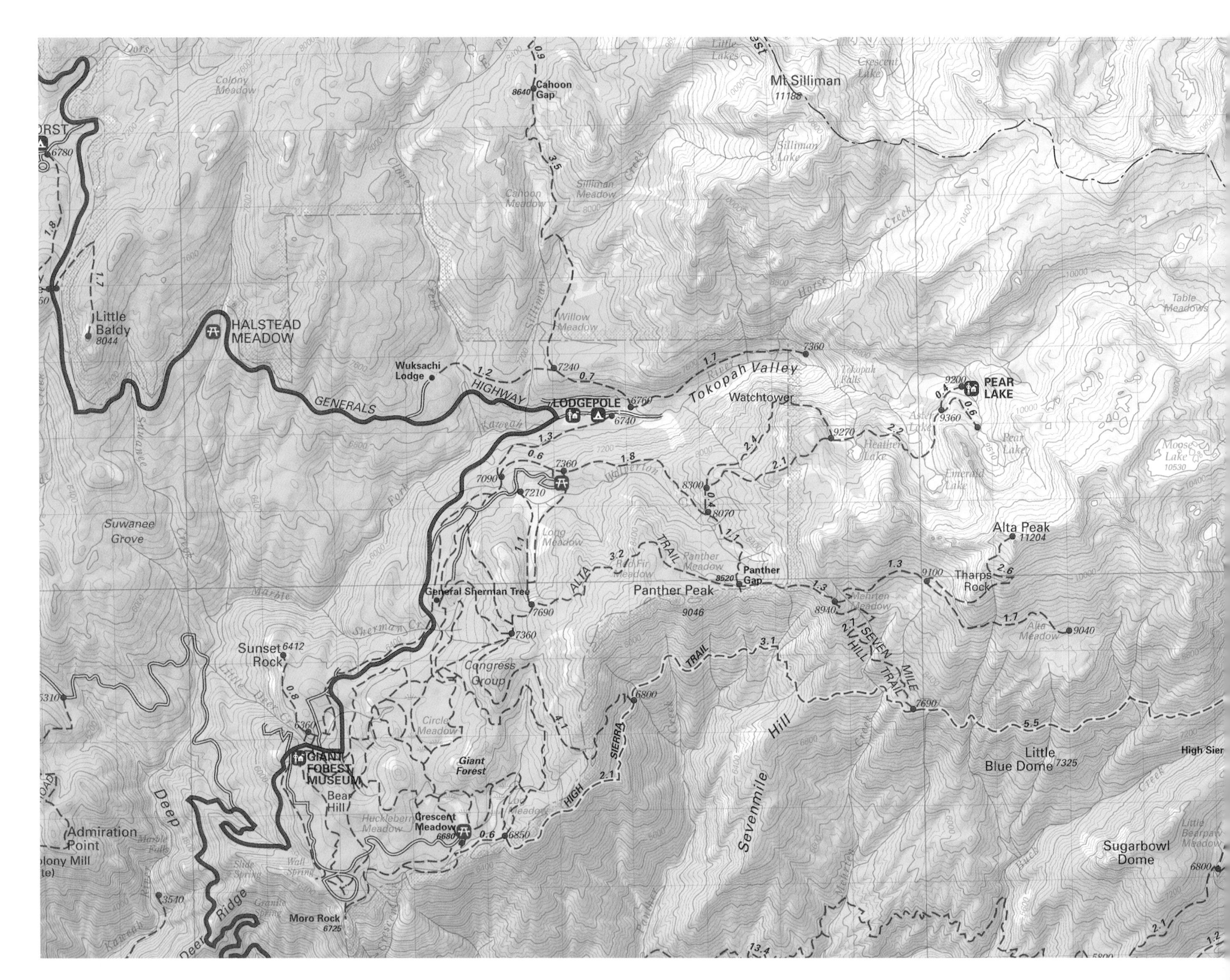

Mt Silliman
11188
Cahoon Gap
8640
Colony Meadow
Silliman Lake
Silliman Meadow
Cahoon Meadow
Willow Meadow
Little Baldy
8044
HALSTEAD MEADOW
Wuksachi Lodge
GENERALS HIGHWAY
LODGEPOLE
Tokopah Valley
Watchtower
PEAR LAKE
Heather Lake
Emerald Lake
Pear Lake
Table Meadows
Moose Lake
10530
Alta Peak
11204
Tharps Rock
Panther Gap
Panther Peak
9046
Mehrten Meadow
Alta Meadow
ALTA TRAIL
Red Fir Meadow
Panther Meadow
Long Meadow
General Sherman Tree
Suwanee Grove
Sunset Rock
6412
Congress Group
Circle Meadow
Giant Forest
GIANT FOREST MUSEUM
Bear Hill
Huckleberry Meadow
Crescent Meadow
HIGH SIERRA TRAIL
SEVEN MILE HILL TRAIL
Sevenmile Hill
Little Blue Dome
7325
Sugarbowl Dome
Admiration Point
Moro Rock
6725
Deep Ridge

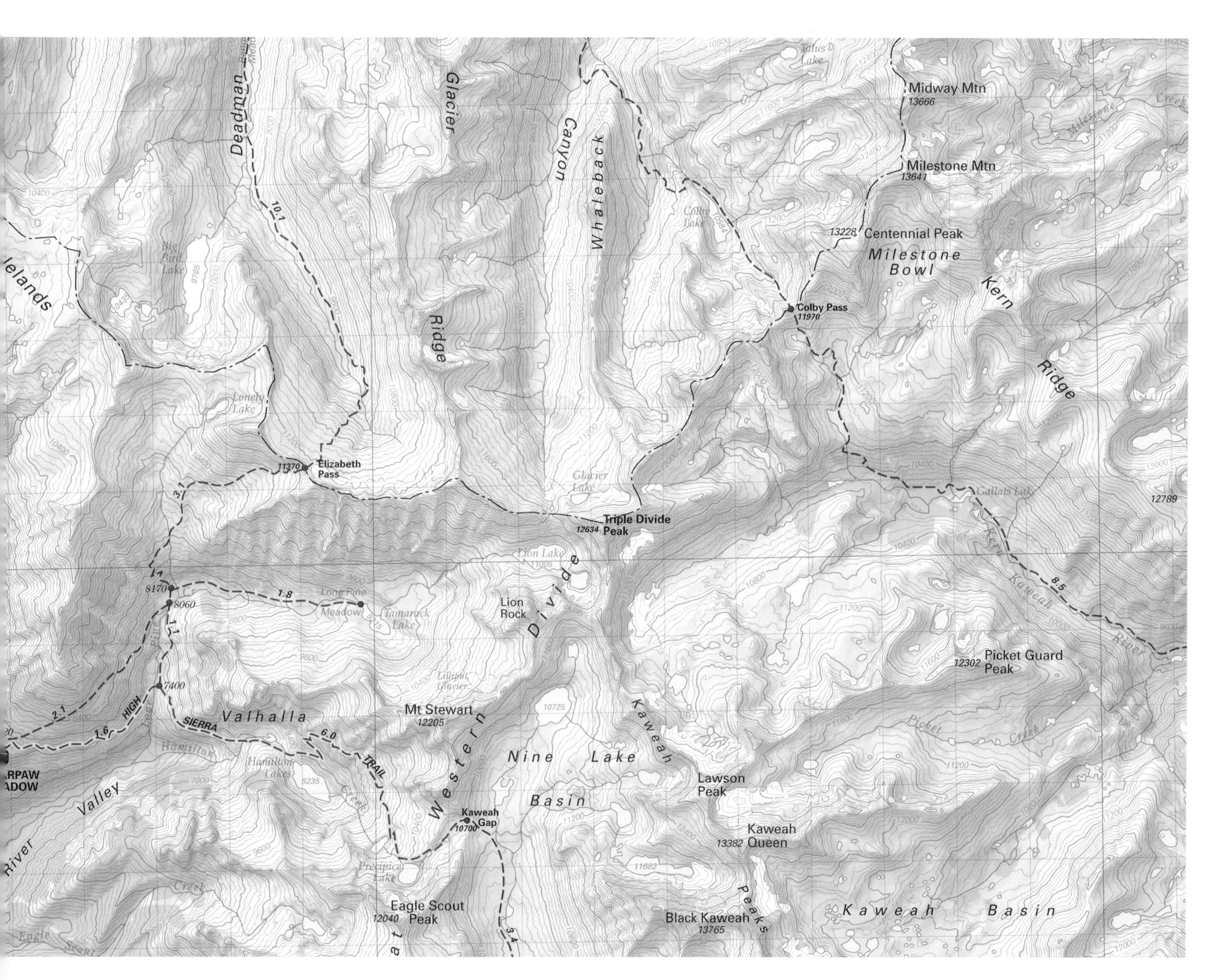
Deadman
Glacier
Canyon
Whaleback
Ridge
Midway Mtn
13666
Milestone Mtn
13641
Centennial Peak
13228
Milestone Bowl
Kern Ridge
Colby Pass
11970
Colby Lake
Big Bird Lake
Lonely Lake
Elizabeth Pass
11370
Glacier Lake
Triple Divide Peak
12634
Gallats Lake
12789
Lion Lake
Lion Rock
Divide
Lone Pine Meadow
Tamarack Lake
8170
8060
7400
1.8
1.1
3.1
10.1
8.5
2.1
1.6
6.0
3.4
Kern Kaweah River
Picket Guard Peak
12302
Lilliput Glacier
Mt Stewart
12205
Valhalla
HIGH SIERRA TRAIL
Hamilton Lakes
Hamilton
Valley
River
Western
Nine Lake Basin
Kaweah Peaks
Lawson Peak
Kaweah Gap
10700
Kaweah Queen
13382
Picket Creek
Precipice Lake
Eagle Scout Peak
12040
Black Kaweah
13765
Kaweah Basin
Eagle Scout
Creek

3.10 MOUNT WHITNEY HIGH COUNTRY TRAIL MAP (PREVIOUS PAGES)

Detailed contours, spot heights at peaks and trail intersections, and hillshading provide a complete base of terrain information for recreation maps. On this segment of the Mount Whitney High Country (California) trail map, roads and trails stand out well in bold red. This map is printed with five ink colors (red, blue, green, brown, black) rather than CMYK process inks for consistent and accurate results on the printing press. Blue is used in a variety of ways on the map. Hydrography—streams, water bodies, marshes—are symbolized and labeled in blue. The UTM grid is blue. A dashed blue line marks the wilderness boundary, and the blue textured band extends into the wilderness area to identify which side of the boundary biking is prohibited. A light blue halo underlies the High Sierra Trail line to emphasize it by blocking the contours and hillshading below. All of these features use blue ink only, making thin lines less troublesome during printing because multiple inks are not registered on the page. The character of lines among the blue features contrast with each other. For example, straight grid lines are not confused with stream lines. Blue features contrast nicely with map elements in other hues but are light enough to remain in the background, maintaining their position as base information.

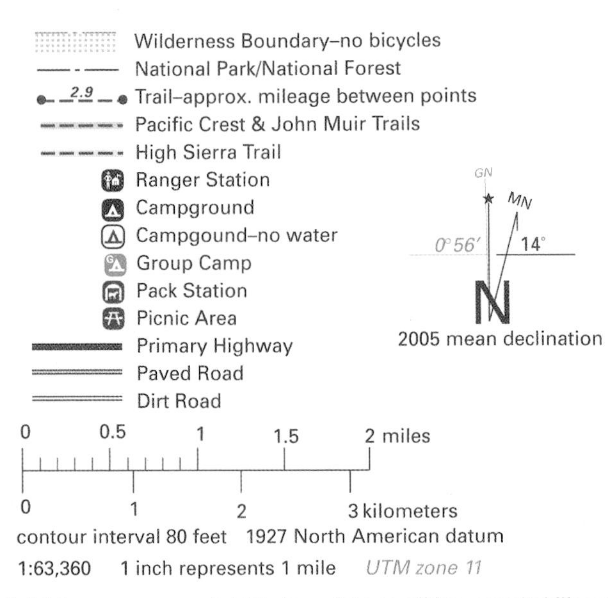

Courtesy of Tom Harrison Maps.

4

SPECIAL-PURPOSE MAPS

INFRASTRUCTURE

Infrastructure maps have a special purpose, as thematic maps do, but they do not share the thematic mapping goal of exploring and communicating overall patterns in a dataset. Infrastructure maps are more focused on allowing users to look up detailed attributes at particular locations with well-chosen symbols. The best infrastructure maps make use of the same visual variables as thematic maps to categorize and order attributes, but they include more intensive labeling, akin to reference maps.

The maps chosen for this chapter inventory a wide range of rules, equipment, and plans. Utility mapping examples include water distribution, sewers, storm drains, and electric and telecommunications services. Property topics include taxes, subdivision plats, zoning, and mineral leases.

Large-scale infrastructure maps are widely used in city management, and they share common base information: parcel boundaries, building footprints, road polygons or curb lines, road labels, and orthophoto background images. These details are meaningful to workers and managers as well as interested citizens, and this suite of detailed base data positions infrastructure features very specifically in relationship to properties. Though most infrastructure maps are large scale, a few that are smaller in scale are included in this chapter—San Francisco-area telecom and Iceland reservoir planning—to show special-topic detail overlaid on more usual reference map information.

The symbols for infrastructure mapping are often domain specific, detailing valve or transformer characteristics, for example. Basic categorization by color hue assists readers in making connections among symbols. For example, the systematic use of red, green, and blue for A, B, and C phases gives an overall structure to the electrical utility map and is useful in the field in identifying equipment. Line colors and casings that echo associated label and halo styles are also characteristic of these maps. For example, green labels name polygons outlined in green on the telecom map, and labels with green halos link to green lines on the electrical utility map. Lightness sequences are useful to show series of zoning categories such as low-to-high-density residential planning and low-to-high flood water planning. Symbol shape is used to group like features and distinguish different types of features, such as square symbols for fuses and triangle shapes for pedestals on the electrical utility map.

In contrast to thematic maps that include minimal labeling and rely on symbol patterns to portray data, infrastructure maps inventory the details of sizes, dates, owners, and contracts with labels. Thus, they often include labels in rigid arrangements not shared in reference mapping. This can be a disconcerting characteristic of the maps when the onus is on the reader to untangle a label from other lines and labels with which it conflicts. While predictability of position is beneficial, this objective competes with the need for legibility. As labeling engines improve and map users bring interactive query tools to the field, design may become more of a priority for these maps.

4.0 TELECOMMUNICATIONS MAP

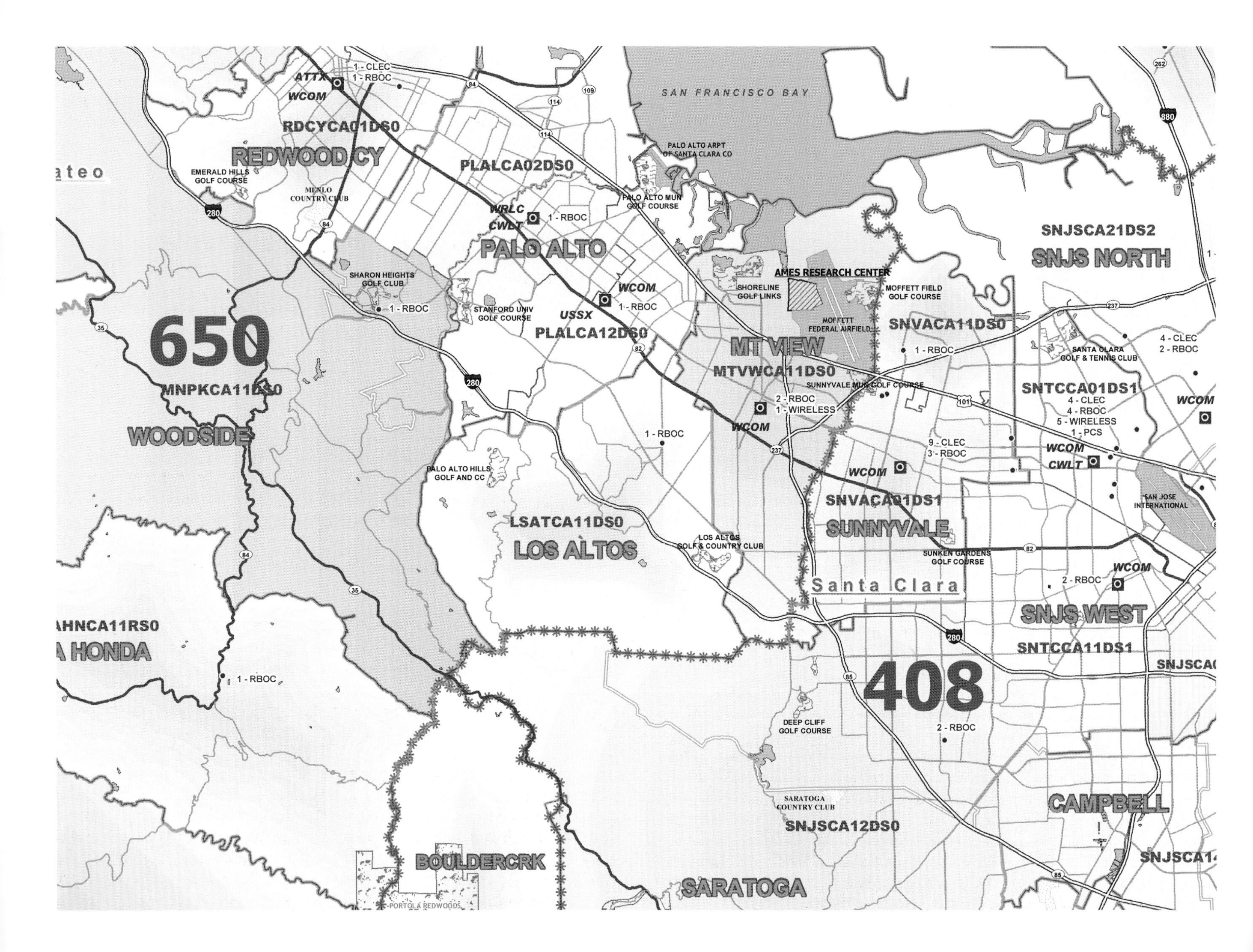

4.0 TELECOMMUNICATIONS MAP

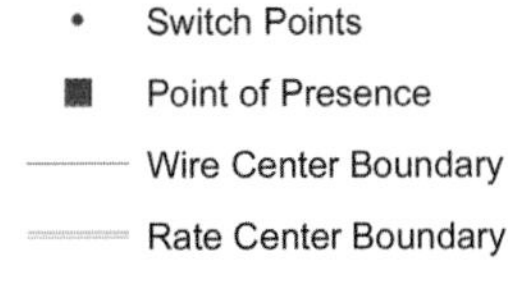

Geographic Data Technology (GDT) data for this map contains a wealth of detail on telecommunications infrastructure in the San Francisco area. Wire centers, rate centers, area code boundaries, switch points, and a base of road detail, parks, and other landmark areas are combined with different emphases in the four map versions that follow.

Courtesy of Tele Atlas.

ORIGINAL DESIGN (Pages 76 and 78)

On the original map by GDT, wire-center areas are bounded by brown lines with corresponding brown labels. These areas also are lightly tinted with radial or gradient fills. Rate-center areas have green lines with larger green labels. Given that smaller wire-center polygons do not necessarily nest within rate-center polygons, white troughs within green rate-center boundaries highlight the lack of coincidence between rate- and wire-center boundaries where they diverge. Asterisk symbols outline the large area-code polygons, allowing rate- and wire-center boundaries to show through gaps between the symbols.

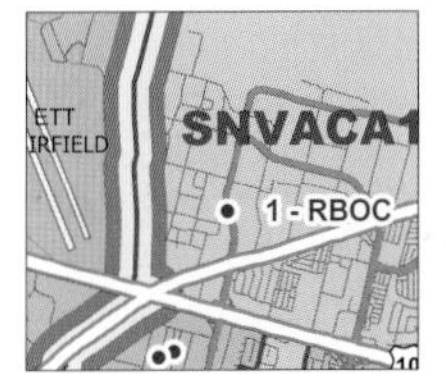

EMPHASIS ON WIRE CENTERS (Page 78)

The road network is more subdued in this design, relying on a hierarchy of line weights in gray with no hue differences. Highways are pulled forward using a cased symbol with a white fill that provides a stark contrast against the background colors The highway shields are rendered in black and white to group them visually with the black-white-gray road network. The broad yellow lines for the rate center boundaries are visually prominent. The brown lines for wire-center boundaries are additionally supported by warm hues filling wire-center areas. Just four hues are the minimum needed to distinguish areas when hues do not represent particular area attributes (hues can be repeated as long as areas are not adjacent). The wide blue area code boundary becomes a triple-layer line of different widths and hues when it coincides with the other two boundary types The brown and yellow lines are seen separately in the lower portion of the map segment where boundaries are not coincident.

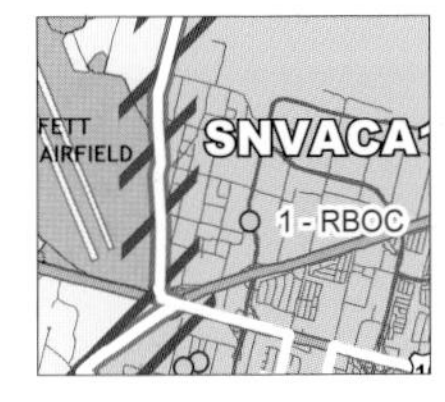

EMPHASIS ON RATE CENTERS (Page 79)

All roads are pushed behind the boundary lines in this design. They are represented with a hierarchy of widths in desaturated red and orange. The red network contrasts with yellow-green-blue hues used on this map to fill rate-center polygons, which are bounded and labeled in green. The smaller wire-center areas are bounded in white which contrasts with the roads and fill colors and is matched with the white labels—this coordinated set of labels and boundaries stands out prominently against the background. Points are less prominent in blues, but since these hues are similar to the hues for rate-center areas, and different from the road hues, they group appropriately with telecom features. Label positions assist in associating labels with their polygons by straddling boundaries that cut across the named polygon. For example, SNJSCA12DS0, a wire-center polygon in the south, crosses three green rate-center boundaries to show that all of these rate centers are in the same wire-center polygon.

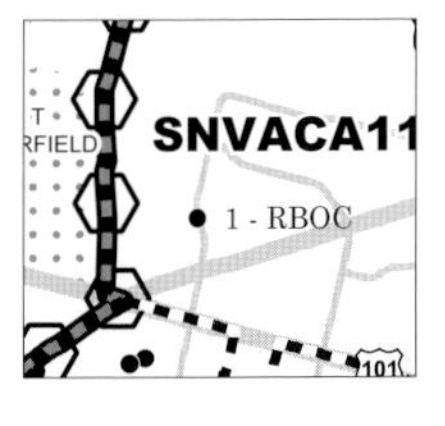

BLACK-AND-WHITE DESIGN (Page 79)

The simpler road network in this design provides visual space for the three boundary types. These boundary styles are coordinated to build logical combinations. Gray wire-center boundaries cased in black combine with black-and-white dashed rate boundaries to build a black-and-gray dash where they run together. The layer order is a key to creating this logic. The white line on the bottom of the set masks the roads so black-and-gray dashing i reserved for only combinations of boundaries (gray roads do not show through black dashes). The layer order is roads on the bottom, white knock-out for boundaries, black lines for boundary casings, gray lines, and black dashes on top. Open patterns are used on the map in the airfield dot pattern, and large hexagons outline the area code boundary. These are matched with an outline text style for area code labels.

REDESIGNS

4A ORIGINAL DESIGN

4B EMPHASIS ON WIRE CENTERS

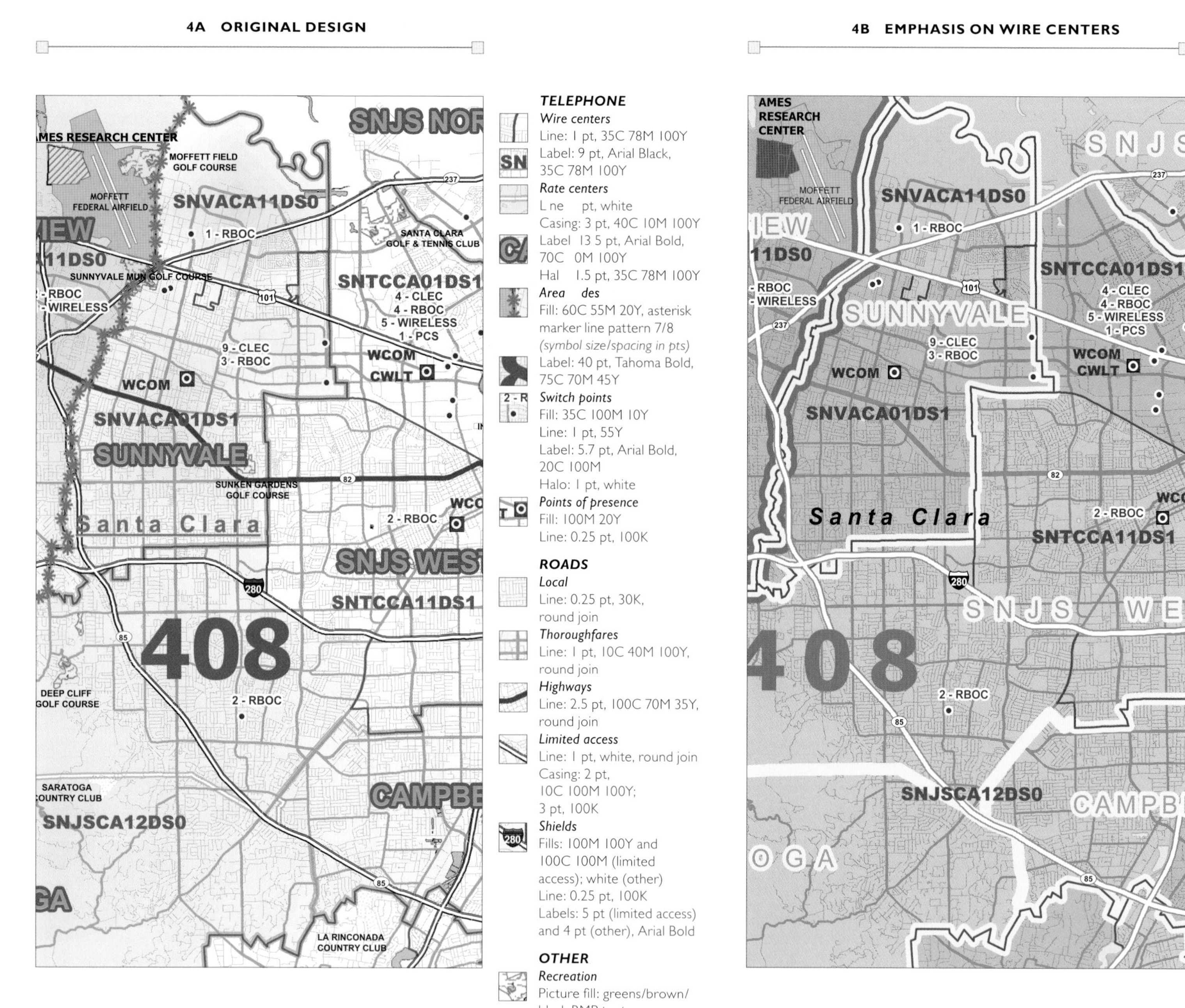

TELEPHONE

Wire centers
Line: 1 pt, 35C 78M 100Y
Label: 9 pt, Arial Black, 35C 78M 100Y

Rate centers
L ne pt, white
Casing: 3 pt, 40C 10M 100Y
Label 13 5 pt, Arial Bold, 70C 0M 100Y
Hal 1.5 pt, 35C 78M 100Y

Area des
Fill: 60C 55M 20Y, asterisk marker line pattern 7/8 *(symbol size/spacing in pts)*
Label: 40 pt, Tahoma Bold, 75C 70M 45Y

Switch points
Fill: 35C 100M 10Y
Line: 1 pt, 55Y
Label: 5.7 pt, Arial Bold, 20C 100M
Halo: 1 pt, white

Points of presence
Fill: 100M 20Y
Line: 0.25 pt, 100K

ROADS

Local
Line: 0.25 pt, 30K, round join

Thoroughfares
Line: 1 pt, 10C 40M 100Y, round join

Highways
Line: 2.5 pt, 100C 70M 35Y, round join

Limited access
Line: 1 pt, white, round join
Casing: 2 pt, 10C 100M 100Y; 3 pt, 100K

Shields
Fills: 100M 100Y and 100C 100M (limited access); white (other)
Line: 0.25 pt, 100K
Labels: 5 pt (limited access) and 4 pt (other), Arial Bold

OTHER

Recreation
Picture fill: greens/brown/ black BMP texture

Ames
Fill: 34C 56M 100Y, 45° line fill pattern 0.5/2.3 *(line width/spacing in pts)*

ArcMap Tips (see pages 161 to 167)

17 Hatched fill	18 Picture fill
19 Gradient fill	38 Multiline labels

TELEPHONE

Wire centers
Fills: 10C 25M; 5C 35M 20Y; 10C 20M 75Y; 5C 20M 45Y
Line: 1 pt, 50C 80M 100Y
Label: 9 pt, Arial Black, 50C 80M 100Y

Rate centers
Line: 6 pt, 55Y
Label: 12 pt, Arial, 50M 100Y, varied character spacing
Halo: 2 pt, 55Y

Area codes
Fill: 12 pt, 70C 50M
Label: 40 pt, Tahoma Bold, 70C 50M, character spacing 40

Switch points
Fill: 20C 100M
Line: 1 pt, white
Label: 5.5 pt, Arial Bold, 20C 100M
Halo: 1 pt, white

Points of presence
Fill: 20C 100M
Line: 0.25 pt, 100K
Label: 6 75 pt Arial Black, 10C 100M 35Y

ROADS

Lines: 0.25, 1.3, and 2.5 pt; 50K, round cap/join

Limited access
Line: 2 pt, white, round cap/join
Casing: 3.5 pt, 50K

Shields
Fills: white and 100K
Line: 0.25 pt, 100K
Labels: 5 pt (limited access) and 4 pt (other shields), Arial Bold

OTHER

Airports
Fills: 30K (base) 25Y (air strip)
Line: 0.35 pt, 60K
Label: 4.5 pt, Arial, 100K

Ames
Fill: 70K
Label: 5.5 pt, Tahoma Bold, 100K

Water
Fill: 20C 5M
Line: 0.5 pt, 50C 15M

ArcMap Tips

20 Four-color fill	27 Label hull
31 Embed font	40 Layer order

4C EMPHASIS ON RATE CENTERS

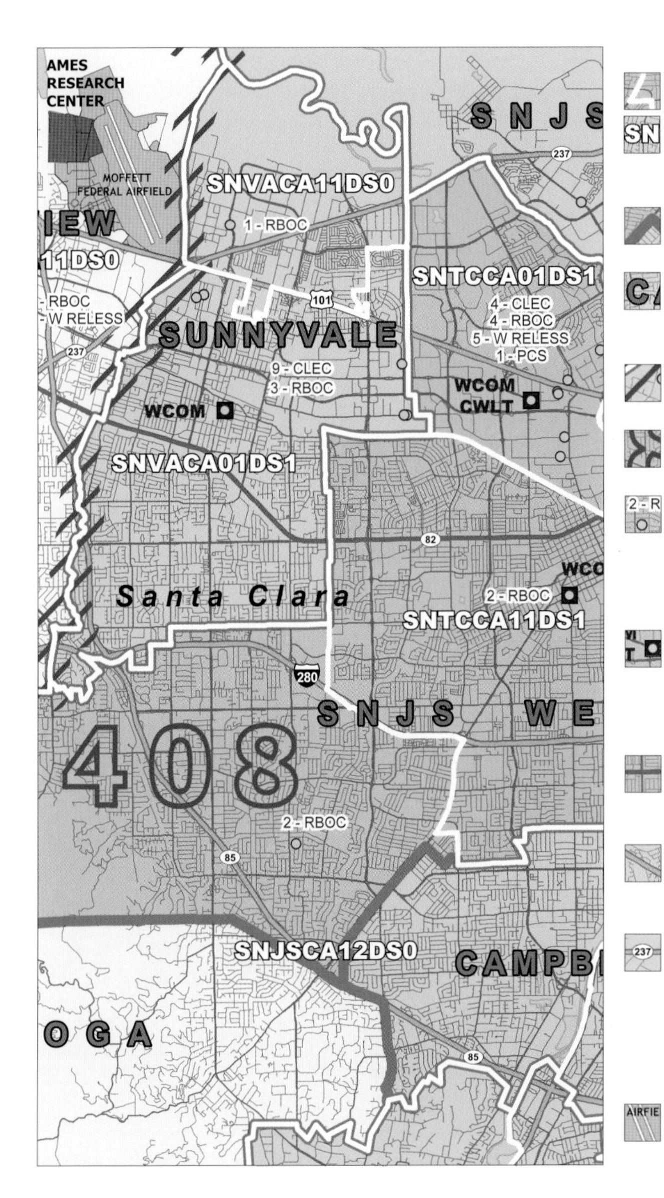

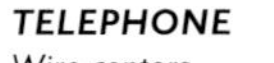

TELEPHONE

Wire centers
Line 2 pt, white
Label: 9 pt, Arial B ack, white
Halo: 0.5 pt, 100K

Rate centers
Fills: 30C 55Y; 30C 20M; 50Y; 35C 30M
Line: 4 pt, 80C 35M 100Y
Label: 12 pt, Arial Bold, 80C 35M 100Y, varied character spacing
Halo: 0.75 pt, 100K

Area codes
Fill: 45° line fill pattern 1/5 *(line width/spacing in pts)*
Label: 40 pt, Tahoma Bold, no fill, character spacing 40
Halo: 2 pt, 65C 85M

Switch points
F ll: 50C
Line: 0.5 pt, 100C 100M
Label: 5.5 pt, Arial, 100C 100M
Halo: 1 pt, white

Points of presence
Fill: 100C 100M
Line: 0.5 pt, 100K
Label: 6.75 pt, Arial Black, 100C 100M

ROADS
Lines: 0.2, 0.5, and 1.8 pt; 20C 80M 70Y, round cap/join

Limited access
Line: 1.5 pt, 35M 80Y, round cap/join
Casing: 2.5 pt, 20C 80M 70Y

Shields
Fills: white and 100K
Line: 0.25 pt, 100K
Labels: 5 pt (limited access) and 4 p (other sh elds), Arial Bold

OTHER

Airports
Fills: 30K (base), 25Y (air strip)
Line: 0.35 pt, 60K
Label: 4.5 pt, Tahoma, 100K

Ames
Fill: 70K
Label: 5.5 pt, Tahoma Bold, 100K

ArcMap Tips

17 Hatched fill	30 Custom symbols
32 Open characters	36 Convert to annotation

4D BLACK-AND-WHITE DESIGN

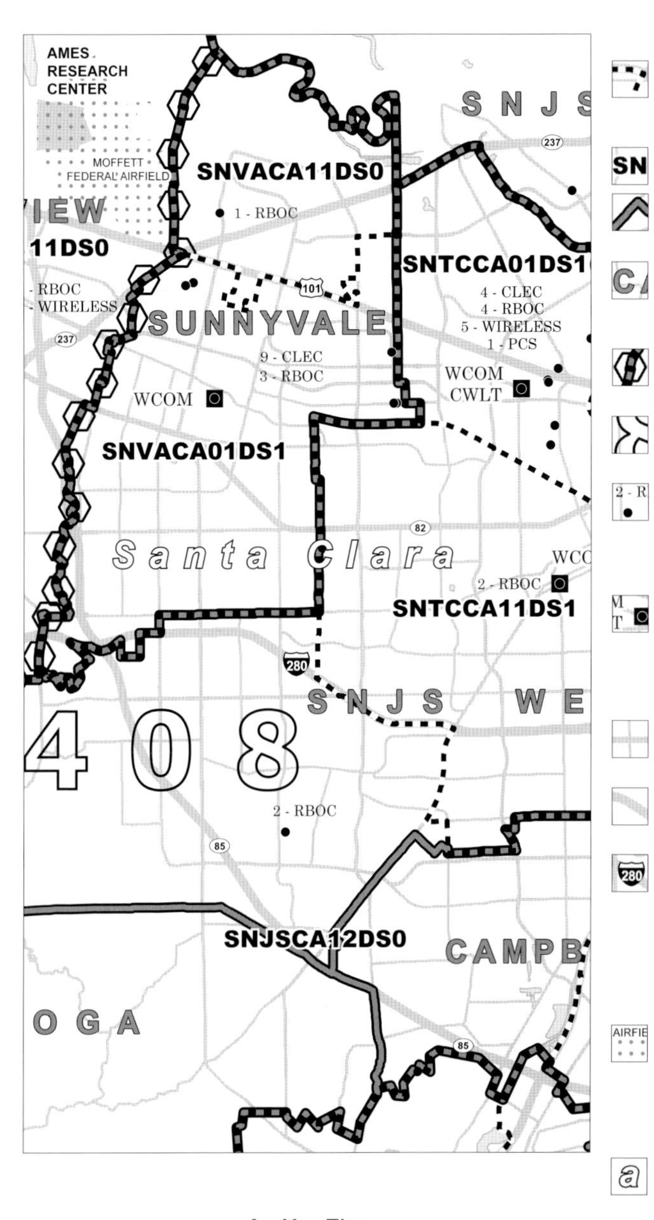

TELEPHONE

Wire centers
Lines: 2 pt, white; 2 pt, 100K, dash pattern 3/3 *(dash/gap in pts)*
Label: 9 pt, Arial Black, 100K
Halo: 0.75 pt, white

Rate centers
Line: 2 pt, 50K
Casing: 4 pt, 100K
Label: 12 pt, Arial Bold, 50K varied character spacing
Halo: 0.35 pt, 100K

Area codes
Line: 20 pt, 100K, hexagon marker line pattern 10/20
Label: 40 pt, Tahoma Bold, no fill, character spacing 70
Halo: 1 pt, 100K

Switch points
Fill: 100K
Line: 0.25 pt, white
Label: 5.5 pt, Century Schoolbook, 100K

Points of presence
Fill: 100K
Label: 6.75 pt, Century Schoolbook, 100K
Halo: 1 pt, white

ROADS
Lines: 1 and 2.5 pt; 20K, round cap/join

Limited access
Line: 3.5 pt, 20K, round cap/join

Shields
Fills: white and 100K
Line: 0.25 pt, 100K
Labels: 5 pt (limited access) and 4 pt (other shields), Arial Bold

OTHER

Airports
Fill: 50K, circle marker fill pattern 1.5/4.5/4.5 *(symbol size/X spacing/ pacing)*
Label: 4.5 pt, Arial, 100K
Halo: 1 pt, white

City
Label: 14 pt, Arial Bold Italic, white
Halo: 0.5 pt, 100K, character spacing 80

ArcMap Tips

5 Marker line	15 Pattern fill
32 Open characters	40 Layer order

4.1 NEW YORK CITY ROAD RESURFACING MAP

Wide, saturated red lines for roads that will soon need resurfacing hold the main message of this map. The symbol sequence goes from red to orange to thin yellow lines for roads most recently resurfaced, thus, needing the least attention in planning. This triple combination of width change, hue transition, and lightness sequence provides a logical visual inventory for city workers. The map design is spare—the base information is limited to city board boundaries and the coastline—but the logic in New York City street naming provides the structure needed to locate streets. The minimal base information on this map reduces distractions from the basic task of seeing where resurfacing work will be needed, allowing for quick prioritizing and scheduling.

Courtesy of Robert Nossa, Bowne Management Systems, Inc., and John Griffin, New York City Department of Transportation.

Community Boards

Protected Streets

Expires within 6 months

Expires within 6 to 12 months

Expires within 2 years

Expires within 3 years

Expires within 4 years

Expires within 5 years

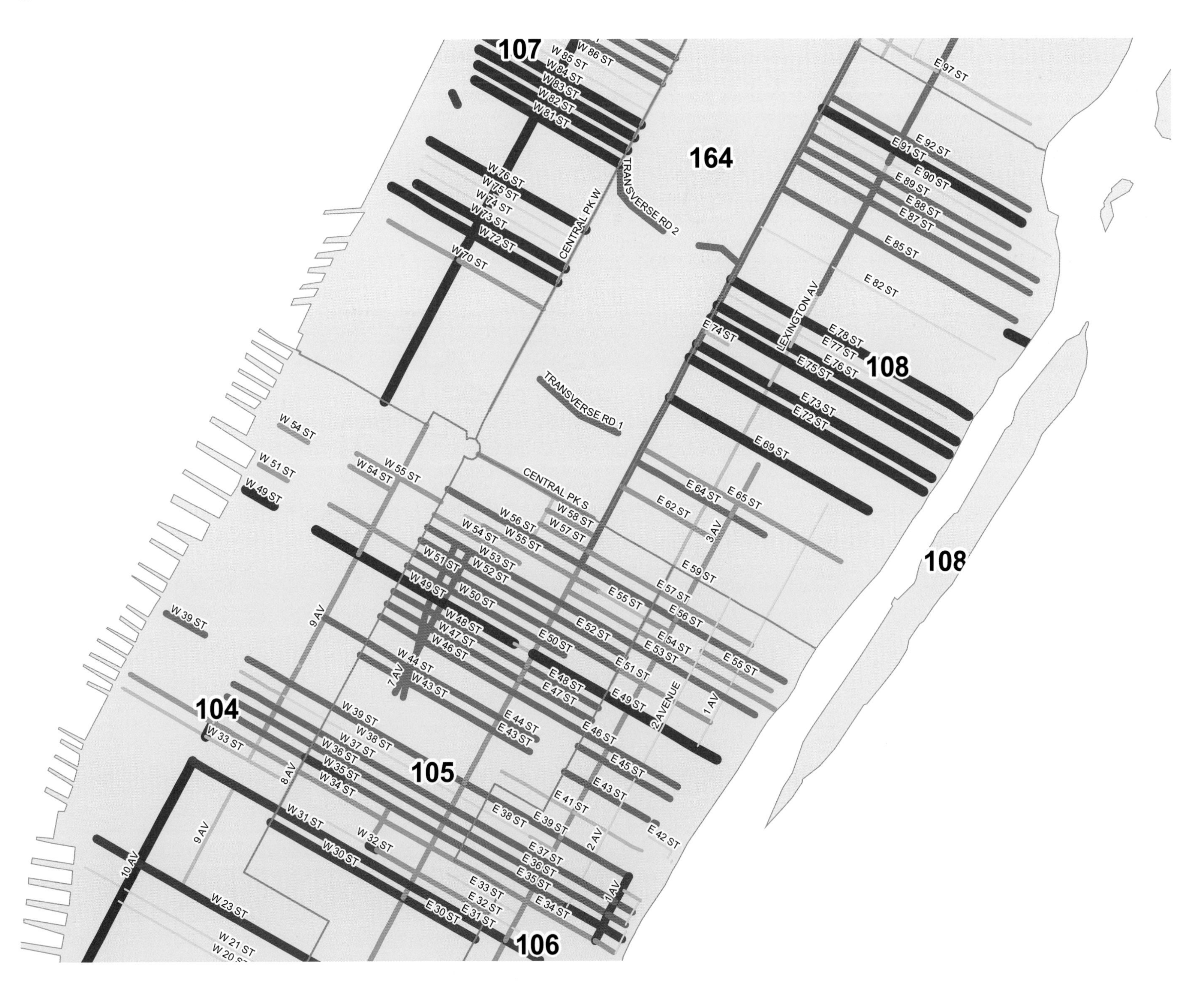
107
164
108
108
104
105
106
W 86 ST
W 85 ST
W 84 ST
W 83 ST
W 82 ST
W 81 ST
W 76 ST
W 75 ST
W 74 ST
W 73 ST
W 72 ST
W 70 ST
CENTRAL PK W
TRANSVERSE RD 2
TRANSVERSE RD 1
E 97 ST
E 92 ST
E 91 ST
E 90 ST
E 89 ST
E 88 ST
E 87 ST
E 85 ST
E 82 ST
LEXINGTON AV
E 74 ST
E 78 ST
E 77 ST
E 76 ST
E 75 ST
E 73 ST
E 72 ST
E 69 ST
W 54 ST
W 51 ST
W 49 ST
W 55 ST
W 54 ST
CENTRAL PK S
E 64 ST
E 65 ST
E 62 ST
W 58 ST
W 57 ST
W 56 ST
W 55 ST
W 54 ST
W 53 ST
W 52 ST
W 51 ST
3 AV
E 59 ST
W 49 ST
W 50 ST
E 57 ST
E 55 ST
E 56 ST
W 39 ST
9 AV
W 48 ST
W 47 ST
W 46 ST
E 52 ST
E 50 ST
E 54 ST
E 53 ST
E 55 ST
W 44 ST
W 43 ST
7 AV
E 51 ST
E 48 ST
E 47 ST
E 49 ST
2 AVENUE
1 AV
W 39 ST
W 38 ST
W 37 ST
W 36 ST
W 35 ST
W 34 ST
W 33 ST
8 AV
E 44 ST
E 43 ST
E 46 ST
E 45 ST
E 43 ST
E 41 ST
E 38 ST
E 39 ST
E 42 ST
W 31 ST
W 32 ST
W 30 ST
9 AV
2 AV
10 AV
E 37 ST
E 36 ST
E 35 ST
E 34 ST
1 AV
E 33 ST
E 32 ST
E 31 ST
E 30 ST
W 23 ST
W 21 ST

4.2 DEDHAM SANDING ROUTES MAP

This simple map shows one of five sanding regions for Dedham, Massachusetts. Each region has similar total road distances to organize truck drivers' snow plow and sanding work. Each region also has a signature hue used for road fills, boundary lines, and gradient bands that blend to each region color. This gradient clearly outlines the region but does not fill it with the hue, so road lines remain in high contrast against the background. Road labels, essential on this map, are well positioned next to roads, so they do not interfere with red casings that highlight roads needing extra attention in bad weather conditions. Red fills also highlight schools and other areas that need prompt clearing. Area feature labels are lighter and larger, giving basic landmark locations, such as parks, without the clutter of building footprints or parcels, which are not needed for the sanding and plow work.

Courtesy of Applied Geographics, Inc.

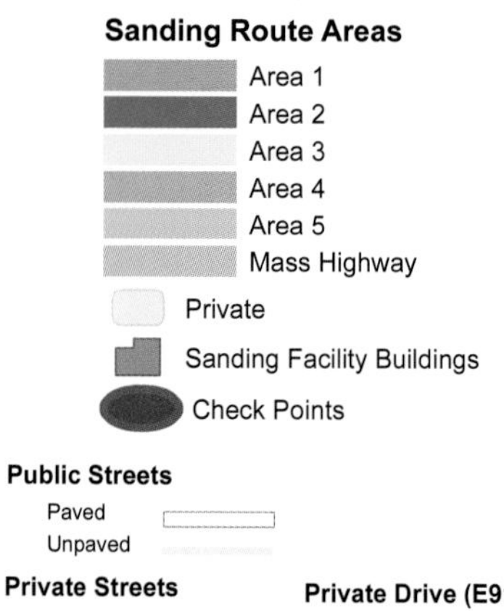

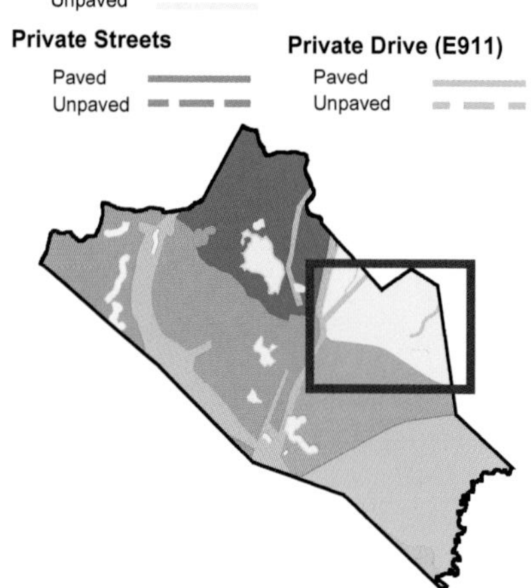

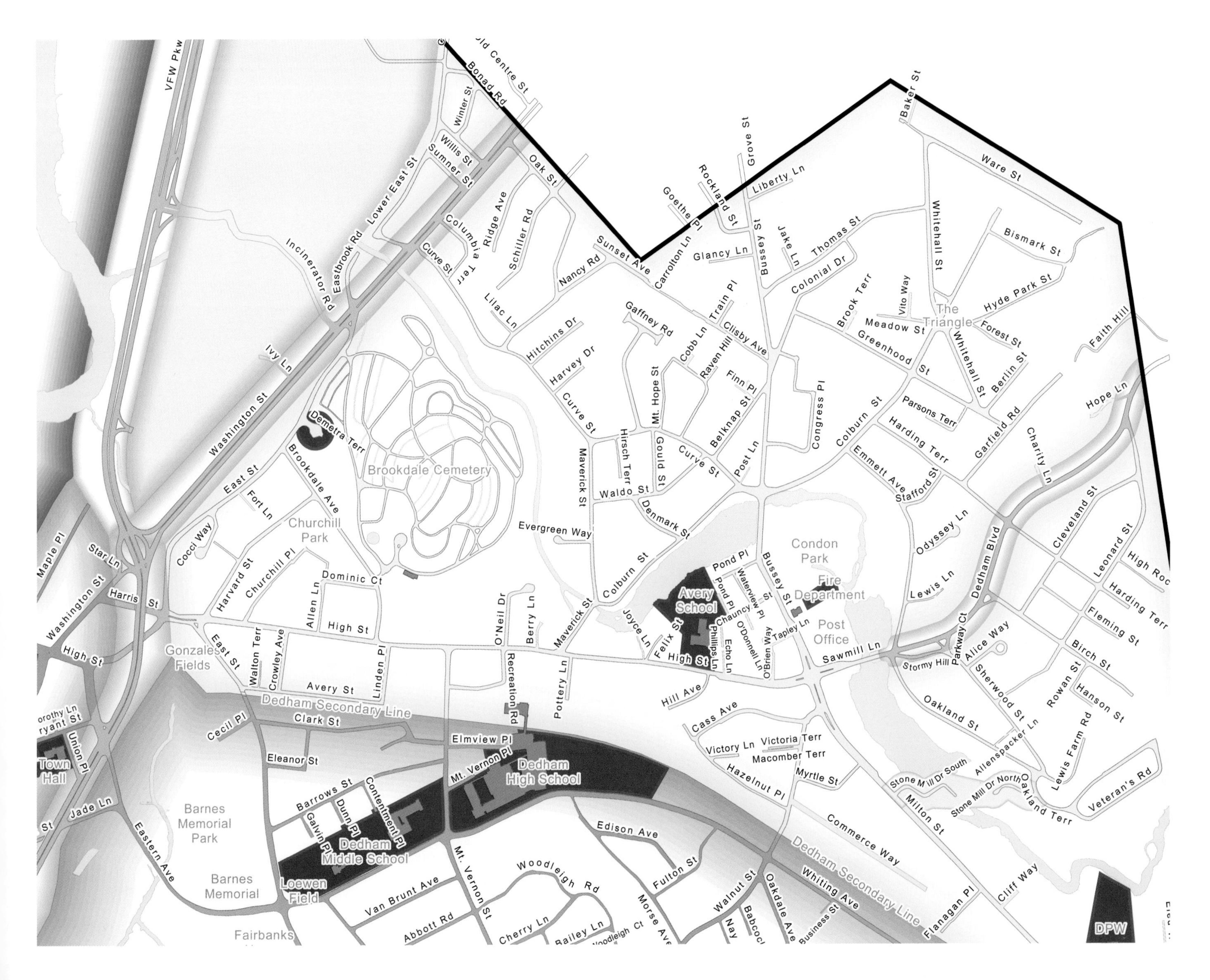
Brookdale Cemetery
Churchill Park
Condon Park
Fire Department
Post Office
Avery School
Dedham High School
Dedham Middle School
Loewen Field
Barnes Memorial Park
Gonzales Fields
Town Hall
The Triangle
DPW
Dedham Secondary Line
Washington St
High St
Bussey St
Sawmill Ln
Colburn St
Maverick St
Eastern Ave
Mt. Vernon St
Oakland St
Milton St
Commerce Way
Whiting Ave
Dedham Blvd
Ware St
Whitehall St
Sunset Ave
Oak St
VFW Pkwy

4.3 DOWNEY UTILITIES MAP

Downey, California, utilities maps show water, sewer, and storm drain systems in blue, green, and red, respectively, with detailed annotation along mains (such as diameter, material, offset, length, and bar code) and at points. The airphoto underlay of buildings, parking areas, and vegetation provide location information augmented with parcel boundaries and curb lines registered with the image. Transparent hues categorize areas; for example, parcels are yellowish and named locations (such as shopping malls) are greenish. These very pale overlays provide a clear overall structure for the image and also reduce the contrast in the photo so it does not hinder legibility of the utilities content, the primary message of the map. The map shown here is a portion of one page from a utilities map book. The map book pages are organized with overlap at each edge, which prevents property from landing only at a page break. Light lines near the page edges mark this overlap.

Courtesy of the City of Downey, California.

Water Distribution System

AC Auto Control Valve
AH Access Manhole
AR Air Release Valve
BF Backflow Device
BL Ball Valve
BO Blowoff Valve
BV Butterfly Valve
CM Compound Meter
CV Check Valve
CV Detector Check Valve
DC Double Check Valve
DE Dead End
EI Emergency Intertie
FC Flow Control
FH Fire Hydrant
FI Fittings
PR Pressure Regulating Station
PS Pump Station
PT Point
PV Plug Valves
RE Reducer
RS Reservoir
SM Service Meter
SP Sampling Point
TM Telemetry
TO Turnout
TP Turbo Meter
TS Cathodic Protection Test Station
TV Tapping Valve
UN Unknown
VA Gate Valve
VT Vault
WE Well

Main Annotation:
OWNER (if not Downey), DIAMETER, MATERIAL, PL OFFSET, LENGTH, BAR CODE

Node Annotation:
Fire Hydrant Number , **Valve Number** , Fitting Number

Abbreviations:

AC - Asbestos Cement
DI - Ductile Iron
CI - Cast Iron
PL - Property Line
CL - Centerline
PVC - Polyvinyl Chloride
CO - County
STL - Steel

Sanitary Collection System

CO Clean out
CV Check Valve
Direction of Flow
DM Manhole Drop
FT Flush Tank
GT Grease Trap
LH Lamp Hole
LS Lift Station
ST Sand Trap
ST Septic Tank
TM Telemetry
UN Unknown
VA Valve
VT Vault
WE Well

Main Annotation:
OWNER (if not Downey), CL OFFSET, LENGTH, DIAMETER, MATERIAL, BARCODE

Manhole Annotation:
RIM ELEVATION, INVERT ELEVATION

Abbreviations:

CL - Centerline
RCP - Reinforced Concrete Pipe
CO - County
VCP - Vitrified Clay Pipe

Storm Drains

CB Catch Basin
CI Curb Intake
DB Debris Basin
Direction of Flow
DW Dry Well
FI Fitting
IT Inlet
JB Junction Box
JM Junction Structure w/Manhole
JS Junction Structure
LS Lift Station
MH Standard Manhole
OT Outlet
PC Private Connection Point
PT Point
TM Telemetry
TS Transition Structure
UN Unknown

Main Annotation:
OWNER (if not Downey), CL OFFSET, LENGTH, DIAMETER, MATERIAL, BAR CODE

Service Lateral Annotation:
LENGTH, DIAMETER

Abbreviations:

CL - Centerline
RCB - Reinforced Concrete Box
CMP - Corrugated Metal Pipe
RCP - Reinforced Concrete Pipe
CO - County

Basemap

Centerline Tie Elevations
Casings
Laterals
Centerlines
Mains
Curb and Sidewalk
Laterals
row
Mains
Parcels
Laterals
Gov. Properties
Major Location Names
Easements

TYPE

Distribution
Reclaim
Transmission

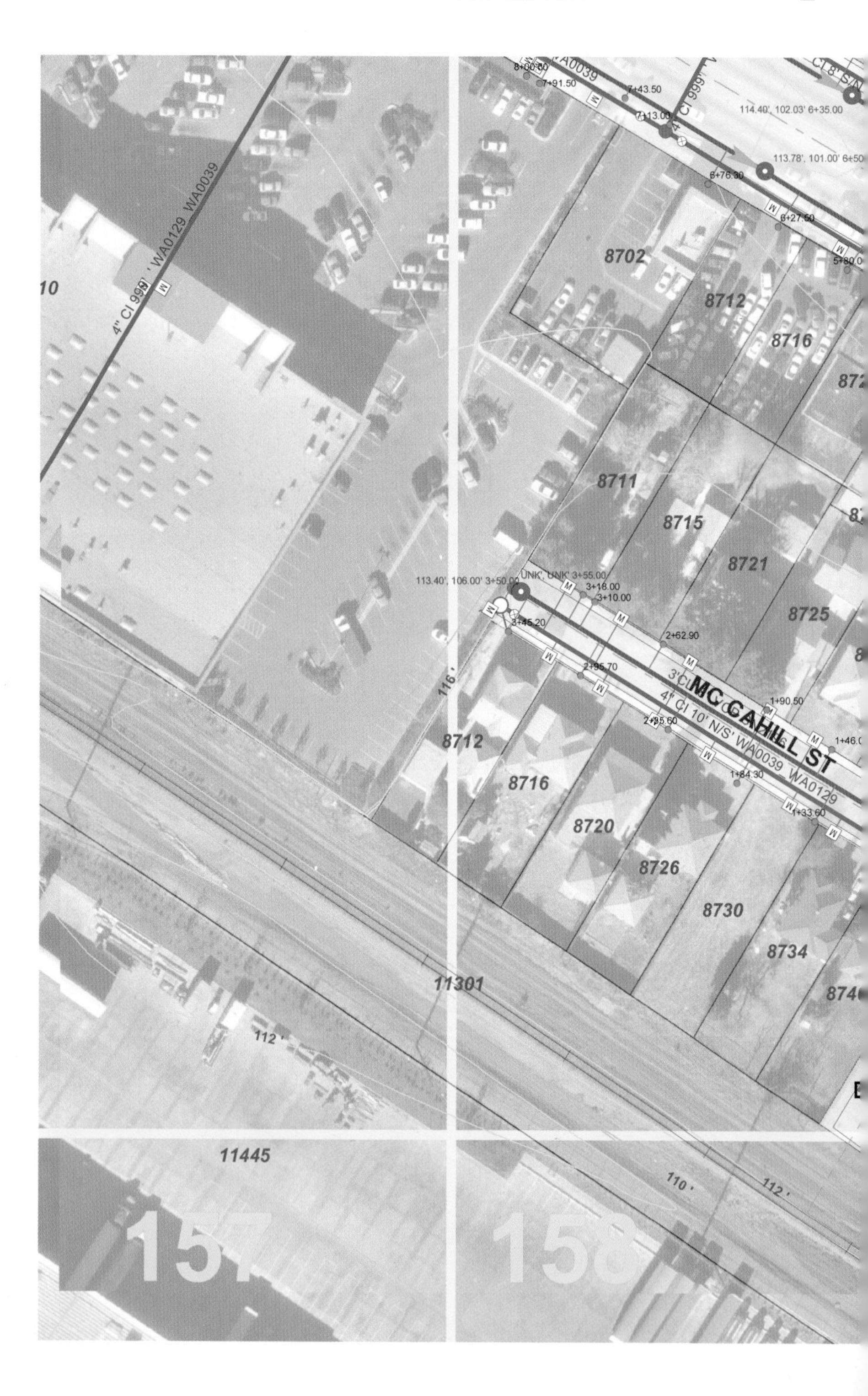

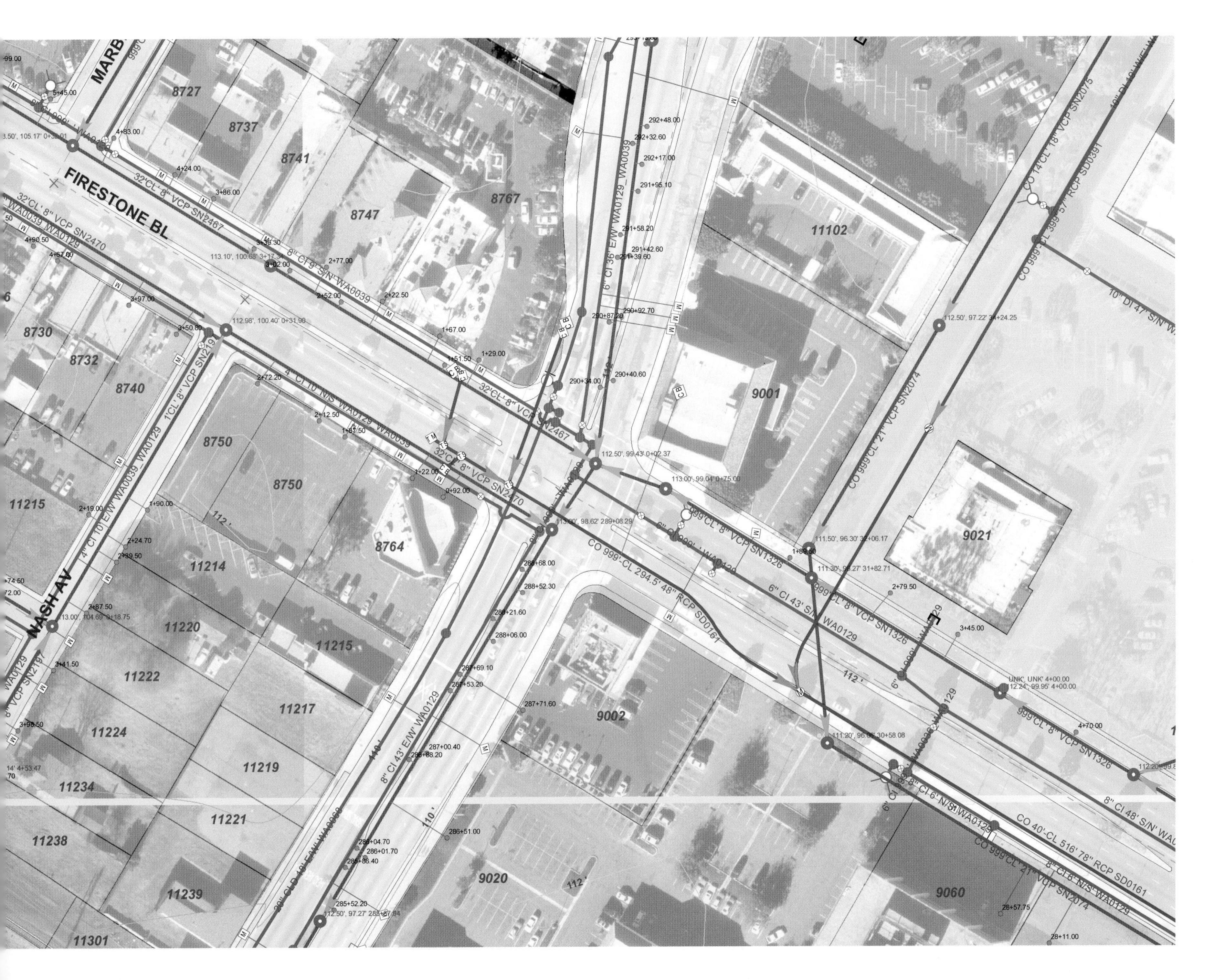
FIRESTONE BL
NASH AV
8727
8737
8741
8747
8767
11102
9001
9021
8730
8732
8740
8750
8750
8764
11215
11214
11220
11215
11222
11217
11224
11219
11234
11221
11238
11239
11301
9002
9020
9060
32'CL 8" VCP SN2467
32'CL' 8" VCP SN2470
4" CI 10' N/S' WA0129 WA0039
CO 999'-CL 294.5' 48" RCP SD0161
6" CI 43' S/N' WA0129
999'CL' 8" VCP SN1326
CO 40'-CL 516' 78" RCP SD0161
8" CI 43' E/W' WA0129
6" CI 36' E/W' WA0129_WA0039
CO 999'CL' 21" VCP SN2074
8" CI 48' S/N' WA0

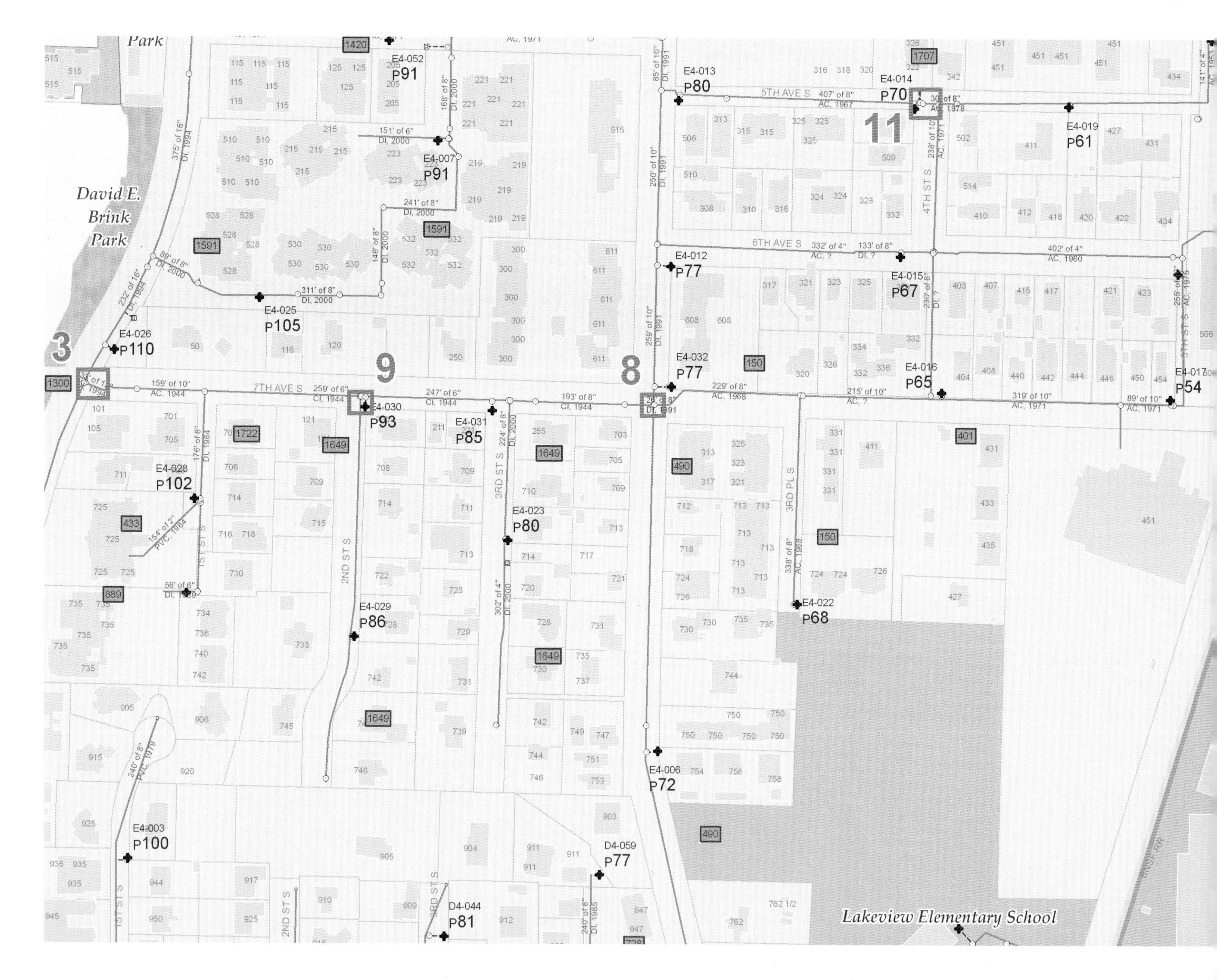
David E. Brink Park
Lakeview Elementary School
5TH AVE S
6TH AVE S
7TH AVE S
1ST ST S
2ND ST S
3RD ST S
3RD PL S
4TH ST S
5TH ST S
BNSF RR
3
9
8
11

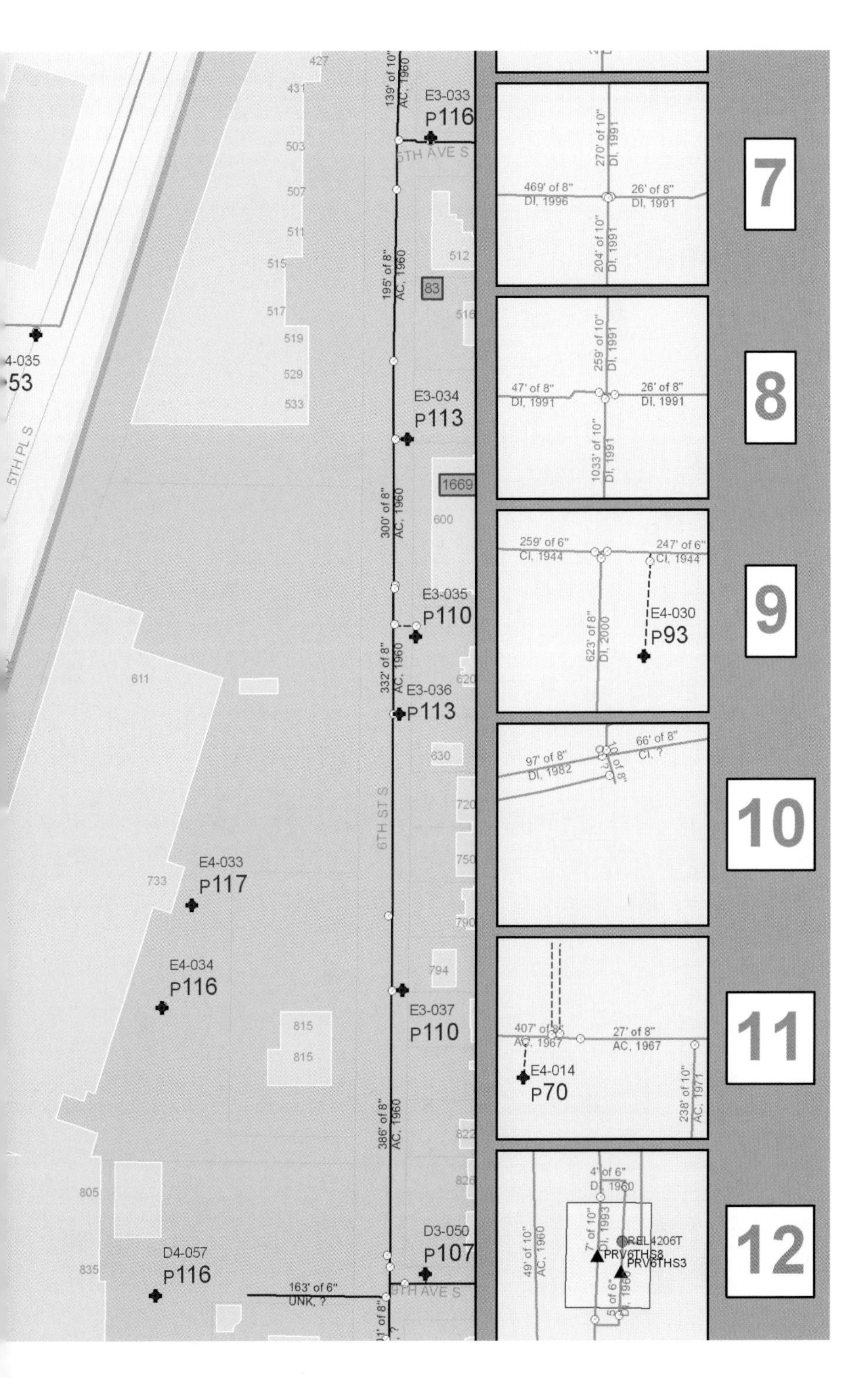

4.4 KIRKLAND WATER UTILITIES MAP

The map book for water utility infrastructure of Kirkland, Washington, includes mains, hydrants, and valves. Specific water utility attributes stand out clearly against a light basemap of gray buildings and zone hues (green and red are used for the zones shown). Parcel lines that are only slightly darker than the background, gray rather than black building numbers, and white outlines on buildings (to ensure definition against varied background hues of similar lightness) all contribute to the visual prominence of the utility details. The orange squares at more complex intersections of mains link to a series of enlarged inset maps in the right margin of the map book page. These large-scale insets provide clarity where it is needed and allow each map book page to show a larger extent.

Courtesy of the City of Kirkland, Washington.

Valve
Zone Valve
Hydrant
Air-Vac
Blow-Off
Pump
285 PZ Main
315 PZ Main
395 PZ Main
450 PZ Main
510 PZ Main
545 PZ Main
590 PZ Main
650 PZ Main
Unknown
Hydrant Lateral
Other Lateral

1301 As-Built Tag Number
Water Service Boundary
285 Pressure Zone
315 Pressure Zone
395 Pressure Zone
450 Pressure Zone
510 Pressure Zone
545 Pressure Zone
590 Pressure Zone
650 Pressure Zone
1/4 Section Grid
Parks
Schools
Lakes

4.5 THE COLONY ELECTRIC UTILITIES MAP

This detailed inventory of electrical utility infrastructure near Dallas, Texas, makes systematic use of hue and shape in organizing a large set of technical symbols. Buildings are reduced to simple rectangles representing service points, thereby reducing clutter and constructing simple relationships with parcels. Squares for transformers, triangles for pedestals, and more detailed shapes for elements such as streetlights make the features easy to identify. Red, green, and blue hues for these basic shapes and for electrical lines are used to categorize equipment by phase (A, B, C), and these hues are also used as light-colored halos around line labels. Short dashes for underground lines do not interfere with line continuity, and the texture brings lines forward visually from the thin solid lines used for street and parcel baselines.

Courtesy of CoServ Electric.

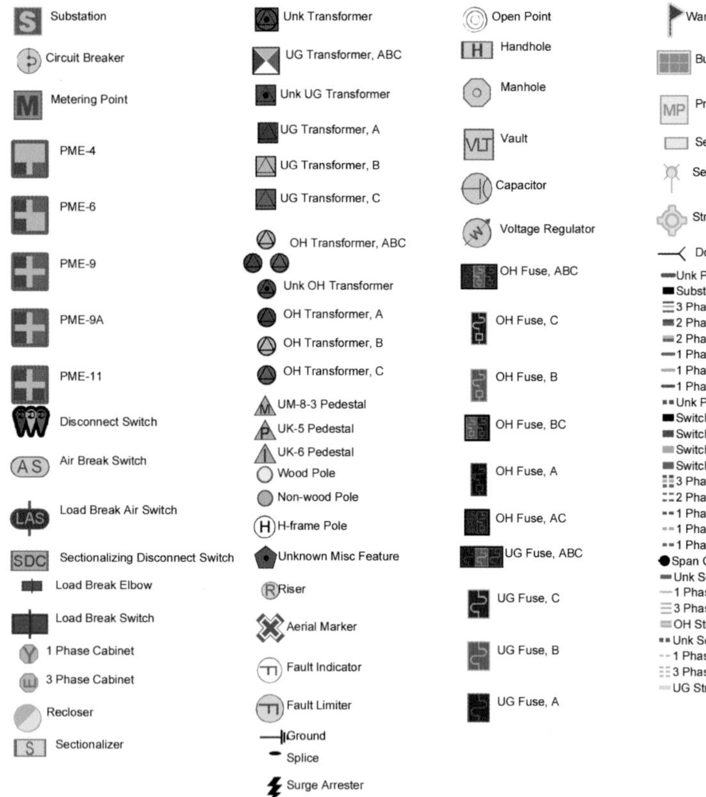

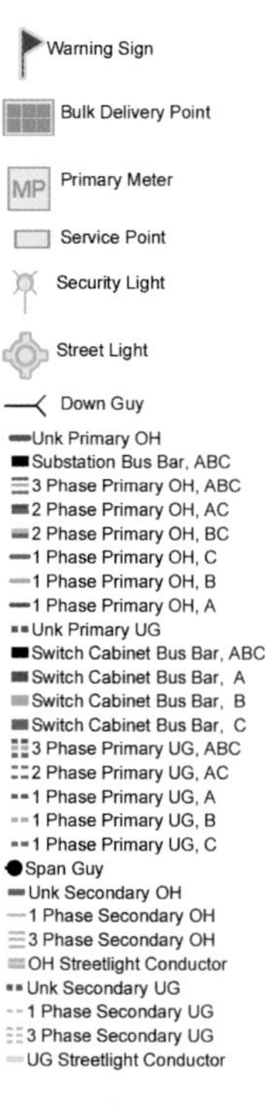

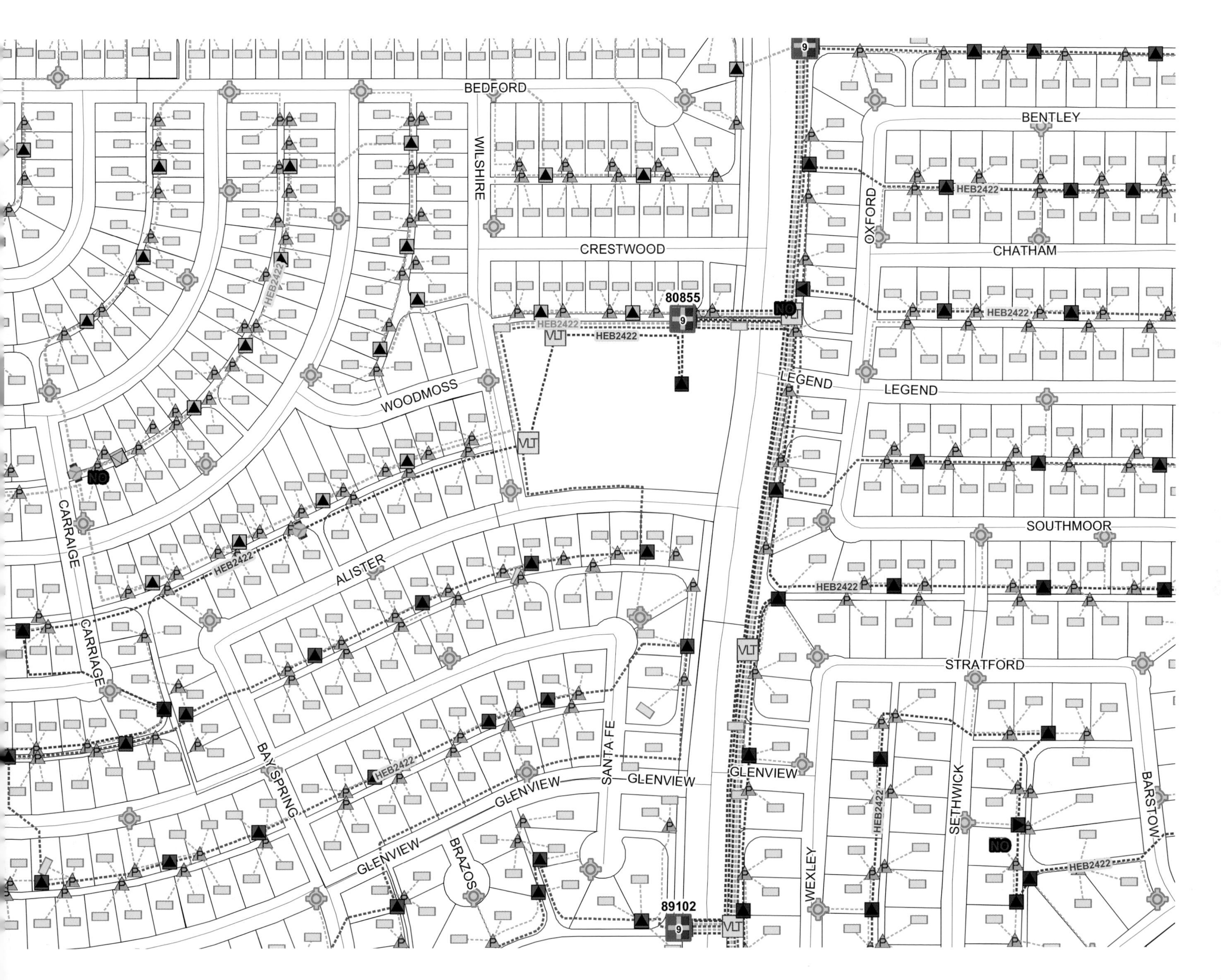
BEDFORD
BENTLEY
WILSHIRE
HEB2422
OXFORD
CRESTWOOD
CHATHAM
80855
NO
VLT
LEGEND
WOODMOSS
CARRAIGE
ALISTER
SOUTHMOOR
STRATFORD
SANTA FE
GLENVIEW
BAY SPRING
BRAZOS
WEXLEY
SETHWICK
BARSTOW
89102

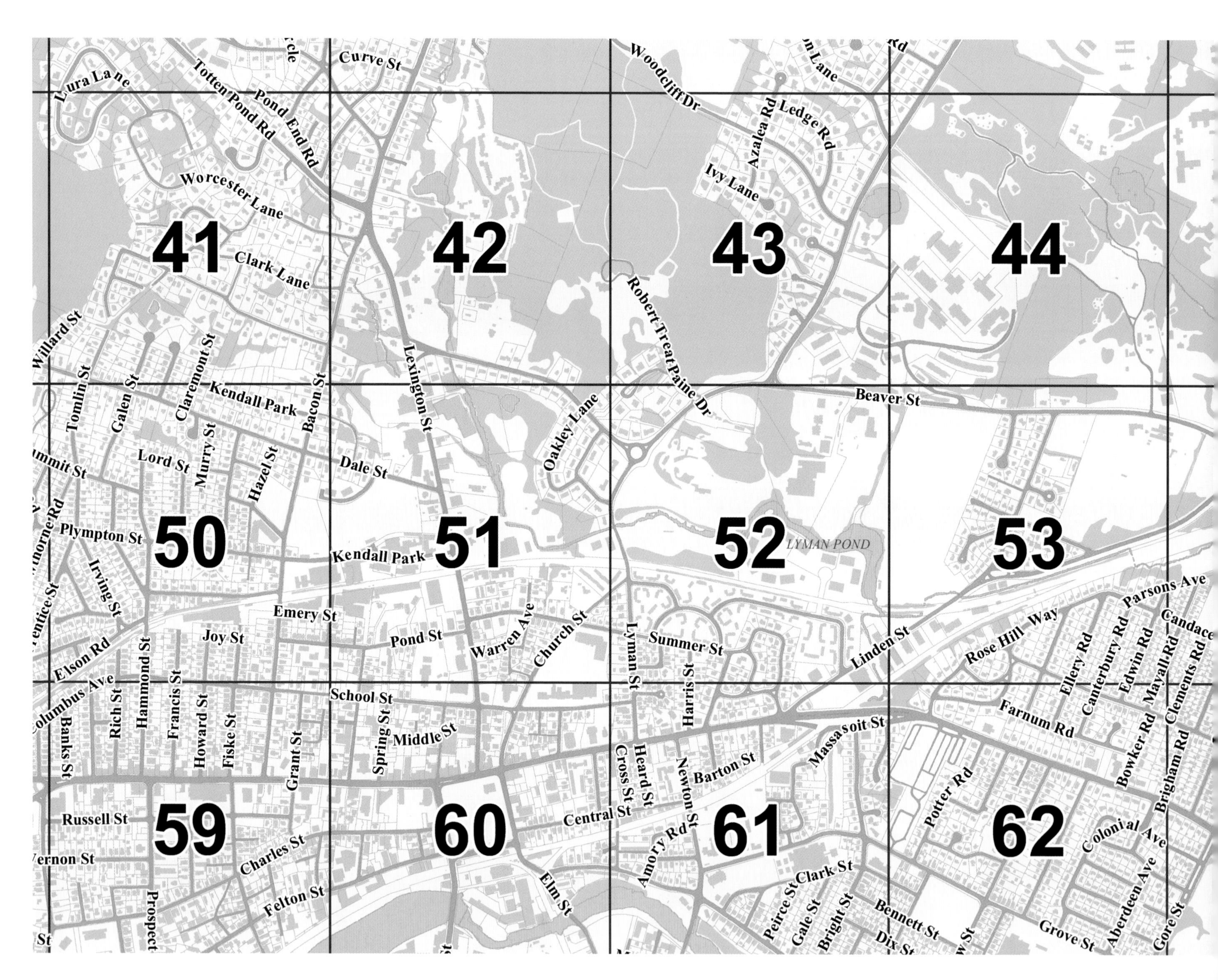
41
42
43
44
50
51
52
53
59
60
61
62
Lura Lane
Totten Pond Rd
Pond End Rd
Curve St
Woodcliff Dr
Ledge Rd
Azalea Rd
Ivy Lane
Worcester Lane
Clark Lane
Willard St
Tomlin St
Galen St
Claremont St
Kendall Park
Bacon St
Lexington St
Robert Treat Paine Dr
Beaver St
Oakley Lane
Lord St
Murry St
Hazel St
Dale St
Plympton St
Irving St
Kendall Park
LYMAN POND
Parsons Ave
Emery St
Joy St
Pond St
Warren Ave
Church St
Lyman St
Summer St
Linden St
Rose Hill Way
Candace
Elson Rd
Columbus Ave
Hammond St
Francis St
Rich St
Banks St
Howard St
Fiske St
School St
Spring St
Middle St
Grant St
Harris St
Ellery Rd
Canterbury Rd
Edwin Rd
Mayall Rd
Clements Rd
Farnum Rd
Massasoit St
Cross St
Heard St
Newton St
Barton St
Potter Rd
Bowker Rd
Brigham Rd
Russell St
Charles St
Central St
Amory Rd
Colonial Ave
Felton St
Elm St
Clark St
Peirce St
Gale St
Bright St
Bennett St
Dix St
Grove St
Aberdeen Ave
Gore St
Prospect

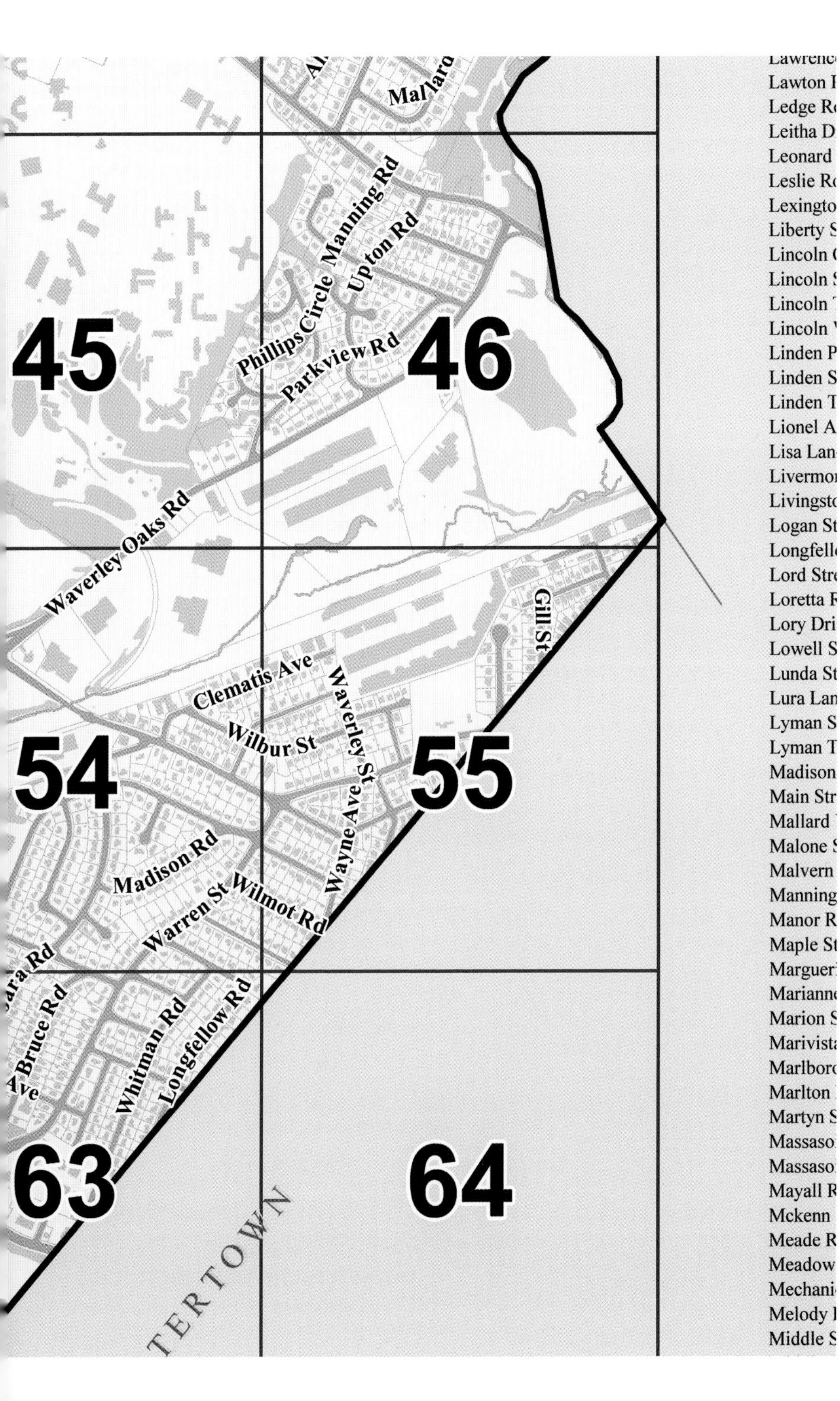

4.6 WALTHAM ASSESSOR'S INDEX MAP

This index map provides an easy reference that points to more detailed tax map pages for the city of Waltham, Massachusetts. The index grid and numbers stand out boldly against the city background, the details of which are reduced to basic urban patterns. To establish the contrast between grid and basemap, light desaturated hues are used for basic features: grays for roads and buildings, blue for water, green for vegetated areas, yellow for other base areas. Faint parcel lines support this pattern but are the lowest contrast elements. The city has a distinctive structure captured well by this design, and street search is also supported by a street index listing. Thus, further graphic detail is not needed to allow the user to find their page of interest. The eighty-one tax map pages referred to by this index each show large-scale parcel boundaries overlaid on orthophotos.

Courtesy of Applied Geographics, Inc.

4.7 WEST LINN PLATS MAP

The map of plats for the city of West Linn, Oregon, emphasizes overlapping area features. A housing subdivision map is an example of a plat. Areas on this map of plats have labels with distinct characteristics: black for plat names and numbers, blue for partition plat numbers, and a decorative font for donation land claim labels. Parcel lines in gray are background and parcels are aggregated by light hues to subdivisions. Numerous hues distinguish plats (pinks, purples, oranges, etc.), but they are similar in lightness and saturation. Labels remain legible over these hues, and plats pull together as a coherent layer of related areas rather than separating into ordered areas.

Courtesy of the City of West Linn, Oregon.

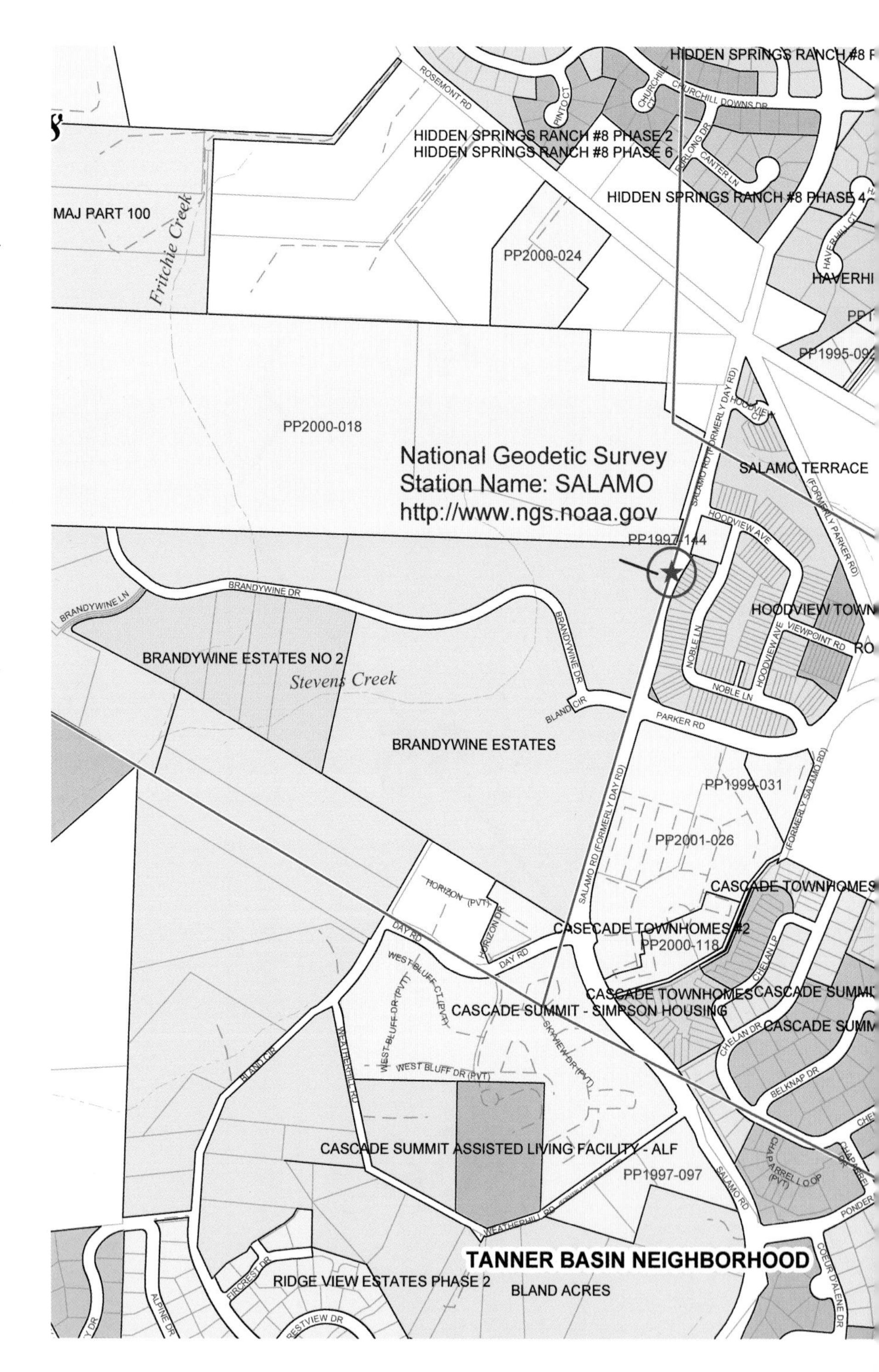

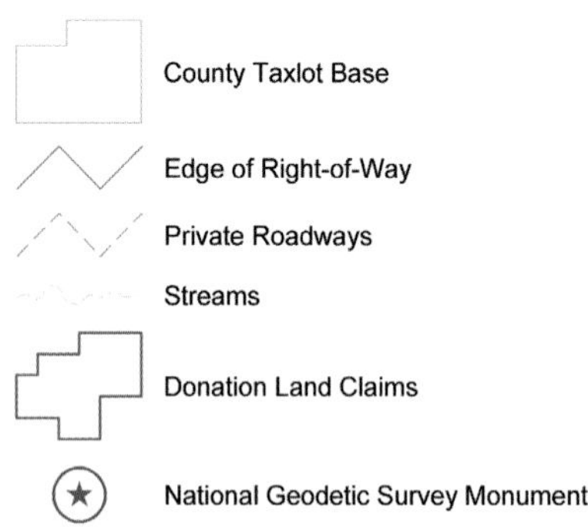

CHRISTY ADDITION (Plat no: 2526) MAP GRID: 3I
CHRISTY COURT (Plat no: 0) MAP GRID: 3I
COLLEGE HILL ESTAES ANNEX #1 (Plat no: 2487) MAP GRID: 3B
COLLEGE HILL ESTATES (Plat no: 2487) MAP GRID: 3B
CRYSTAL TERRACE (Plat no: 3288) MAP GRID: 4H
CUSHMANS SUBDIVISION (Plat no: 828) MAP GRID: 7E
DAVENPORT (Plat no: 2154) MAP GRID: 6F
DAVID TRACTS (Plat no: 1859) MAP GRID: 3H
DILLOW TRACTS (Plat no: 698) MAP GRID: 6E
DODGE GROVE (Plat no: 865) MAP GRID: 3H
EASTMAN REPLAT (Plat no: 1466) MAP GRID: 2H
ERLAND ADDITION (Plat no: 1633) MAP GRID: 6C
FARMSTEAD (Plat no: 2962) MAP GRID: 2H
FARRVIEW (Plat no: 1574) MAP GRID: 2H
FARRVISTA (Plat no: 1851) MAP GRID: 2G
FIELDVIEW ESTATES (Plat no: 2339) MAP GRID: 2G
FIR VIEW (Plat no: 3081) MAP GRID: 4H
FLORENDOS HIDEAWAY (Plat no: 3800) MAP GRID: 4G
FOWLERS OAK VIEW ESTATES (Plat no: 2736) MAP GRID: 6G
FRITCHIES ADDITION (Plat no: 1418) MAP GRID: 2H
GALLERY PLACE (Plat no: 0) MAP GRID: 3C
GLEN GLENN (Plat no: 2087) MAP GRID: 4B
GLENDORRA ADDITION (Plat no: 1853) MAP GRID: 2G
GLENESK (Plat no: 584) MAP GRID: 5G
GRAND VIEW (Plat no: 3575) MAP GRID: 4G
GREEN HILLS ESTATES #2 (Plat no: 2050) MAP GRID: 3G
GREEN HILLS ESTATES #3 (Plat no: 2141) MAP GRID: 3G
GREEN HILLS ESTATES #4 (Plat no: 2350) MAP GRID: 3G
GREEN HILLS ESTATES (Plat no: 1983) MAP GRID: 3G
HAVERHILL (Plat no: 2770) MAP GRID: 4E
HAVERHILL ESTATES (Plat no: 3044) MAP GRID: 4E
HIDDEN SPRINGS RANCH #1 (Plat no: 1933) MAP GRID: 5D
HIDDEN SPRINGS RANCH #2 (Plat no: 2108) MAP GRID: 5D
HIDDEN SPRINGS RANCH #3 (Plat no: 2151) MAP GRID: 4D
HIDDEN SPRINGS RANCH #4 (Plat no: 2270) MAP GRID: 4E

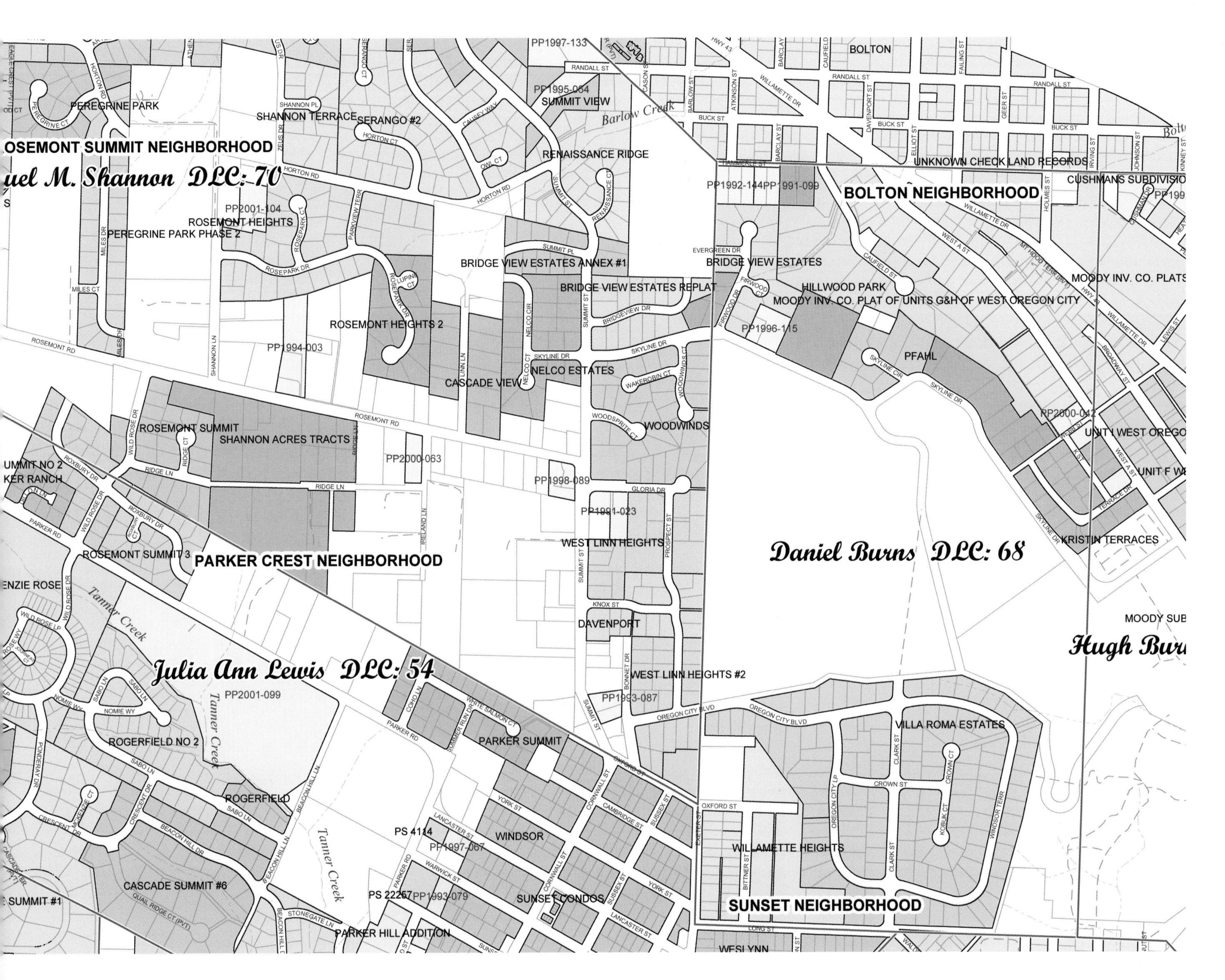
OSEMONT SUMMIT NEIGHBORHOOD
uel M. Shannon DLC: 70
BOLTON NEIGHBORHOOD
BOLTON
PARKER CREST NEIGHBORHOOD
Daniel Burns DLC: 68
Julia Ann Lewis DLC: 54
Hugh Bur
SUNSET NEIGHBORHOOD
PEREGRINE PARK
SHANNON TERRACE
SERANGO #2
SUMMIT VIEW
RENAISSANCE RIDGE
Barlow Creek
UNKNOWN CHECK LAND RECORDS
ROSEMONT HEIGHTS
PEREGRINE PARK PHASE 2
BRIDGE VIEW ESTATES ANNEX #1
BRIDGE VIEW ESTATES
BRIDGE VIEW ESTATES REPLAT
HILLWOOD PARK
MOODY INV. CO. PLAT OF UNITS G&H OF WEST OREGON CITY
MOODY INV. CO. PLATS
ROSEMONT HEIGHTS 2
NELCO ESTATES
CASCADE VIEW
PFAHL
ROSEMONT SUMMIT
SHANNON ACRES TRACTS
WOODWINDS
ROSEMONT SUMMIT 3
WEST LINN HEIGHTS
KRISTIN TERRACES
DAVENPORT
WEST LINN HEIGHTS #2
Tanner Creek
ROGERFIELD NO 2
PARKER SUMMIT
VILLA ROMA ESTATES
ROGERFIELD
WINDSOR
WILLAMETTE HEIGHTS
CASCADE SUMMIT #6
SUNSET CONDOS
PARKER HILL ADDITION
MOODY SUB

4.8 SALINAS ZONING MAP

In contrast to the muted area fills on previous maps, saturated hues code the primary message for this map of Salinas, California—zoning by district. Lightness sequences are also used to connect zoning categories that are related, such as low-to-high-density residential areas that range from yellow through oranges to dark brown in an ordered sequence. Other basic hue/zoning associations used on this and many other zoning maps include: greens for agriculture, parks, and open space; pinks and reds for commercial; and purples for industrial. Haloed labels within areas provide redundant zoning designations to ensure the zoning categories are unambiguously communicated, an important objective since they determine regulations and property development standards that apply to parcels. The light black-and-white skeleton for the road infrastructure provides a suitable base that does not compete with the zoning colors or the bolder black boundaries for overlay districts where special regulations apply.

Courtesy of the City of Salinas, California.

Overlay Legend

Gateway Overlay Districts:

1. West Boronda Road @ U.S. 101
2. North Main Street @ U.S. 101
3. West Market Street @ Davis Road
4. South Main Street @ Blanco Road
5. Sanborn Road @ U.S. 101

Focused Growth Overlay Districts:

6. FG 1 Laurel Drive at North Main Street
7. FG 2 North Main Street/Soledad Street
8. FG 3 South Main Street
9. FG 4 Abbott Street
10. FG 5 East Alisal Street/East Market Street

Specific Plan Overlay Districts:

11. SP-1 Harden Ranch
12. SP-2 Williams Ranch
13. SP-3 Westridge Center
14. SP-4 Salinas Auto Center
15. SP-5 Mountain Valley
16. SP-6 Boronda Crossing

Other:

17. East Romie Lane Corridor Overlay District
18. Central City Overlay District (see map inset)

Zoning District Legend

A	Agriculture
R-L-5.5	Residential Low Density
R-M-3.6	Residential Medium Density
R-M-2.9	Residential Medium Density
R-H-2.1	Residential High Density
R-H-1.8	Residential High Density
CO/R	Commercial Office / Residential
CO	Commercial Office
CR	Commercial Retail
CT	Commercial Thoroughfare
MX	Mixed Use
MAF	Mixed Arterial Frontage
IGC	Industrial - General Commercial
IBP	Industrial - Business Park
IG	Industrial - General
PS	Public / Semipublic
P	Parks
OS	Open Space

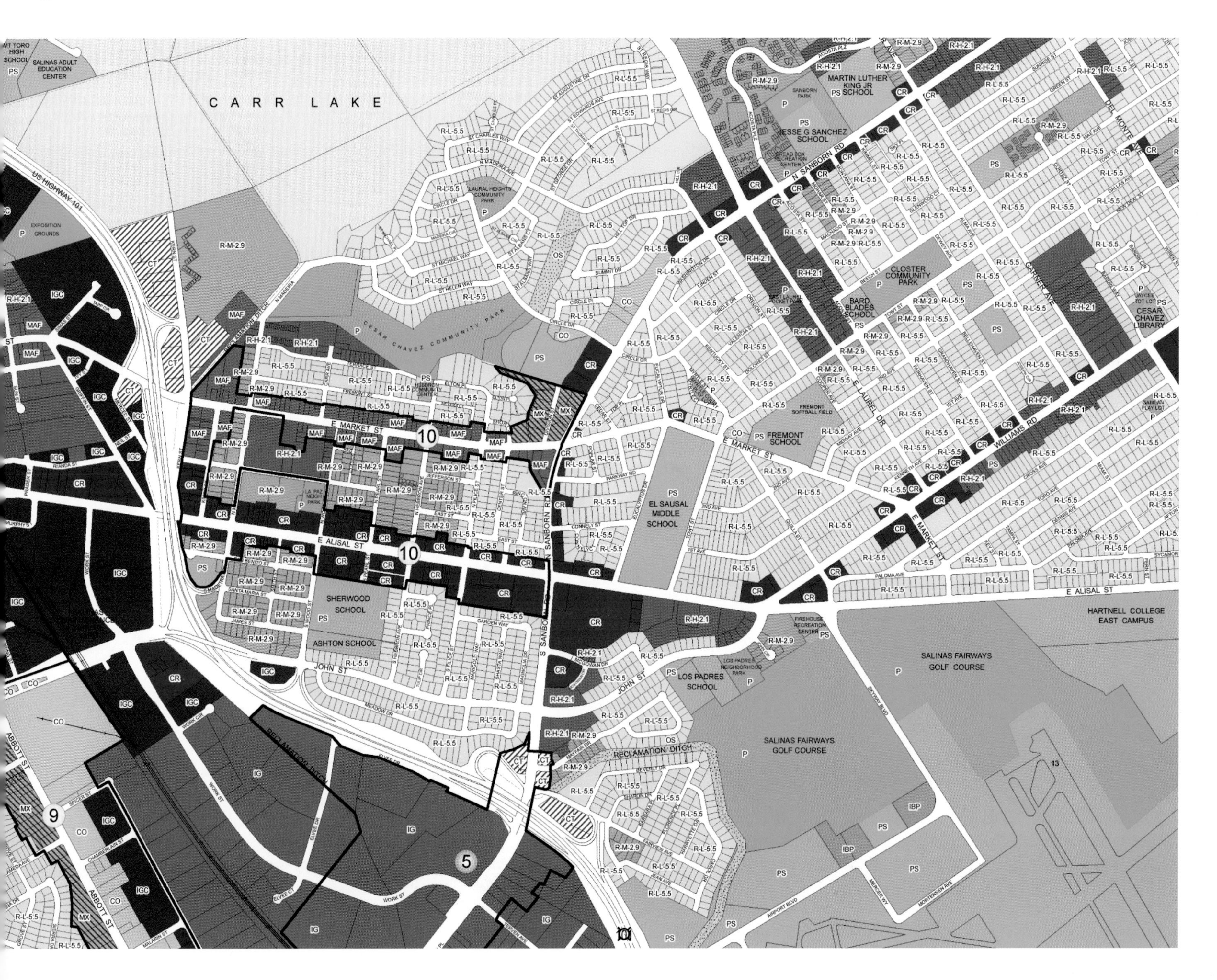
CARR LAKE
MT TORO HIGH SCHOOL
SALINAS ADULT EDUCATION CENTER
US HIGHWAY 101
EXPOSITION GROUNDS
CESAR CHAVEZ COMMUNITY PARK
LAURAL HEIGHTS COMMUNITY PARK
MARTIN LUTHER KING JR SCHOOL
JESSE G SANCHEZ SCHOOL
CLOSTER COMMUNITY PARK
BARD BLADES SCHOOL
CESAR CHAVEZ LIBRARY
FREMONT SCHOOL
FREMONT SOFTBALL FIELD
EL SAUSAL MIDDLE SCHOOL
SHERWOOD SCHOOL
ASHTON SCHOOL
LOS PADRES SCHOOL
LOS PADRES NEIGHBORHOOD PARK
FIREHOUSE RECREATION CENTER
SALINAS FAIRWAYS GOLF COURSE
HARTNELL COLLEGE EAST CAMPUS
E MARKET ST
E ALISAL ST
N SANBORN RD
S SANBORN RD
E LAUREL DR
WILLIAMS RD
GARNER AVE
DEL MONTE AVE
JOHN ST
WORK ST
ABBOTT ST
AIRPORT BLVD
RECLAMATION DITCH

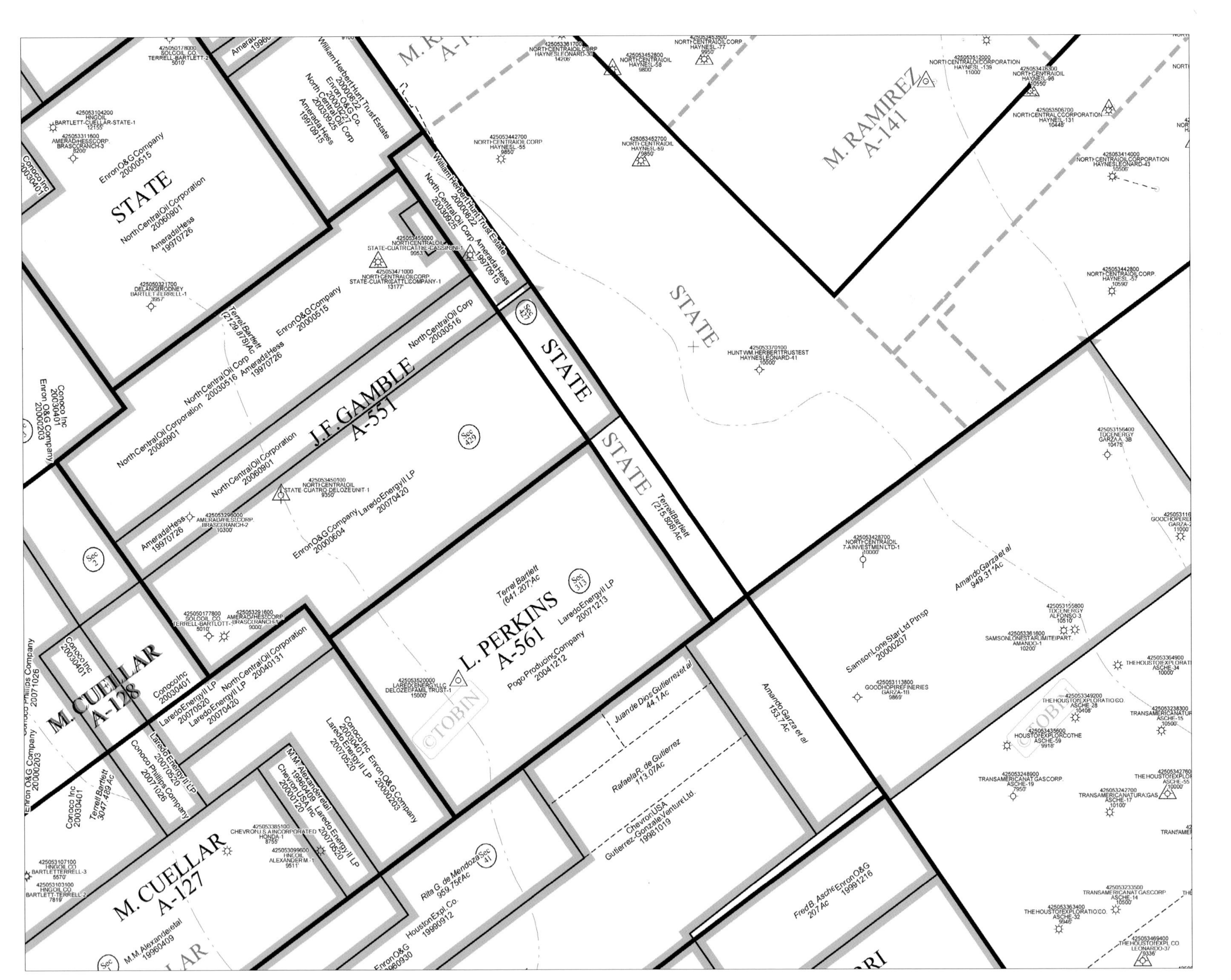
STATE
Enron O&G Company 20000515
North Central Oil Corporation 20060901
Amerada Hess 19970726
M. RAMIREZ
A-141
STATE
STATE
STATE
J.F. GAMBLE
A-551
North Central Oil Corporation 20060901
North Central Oil Corp 20030516
Amerada Hess 19970726
Enron O&G Company 20000604
Laredo Energy II LP 20070420
Terrel Bartlett (641.207Ac
L. PERKINS
A-561
Laredo Energy II LP 20071213
Pogo Producing Company 20041212
Amando Garza et al 949.31Ac
Samson Lone Star Ltd Ptnsp 20000207
M. CUELLAR
A-128
M. CUELLAR
A-127
Juan de Dios Gutierrez et al 44.1Ac
Rafaela R. de Gutierrez 113.07Ac
Chevron USA Gutierrez-Gonzale Venture Ltd. 19981019
Amando Garza et al 153.7Ac
Rita G. de Mendoza 959.75Ac
Houston Expl. Co. 19990912
Fred B. Asche 207Ac
Enron O&G 19981216
©TOBIN

4.9 TEXAS LEASE SUMMARY MAP

Good black-and-white design requires precise organization of line weight, pattern, lightness, and label characteristics. In this example by Tobin International, mineral lease information for a portion of Zapata County, Texas, rests on a base of legal boundaries. Bold and black lines for abstract boundaries contrast with broad light lines that are offset to the interior of lease areas. Two types of dashed lines show background information on surveys and ownership. Large serif fonts for grantees and smaller sans serif fonts distinguish owner details (in italics) and lease details (in nonitalics). This systematic use of lines and label styles allows readers to comprehend groups of features, understand how areas nest and overlap, or focus on specific details at locations of interest.

Courtesy of P2 Energy Solutions, L.P.

4.10 HARRIS COUNTY FLOODPLAIN MAP

Three lightness levels of transparent blue show three nested floodplain extents, resting on a base network of lightly colored streets and parcels. The brown parcel lines turn to gray when overlaid by the transparent floodplain blues. Parcel lines partly define the street network of Harris County, Texas. Gray road centerlines bolster this road network and carry the road labels. The centerlines provide minimal, but adequate, contrast with the darker floodplain blues to define roads within parcels affected by floods without becoming a distraction from the main objective; that is, quickly identifying which parcels will be inundated. In addition to the floodplain areas, supporting information on detention basins is shown (for example, regional basins are symbolized with red outlines).

Courtesy of Harris County Flood Control District, Texas.

Legend

- Airport
- City Hall and/or Clerk Office
- Colleges & Universites
- Community Center
- Convention Center and/or Training Facility
- Court
- Elementary School
- Fire Protection
- Harris County Office
- High School
- Hospital
- Intermediate School
- Library
- Middle School
- Police Protection
- Post Office
- Primary School
- Private High School
- Private School

Drainage Network

- Open
- Storm Sewer
- Historical

HC Precincts

- PCT BNDY

Right of Way

Grantee

- H.C.F.C.D.
- Other

Interest

- EASEMENT
- FEE

Detention Basin

- Development Basin
- Flood Plain Preservation
- Proposed Basin
- Regional Basin
- Site Spacific Basin
- Trust for Public Land
- Wetland Mitigation Bank

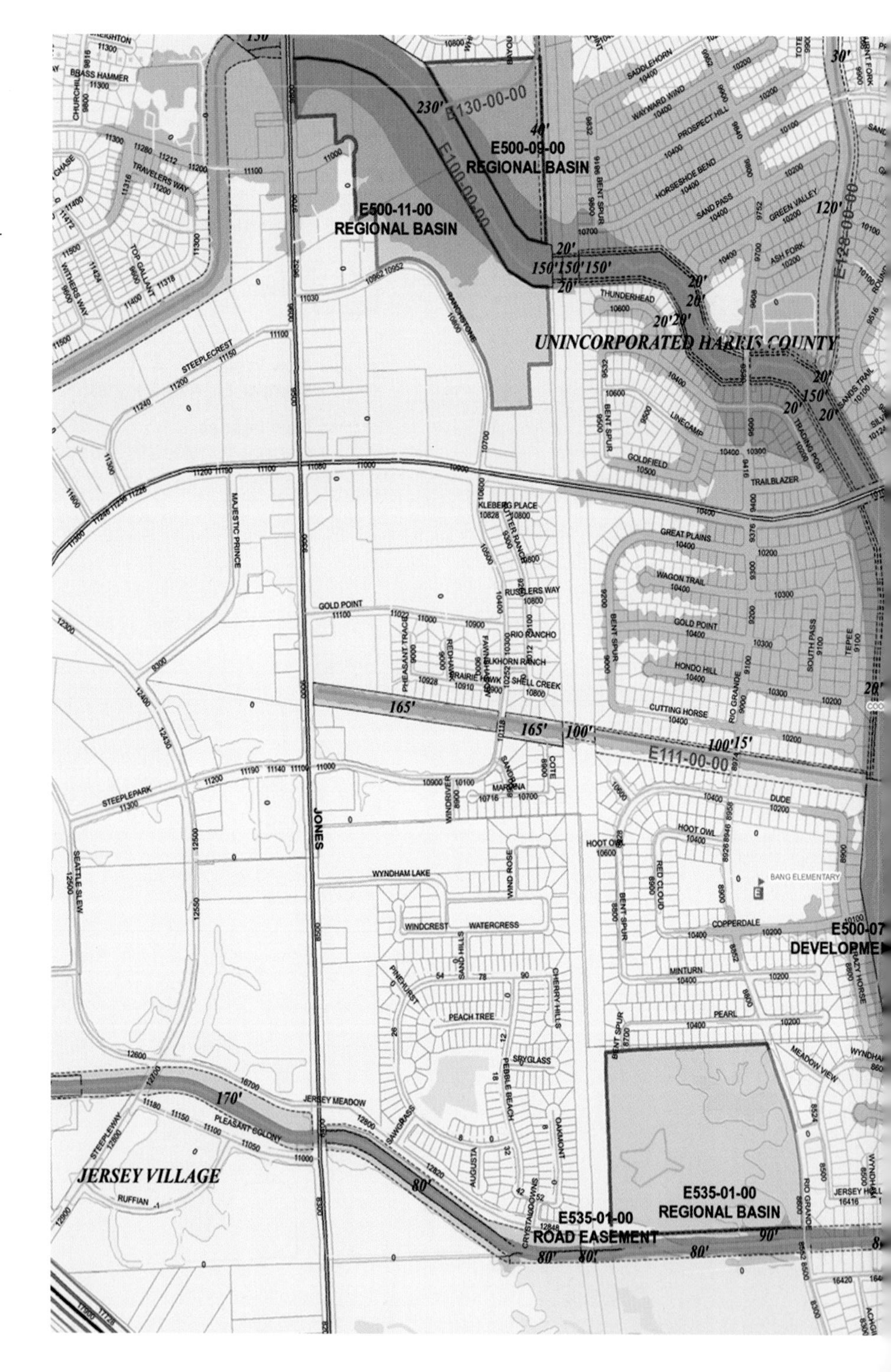

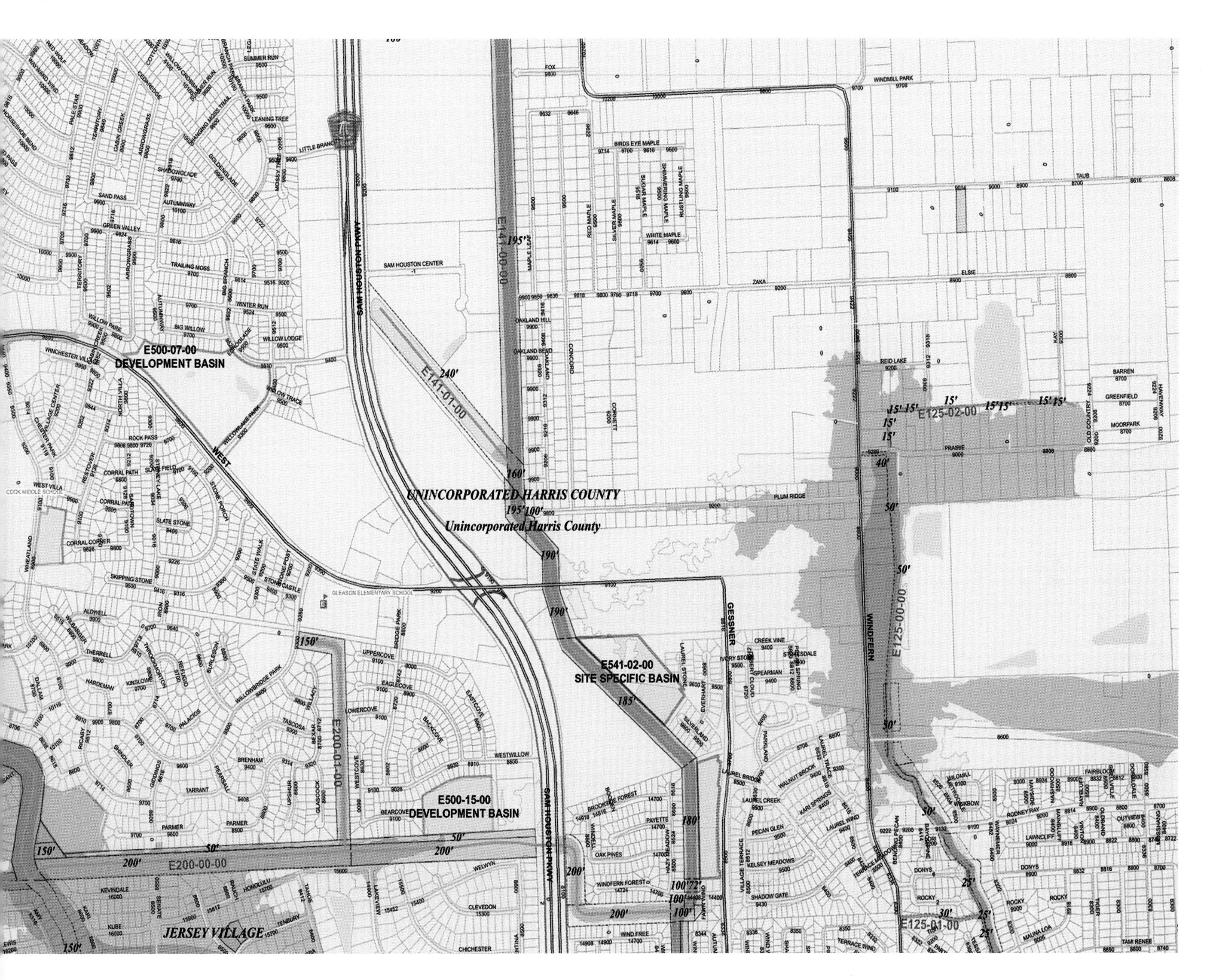
E500-07-00
DEVELOPMENT BASIN
E141-00-00
E141-01-00
UNINCORPORATED HARRIS COUNTY
Unincorporated Harris County
E125-02-00
E125-00-00
E541-02-00
SITE SPECIFIC BASIN
E200-01-00
E500-15-00
DEVELOPMENT BASIN
E200-00-00
E125-01-00
JERSEY VILLAGE
SAM HOUSTON PKWY
GESSNER
WINDFERN
WEST
COOK MIDDLE SCHOOL
GLEASON ELEMENTARY SCHOOL
SAM HOUSTON CENTER

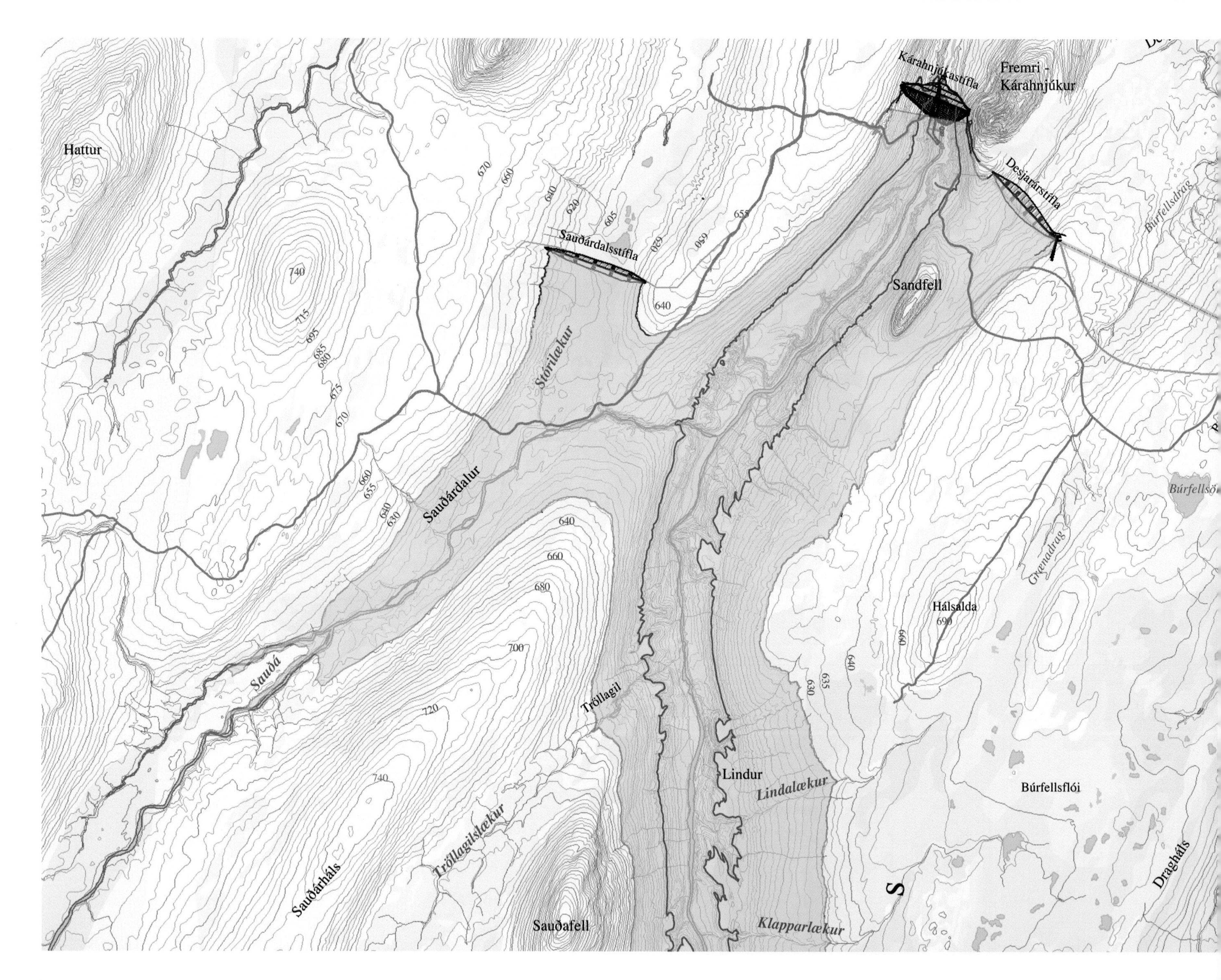
Hattur
Kárahnjúkastífla
Fremri - Kárahnjúkur
Desjarárstífla
Búrfellsdrag
Sauðárdalsstífla
Sandfell
Stórilækur
Sauðárdalur
Búrfellsó
Grænadrag
Hálsalda
Sauðá
Tröllagil
Lindur
Lindalækur
Búrfellsflói
Tröllagilslækur
Sauðárháls
Sauðafell
Klapparlækur
Draghals
S

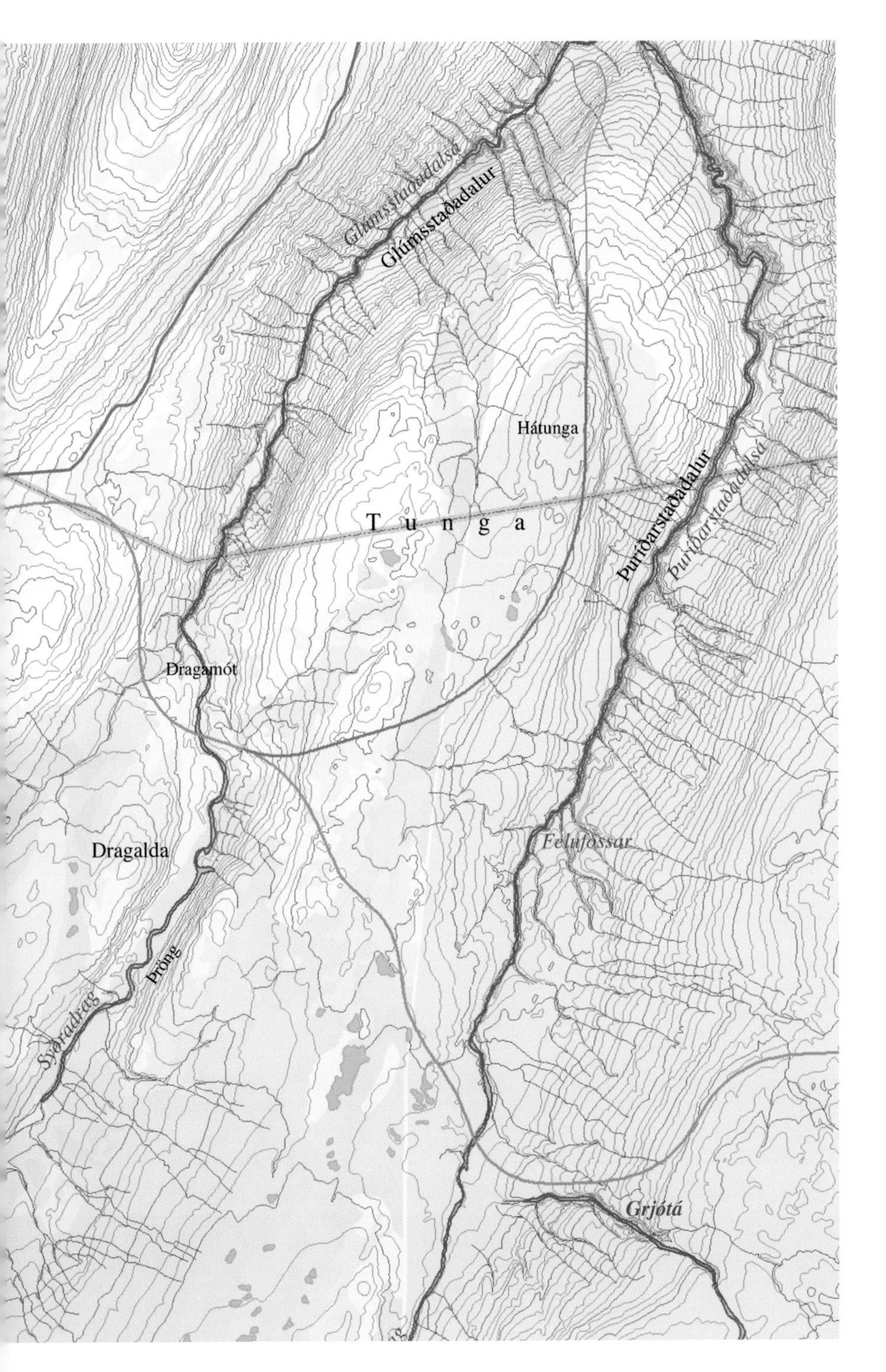

4.11 ICELAND HYDROELECTRIC PROJECT MAP

In this example, map readers see the tributaries and landforms that will be flooded through the transparent blue of the planned Hálslón Reservoir in Iceland. The full extent of the reservoir is shown at the 625m level, and a gray isoline breaks the inundation at an intermediate 550m level. Ladders of contour labels sweeping up slopes provide landform detail. The labels were created by adding annotation to a raster rendition of the contours; the slow processing of intensive contour labeling sometimes makes this type of work-around a cost-effective solution. Multiple representations of the glacial river braids are seen in the hydrography overlays. Very light land-cover variations create background texture across the terrain that is not flooded.

Courtesy of The National Power Company of Iceland.

Skýringar

Hálslón 625 m y.s.

Fyrirhugaðar stíflur

Vegir og vegslóðar

Göng

5 m hæðarlínur

4.12 ANTARCTIC REGION MAP

Nations' claims on territory in Antarctica are overlapping and partly undefined, and this map provides a reference for the mix of national research stations within overlapping wedges that converge on the pole. The azimuthal equal-area map projection centered on the pole is key to understanding these claims as defined by latitude and longitude ranges, and it allows accurate comparison of areas that each country claims.

Indistinct boundaries are rendered in a variety of ways on the map. The average minimum extent of sea ice is shown with a broken texture of white ice against light cyan water. The Antarctic Convergence, where cold Antarctic water sinks beneath warmer waters to the south, is a blurred band of darker cyan that appropriately marks this shifting hydrology zone. Positioning labels over areas without particular boundaries is another simple way of identifying indefinite areas, such as Queen Maud Land and the adjacent Weddell Sea. The ragged boundary at the top and bottom of the Norwegian claim contrasts with the smooth arcs of other claims to highlight its undefined limit, which is different in character than the other claims.

Courtesy of CIA—The World Factbook.

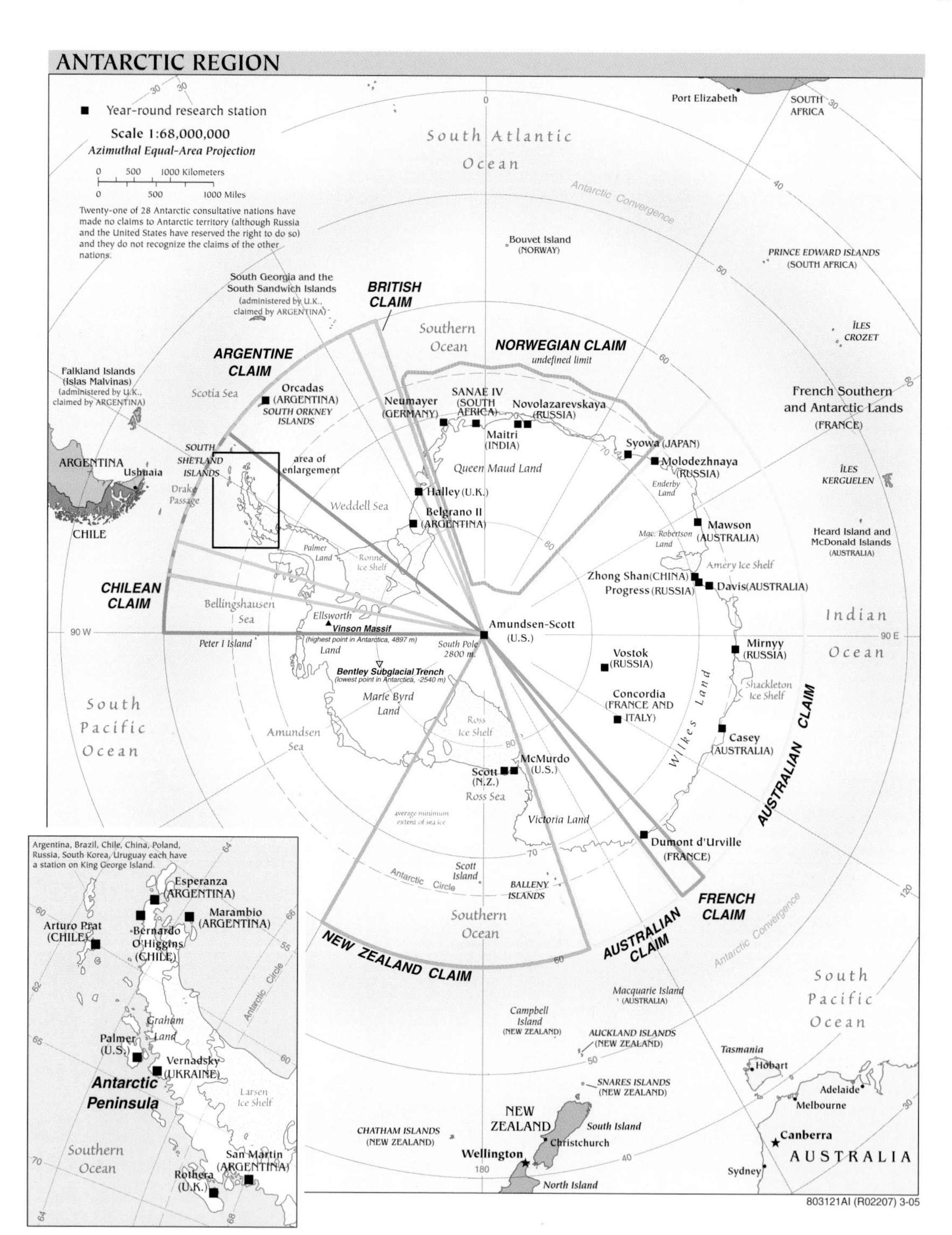

5

THEMATIC MAPS *CATEGORICAL*

People reading categorical thematic maps should gain an understanding of overall patterns and relationships among different kinds of geographic phenomena. The main objective of a categorical map is not to present an inventory of individual features, as is the case with many special-purpose and reference maps. Base information on thematic maps helps map users compare characteristics of their home or a place of particular concern to the patterns evident in the map theme. Lines, points, labels, and terrain in the base information are represented lighter, smaller, and less densely than on reference maps. The maps chosen for this chapter present mostly qualitative information, though a few the multivariate maps include quantitative data.

Hue is an indispensable design tool when mapping categorical data, but shape and pattern are also important. Maps selected for this chapter show a wide range of area hues that are also differentiated by lightness, saturation, and pattern. These additional visual variations are essential to map reading when there are many categories on a map. Two maps in this chapter show linear features that are categorized primarily by hue. Other chapters include maps with point features categorized by hue and shape. For example, the perchlorate map in the next chapter includes point locations symbolized with simple geometric shapes (circles, squares, triangles, and stars), and the infrastructure maps in the preceding chapter include more elaborately shaped symbols.

Pattern and shading are often embedded in transparent hues. The colors resulting from a transparent hue overlaid on an image, hillshading, or other symbols may look quite different than the pure color without transparency. Because of this difference, it is important to construct a legend using small tiles taken directly from the map (or re-created to mimic the map), especially if there are many legend categories. Organizing the legend to make the relationships between categories clear is another quality of an excellent categorical map. Similar categories, such as industrial and extractive land uses on the San Diego area map, are adjacent in the legend and represented in the same hue, but with differences in lightness. The legend colors are also combined with hillshading, as they are on the map, for an accurate representation of map symbols.

A common design choice for categorical maps is to forego using an outline when choosing colors for polygons. While an outline delineates a polygon boundary, the trade-off is it takes up valuable space. Many polygons are so small that an outline would encroach on the small color patch that defines it, making it unreadable. Using no outlines requires adjacent areas be represented by adequately contrasting colors in order to see edges of polygons; differences in lightness help make those edges visible. A high-contrast mesh of outlines may also prevent readers from seeing overall patterns in the area colors, or interfere with the lines used for base information in a design.

5.0 BORDER REGION PLANNED LAND-USE MAP

5.0 BORDER REGION PLANNED LAND-USE MAP

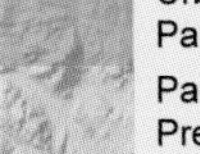

The SANDAG map, San Diego/Baja California Border Region Planned Land Use, *is compiled from local land-use and transportation plans from both sides of the international border. This compilation gives us a complete view of the region. The following redesigns are mainly differentiated by color and pattern choices.*

Courtesy of San Diego Association of Governments.

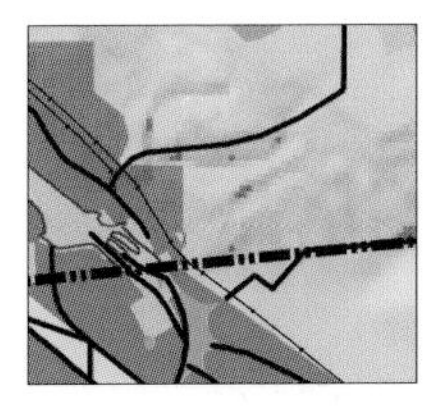

ORIGINAL DESIGN (Pages 104 and 106)

The original design uses saturated hues for land use. Two lightness levels of a hue are used to link related categories, such as light and dark green for urban recreation and parks/open space. Colors are overlaid on hillshading to provide an understanding of landforms that influence the distribution of land use in the region. Labels, roads, and boundary lines are black; however, the land-use polygons are not outlined, so blocks of color abut each other without bounding lines.

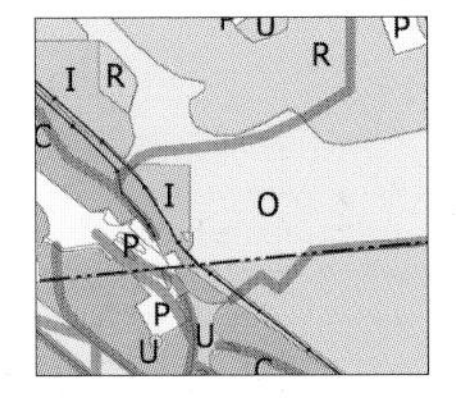

MUTED HUES WITH OUTLINES (Page 106)

Each polygon is outlined and labeled in this map version. Hues are lighter and less saturated. This combination of lightness and labels allows the map to remain readable even as a black-and-white print or photocopy. Roads are wide, transparent gray lines that overlay polygons and some land-use labels. Black place-names have a light, thin halo around letters to retain readability against roads and polygon outlines.

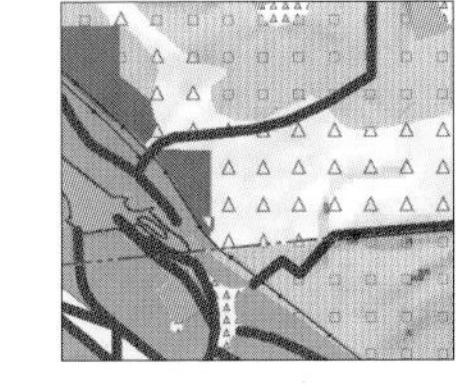

TWO INKS WITH PATTERN (Page 107)

Only two ink colors are needed to print this map: magenta and black. With only two hues to work with, varying lightness levels of hue and area patterns assist in making the map readable. Triangles, with different sizes and spacing, represent related classes of open lands and parks, and squares represent residential and low-density land. Roads are a bold mix of both magenta and black. An advantage of this simple color palette is that another middle-lightness ink color could replace the magenta to produce a similar map that coordinates with particular design content in an inexpensive publication.

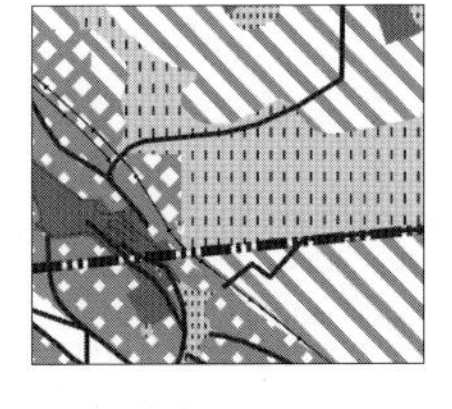

BLACK-AND-WHITE DESIGN (Page 107)

The black-and-white design makes full use of a coordinated set of patterns. Low-density areas are represented by a diagonal hatch of lines; the related residential class is represented by thicker lines with the same spacing as the low-density area hatching. This spacing is continued for commercial and industrial areas, but a second set of diagonals are overlaid for cross-hatch patterns in two line widths. Open land is a contrasting texture of ticks, with denser, horizontally aligned ticks for parks. Areas without patterns—white for water and dark gray for public land—complete the set. Thin black road lines contrast with all of these symbols to provide base information.

REDESIGNS

5A ORIGINAL DESIGN

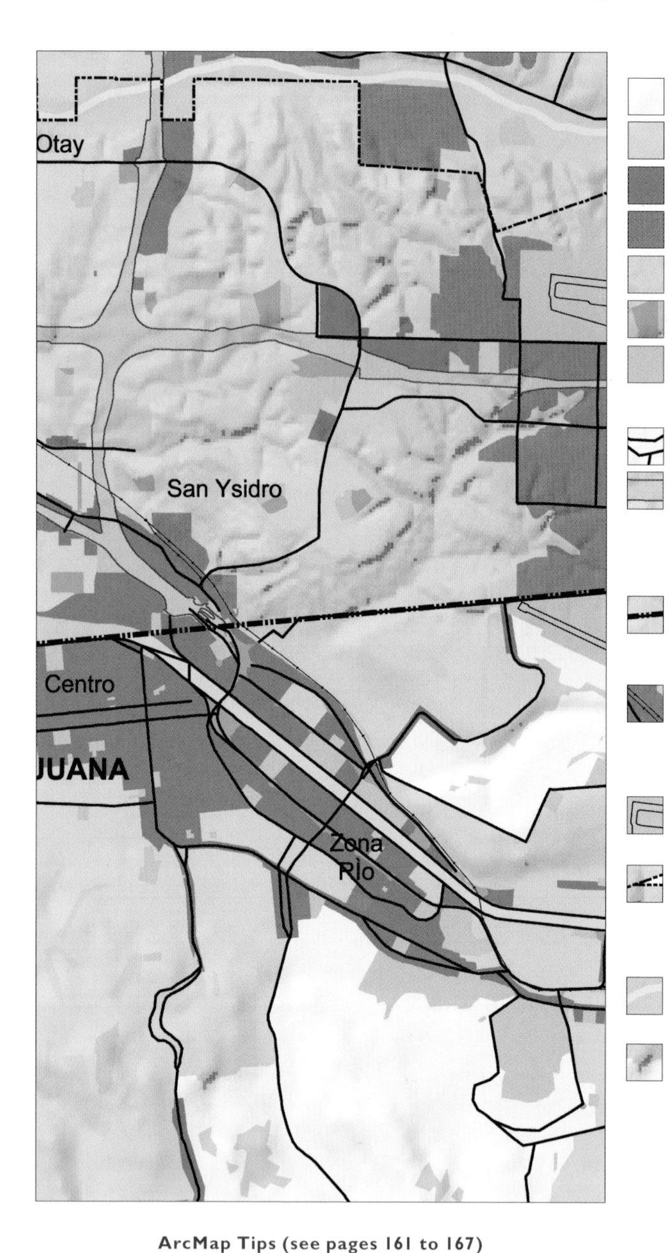

LAND USE

Low density
Fill: 14Y

Residential
Fill: 2C 11M 100Y

Commercial
Fill: 95M

Industrial
Fill: 40C 75M

Park/open
Fill: 36C 53Y

Urban recreation
Fill: 75C 7M 87Y

Public
Fill: 22K

ROADS

Line: 1 pt, 100K, round cap/join

Highways
Fill: 22K
Line: 0.5 pt, 100K, round cap/join

OTHER

Border
Line: 2 pt, 100K, dash pattern 6/1/1/1/1/1 *(dash/gap in pts)*

Railways
Lines: 0.25 pt, 100K; 0.8 pt, 100K, hash line pattern 0.5/12 *(hash width/spacing in pts)*

Airports
Line: 0.5 pt, 100K, round cap/join

City
Line: 1 pt, 100K, dash pattern 5/1/1.5/1/1.5/1
Labels: 11 pt, Arial Bold, 100K; 9 pt, Arial, 100K

Water
Fill: 37C 3Y
Line: 37C 3Y

Hillshade
70% transparent overlay

ArcMap Tips (see pages 161 to 167)

1 Dashed line	22 Transparent fill
23 Transparent legend	

5B MUTED HUES WITH OUTLINES

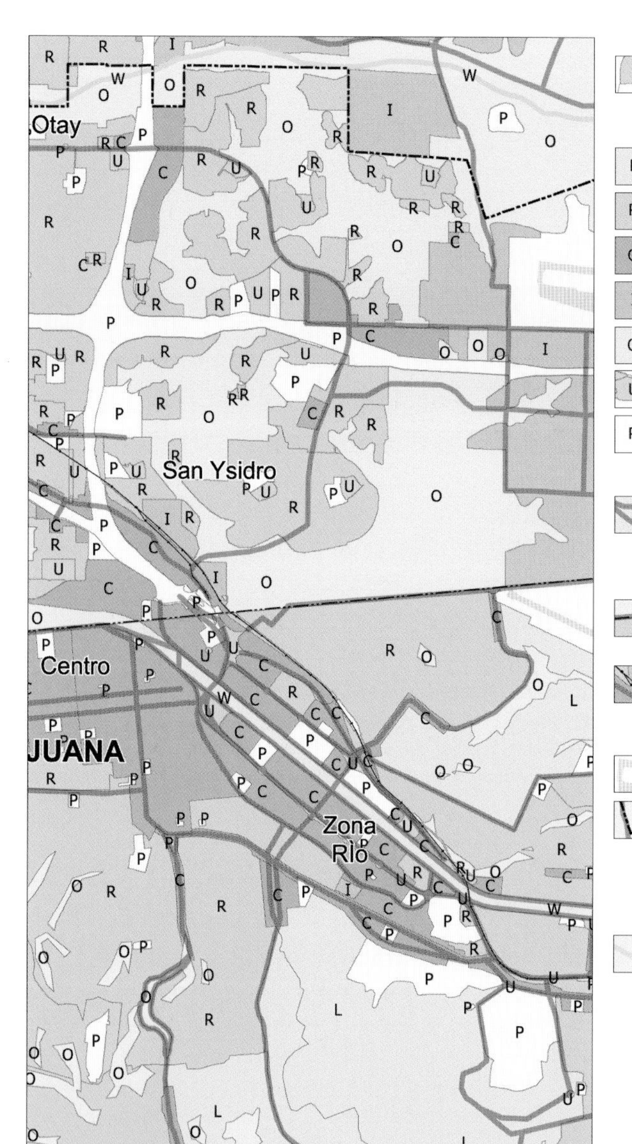

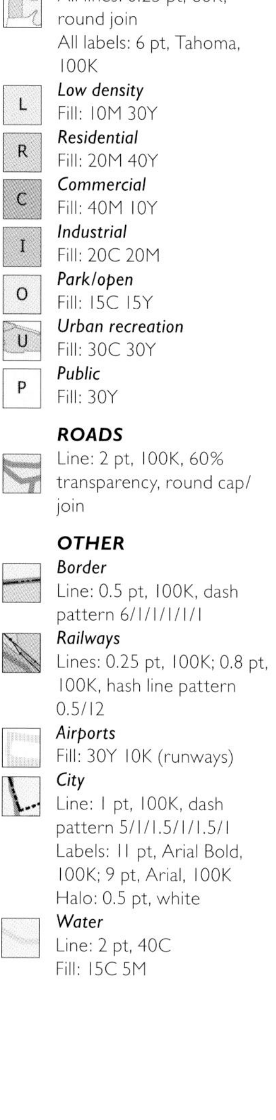

LAND USE

All lines: 0.25 pt, 60K, round join
All labels: 6 pt, Tahoma, 100K

L *Low density*
Fill: 10M 30Y

R *Residential*
Fill: 20M 40Y

C *Commercial*
Fill: 40M 10Y

I *Industrial*
Fill: 20C 20M

O *Park/open*
Fill: 15C 15Y

U *Urban recreation*
Fill: 30C 30Y

P *Public*
Fill: 30Y

ROADS

Line: 2 pt, 100K, 60% transparency, round cap/join

OTHER

Border
Line: 0.5 pt, 100K, dash pattern 6/1/1/1/1/1

Railways
Lines: 0.25 pt, 100K; 0.8 pt, 100K, hash line pattern 0.5/12

Airports
Fill: 30Y 10K (runways)

City
Line: 1 pt, 100K, dash pattern 5/1/1.5/1/1.5/1
Labels: 11 pt, Arial Bold, 100K; 9 pt, Arial, 100K
Halo: 0.5 pt, white

Water
Line: 2 pt, 40C
Fill: 15C 5M

ArcMap Tips

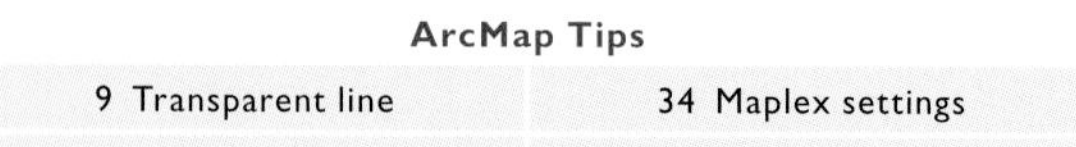

9 Transparent line	34 Maplex settings

5C TWO INKS WITH PATTERNS

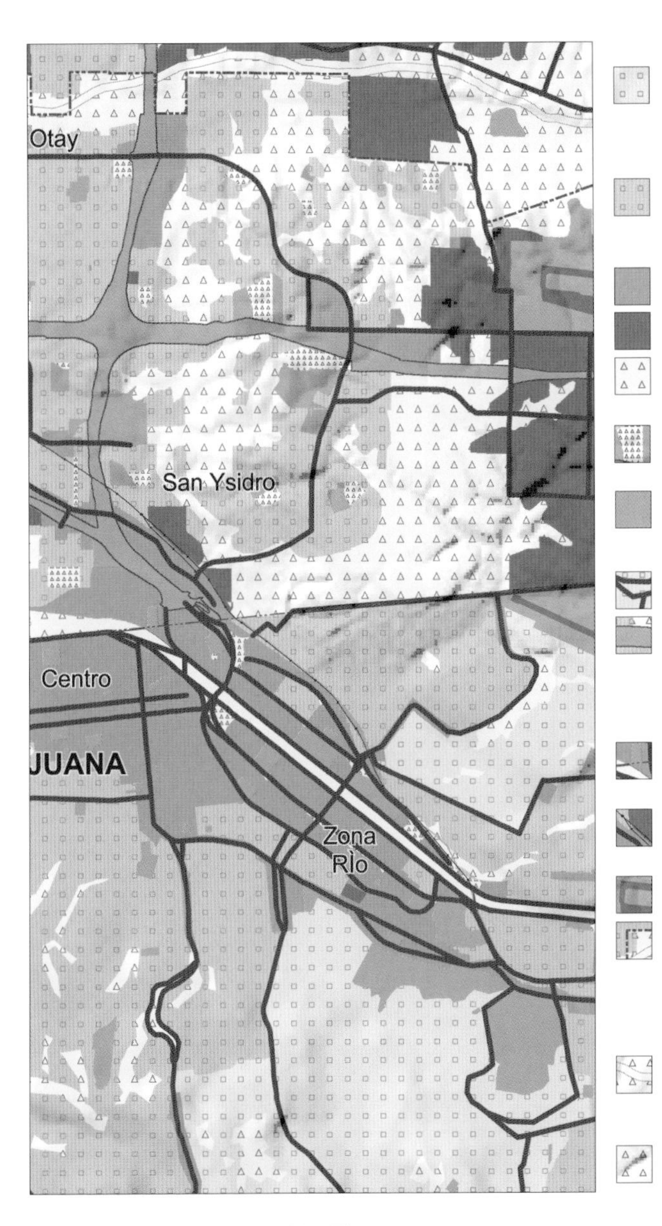

LAND USE

Low density
Multilayer fill: 70K, square marker fill pattern 2/7 *(symbol size/spacing in pts)*; 7M 3K base fill

Residential
Multilayer fill: 70K, square marker fill pattern 2/7; 20M 10K base fill

Commercial
Fill: 50K

Industrial
Fill: 80K

Park/open
Fill: 100K, triangle marker fill pattern 2.5/7

Urban recreation
Fill: 100K, triangle marker fill pattern 1.5/2.5

Public
Fill: 60M

ROADS
Line: 2 pt, 100M 50K, round cap/join

Highways
Fill: 60M
Line: 0.5 pt, 100M 50K, round cap/join

OTHER

Border
Line: 0.5 pt, 100K, dash pattern 6/1/1/1/1/1

Railways
Line: 0.25 pt, 100K, hash line pattern 1/11

Airports
Fill: 100M (runways)

City
Line: 1 pt, 100K, dash pattern 5/1/1.5/1/1.5/1
Labels: 11 pt, Arial Bold, 100K; 9 pt, Arial, 100K
Halo: 0.5 pt, white

Water
Fill: white
Line: 2.9 pt, white
Casing: 3.2 pt, 100K

Hillshade
70% transparent overlay

ArcMap Tips

15 Pattern fill	26 Halo
39 Spot color	

5D BLACK-AND-WHITE DESIGN

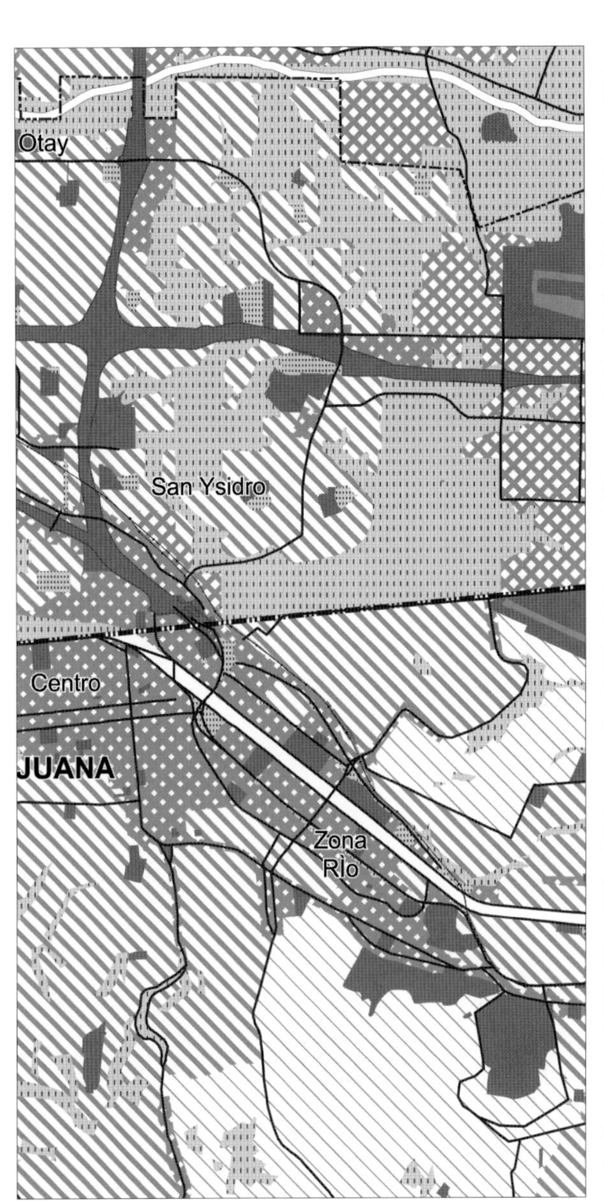

LAND USE

Low density
Fill: 50K, -45° line fill pattern 0.5/7 *(line width/spacing in pts)*

Residential
Fill: 50K, -45° line fill pattern 2/7

Commercial
Multilayer fill: 50K, 45° and -45° line fill patterns 3/7

Industrial
Multilayer fill: 50K, 45° and -45° line fill patterns 2/7

Park/open
Multilayer fill: 100K, rectangle marker fill pattern 1.7/4 (vertical), 0.5/2.5 (horizontal); 20K base fill

Urban recreation
Multilayer fill: 100K, rectangle marker fill pattern 0.9/2 (vertical), 0.5/1.3 (horizontal); 20K base fill

Public
Fill: 70K

ROADS
Line: 1 pt, 100K, round join

Highways
Fill: 70K
Line: 0.5 pt, 85K, round cap/join

OTHER

Border
Line: 2 pt, 100K, dash pattern 6/1/1/1/1/1

Railways
Line: 0.25 pt, 100K, hash line pattern 1/11

Airports
Fill: 50K (runways)

City
Line: 1 pt, 100K, dash pattern 5/1/1.5/1/1.5/1
Labels: 9 pt, Arial, 100K; 11 pt, Arial Bold, 100K
Halo: 0.5 pt, white

Water
Fill: white
Line: 2.9 pt, white
Casing: 3.2 pt, 100K

ArcMap Tips

3 Cased line	15 Pattern fill
17 Hatched fill	

5.1 HONG KONG LAND-USE MAP

This small-scale (1:170,000) land-use map of Hong Kong shows broad spatial patterns to support planning and monitor sustainability. Vegetation categories such as grassland and woodland are greens that contrast with the darker and more varied hues of built-up areas (residential, commercial, industrial, and institutional). Agriculture falls between these two, and is represented by a dark green that ties in visually with the built-up areas. Light roads in yellow also contrast with the dark built-up areas, producing a city block structure where they come together. Vacant land in white and barren land in gray both effectively evoke an absence of land use and land cover.

Courtesy of Planning Department, Hong Kong Special Adminstrative Region.

后海灣
（深圳灣）
DEEP BAY
(Shenzhen Wan)
粉嶺
Fanling
汀角
Ting Kok
船灣淡水湖
PLOVER COVE RESERVOIR
赤門 TOLO CHANNEL (Chek Mu
天水圍
Tin Shui Wai
大埔
Tai Po
吐露港
TOLO HARBOUR
錦田
Kam Tin
Yuen Long
Shek Kong
Ma Liu Shui
馬鞍山
Ma On
Shan
火炭
Fo Tan
北潭涌
Pak Tam
Chung
屯門
Tuen Mun
西貢
Sai Kung
Sha Tin
荃灣
Tsuen Wan
深井
Sham Tseng
葵涌
Kwai
Chung
滘西洲
Kau Sai Chau
馬灣
Ma Wan
牛尾海
PORT SHELTER
(Ngau Mei Hoi)
青衣
Tsing Yi
竹篙灣
Penny's Bay
(Chok Ko
Wan)
九龍
Kowloon
將軍澳
Tseung Kwan O
愉景灣
Discovery
Bay
VICTORIA HARBOUR
東涌
Tung Chung
坪洲
Peng Chau
梅窩
Mui Wo
山
Island
香港島
Hong Kong Island
貝澳
Pui O

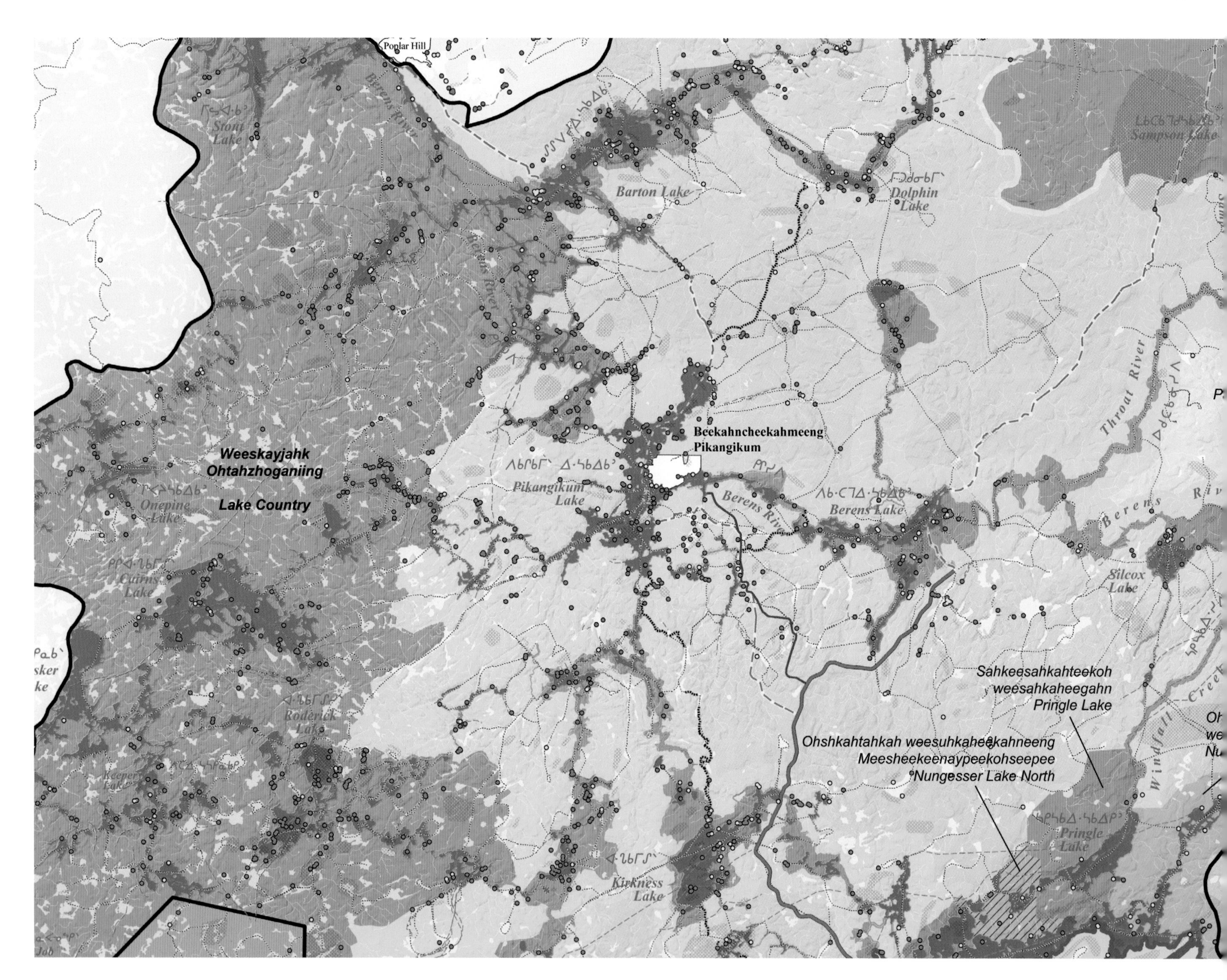
Poplar Hill
Berens River
Stout Lake
Barton Lake
Dolphin Lake
Sampson Lake
Beekahncheekahmeeng
Pikangikum
Weeskayjahk
Ohtahzhoganiing
Lake Country
Pikangikum Lake
Onepine Lake
Berens Lake
Throat River
Berens River
Cairns Lake
Silcox Lake
Roderick Lake
Keeper Lake
Sahkeesahkahteekoh
weesahkaheegahn
Pringle Lake
Ohshkahtahkah weesuhkaheekahneeng
Meesheekeenaypeekohseepee
Nungesser Lake North
Windfall Creek
Kirkness Lake

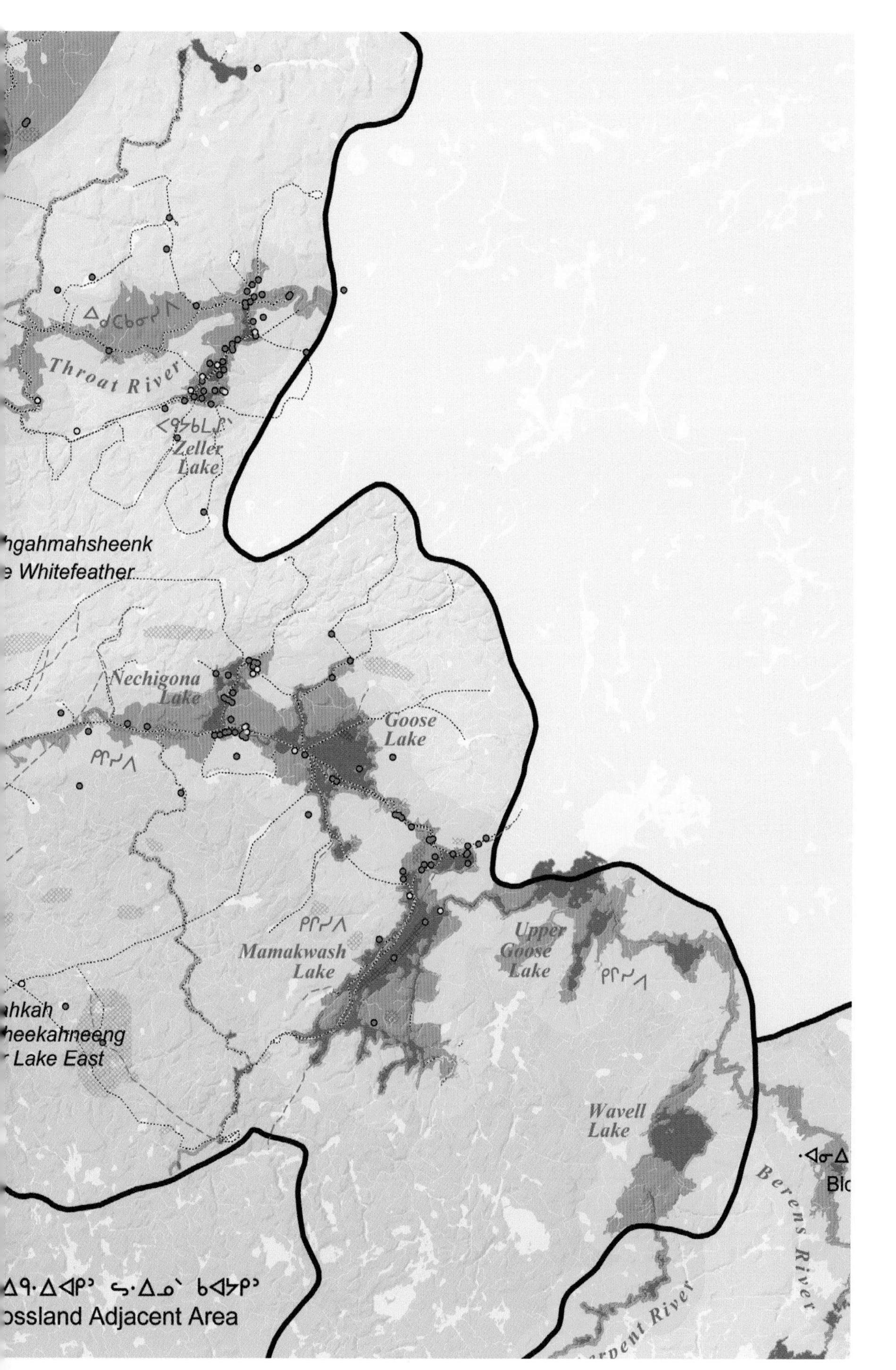

5.2 WHITEFEATHER FOREST AREA LAND-USE MAP

The Ontario government and Pikangikum people worked together to develop the *Whitefeather Forest and Adjacent Land Areas Land-Use Strategy* illustrated by this map. Saturated hues emphasize enhanced management areas and contrast them with waterways and protected areas in saturated purple-blue and green. Remote and general use areas are desaturated colors that are less prominent. A diagonal line overlay across these colors marks areas protected from forestry. Cross-hatched overlays and point locations indicate ecological and cultural features that are detailed on larger-scale maps within a map series created for the project. The map design promotes a balance between protecting culturally significant areas and identifying economic opportunities for forestry and tourism. Pikangikum and English labels ensure this map can be used by multiple communities.

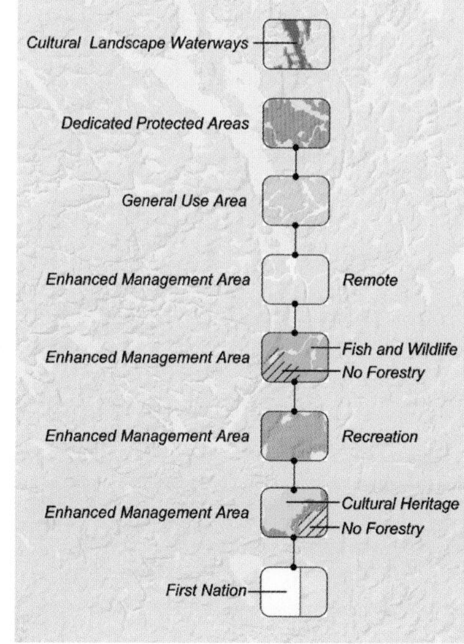

5.3 TAHOE FOREST DYNAMICS MAP

In this map changes in forest cover from 1940 to 2002 are symbolized for a portion of the southern Lake Tahoe Basin. Two greens identify areas where forest density increased and forest expanded into grassland, shrubland, or wetlands. In contrast, warm colors represent areas where forest density decreased (orange), forest transitioned to other vegetation (yellow), or vegetation died (brown). The light yellow and dark brown for tree loss have strong lightness contrast with other vegetation-change colors. Terrain shading and a simple street network provide location information and insight into potential causes of forest change. Subdued elevation tints ranging from teal to brown distinguish low- and high-elevation areas of forest change. The map supports USGS analysis of land management and forest succession during a period of intensive urbanization.

Courtesy of U.S. Geological Survey.

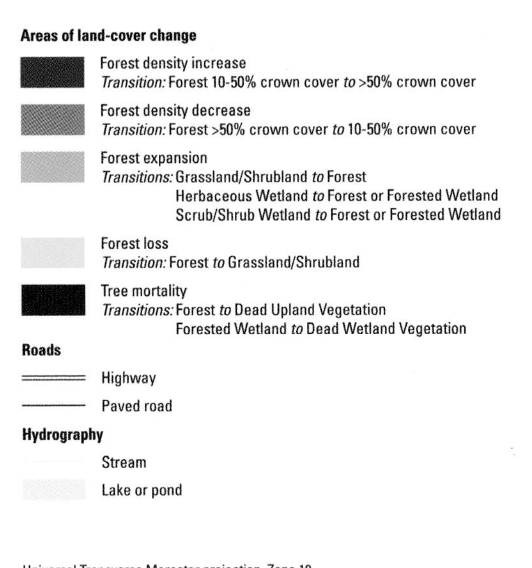

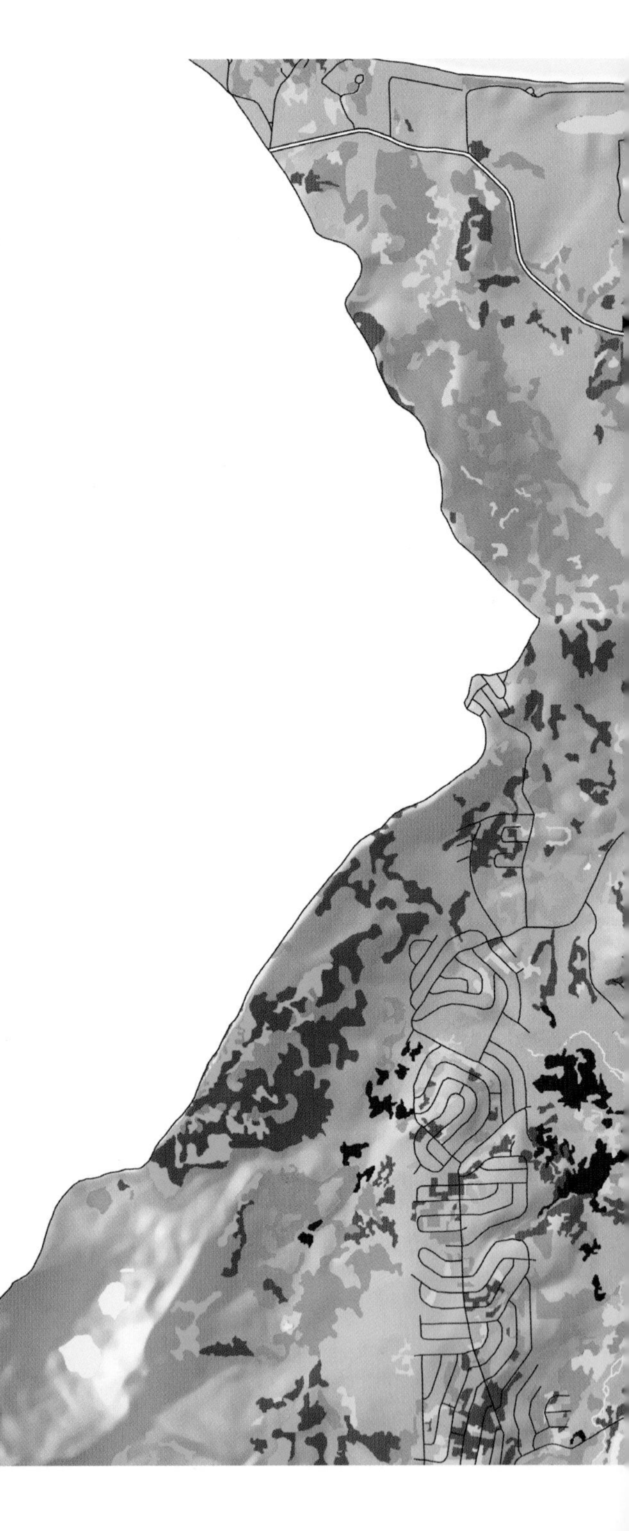

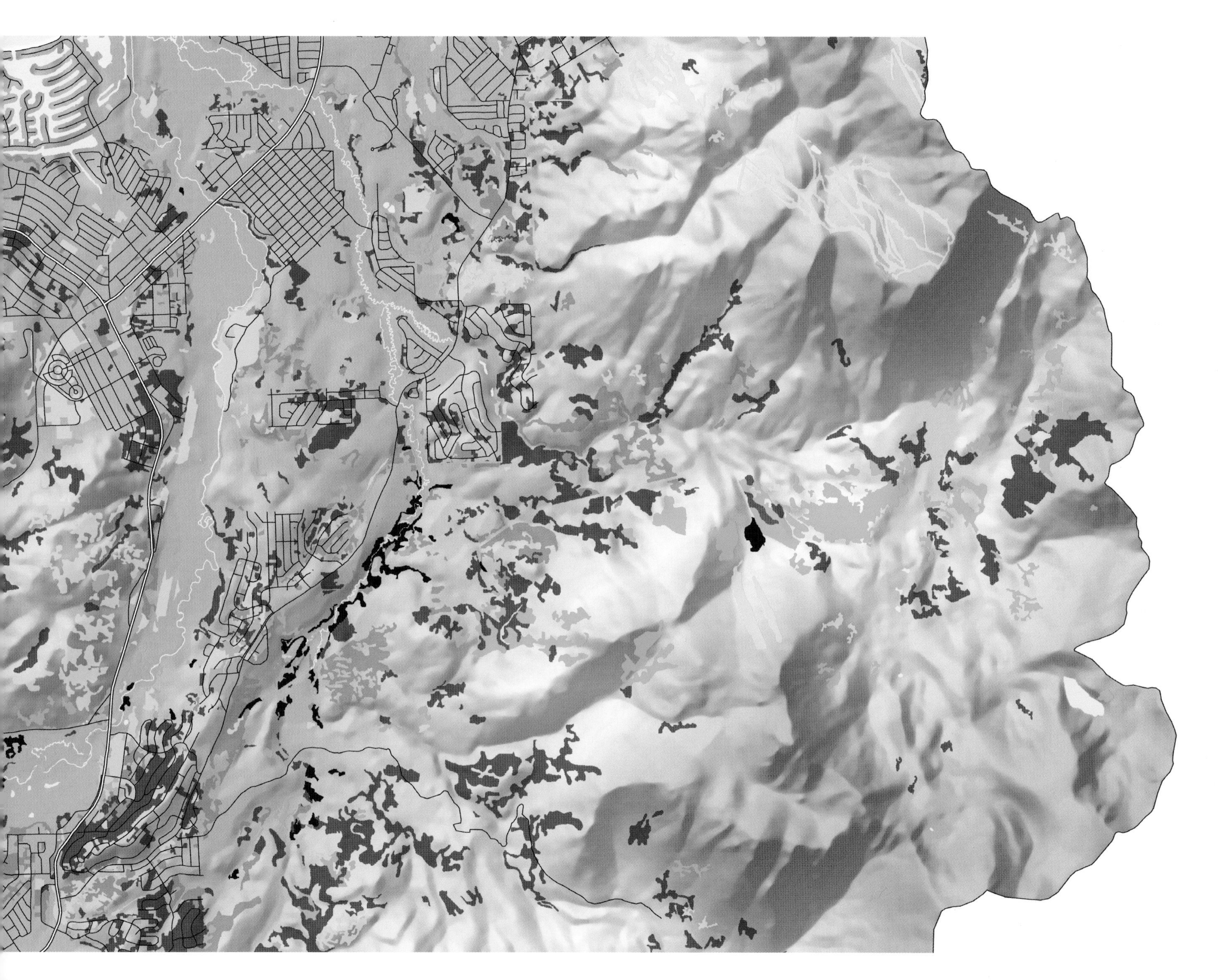

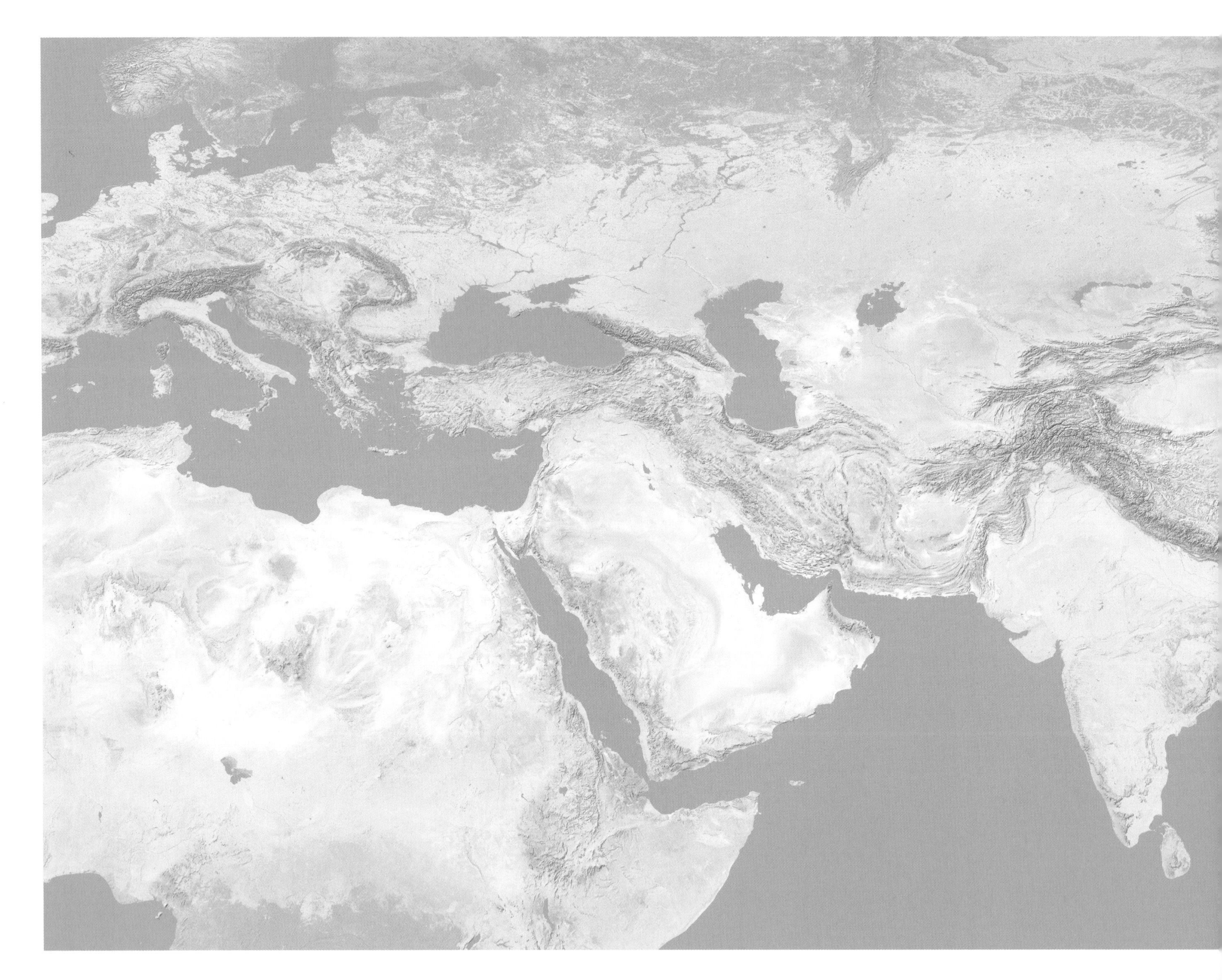

5.4 NATURAL EARTH BASEMAP

This map combines shaded relief with a subdued color palette for general land-cover types and with land-surface textures from NASA satellite imagery. The mapmaker's combination of three types of data produces a more realistic looking portrait of the natural surface than a single image type does. General hue categories include oranges for desert, yellow for herbaceous vegetation, greens for savannah and forest, and neutrals for highland, tundra, and glaciers, producing a natural-looking view of the earth. These classes are not represented with flat areas of color, though; they are gently mottled with the textures of the surface from imagery while retaining the seven vegetation classes. For example, the desert oranges range from reddish to yellowish with different rock and sand types. Gray shaded relief overlays these general color categories as well, so highlands look raised and basins nestle between ranges, showing the geomorphic controls for contrasting vegetation types. The map functions as a stand-alone image or as a base for other information.

Courtesy of Tom Patterson, www.shadedrelief.com.

5.5 AFRICA CHIMPANZEE CONSERVATION MAPS

This series of maps organizes chimpanzee priority areas and species ranges, protected areas, and ecosystem conservation priorities to support Conservation International's analysis for West Africa. The maps each have a simple message and combine ordered priorities, represented by color series, with other area representations. The light diagonal lines in the second map represent general chimpanzee ranges. These areas do not have outlines, thus giving the sense they are approximate areas. The pink lines that arc over west and east portions of the maps sketch the ranges of two different species, *pan troglodytes verus* and *vellerosus.* The labels are built into these sketch lines and arc with them, creating an easy-to-read regionalization that does not clutter the map with additional area symbols. Labels for each priority area are positioned with care in the top map. They sit inside larger areas, next to smaller areas, and arc down to horizontally arranged areas, avoiding the need for leader lines. Each label is closer to the area it labels than any other area, a key characteristic of successful map labeling.

Courtesy of Mark Denil, Conservation International.

Chimpanzee Priority Areas

Chimpanzee Priority Areas

- Extremely Important Areas
- Very Important Areas
- Areas Needing Surveys

Chimpanzee Ranges

- chimpanzee species range

Data:
Butynski, T.M. "Africa's great apes" pp. 3-56, *Great apes and humans: The ethics of coexistence*, (Smithsonian Institution Press, Washington D.C. 2001)

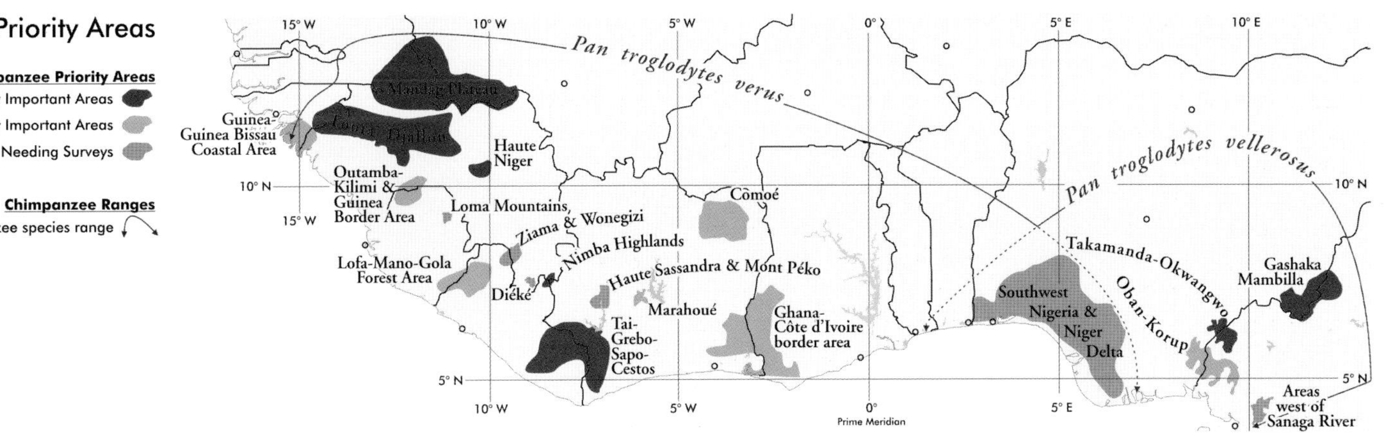

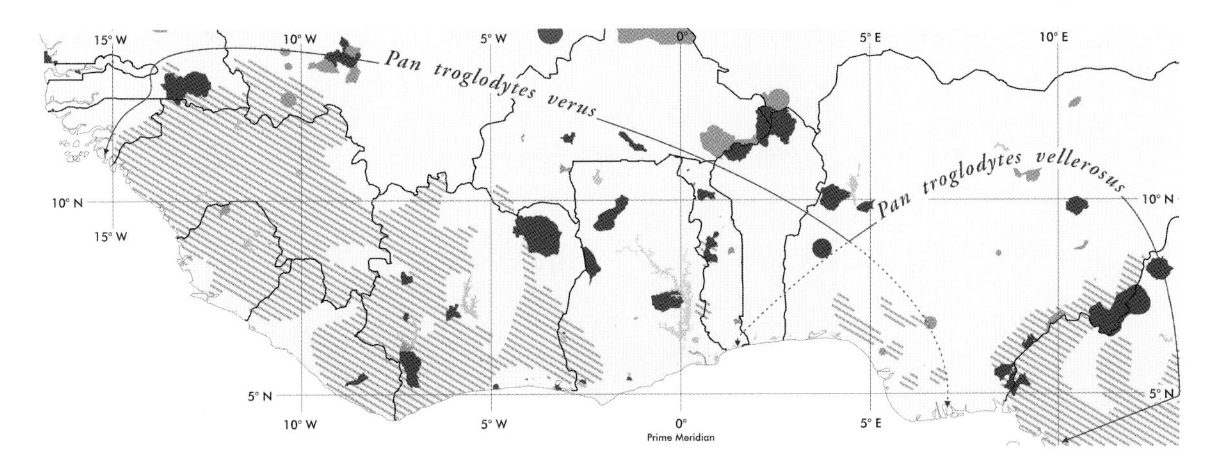

Chimpanzee Ranges and Conservation Protected Areas

Chimpanzee Ranges

- chimpanzee range
- chimpanzee species range

data:
Butynski, T.M. "Africa's great apes" pp. 3-56, *Great apes and humans: The ethics of coexistence*, (Smithsonian Institution Press, Washington D.C. 2001)

IUCN Protected Areas Management Categories

- II - national park
- III - natural monument
- IV - habitat/species management area
- VI - managed resource protected area

data:
UNEP - WCMC world protected areas

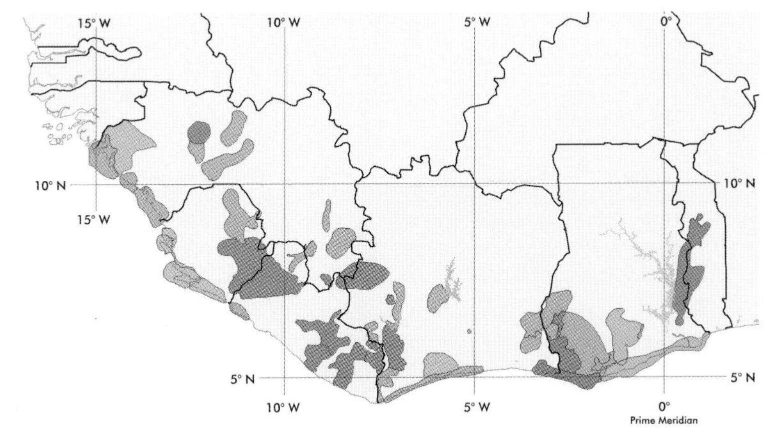

Biodiversity Conservation Priorities Upper Guinea Forest

Forest and Aquatic Ecosystem Priority Areas

- Exceptionally High Priority
- Very High Priority
- High Priority

data:
Conservation Priority Setting Workshop
From the Mountains to the Sea:
Biodiversity Connections from Guinea to Togo
1999

100 0 100 200 300 400 500 600 700
kilometers

scale: 1/20 million
projection: Equal Area Cylindrical
data: **Digital Chart of the World**
cartography: M.Denil
these maps were produced by the Conservation Mapping Program, GIS & Mapping Laboratory of the Center for Applied Biodiversity Science at Conservation International

CENTER FOR APPLIED BIODIVERSITY SCIENCE
AT CONSERVATION INTERNATIONAL

AFRICA
area of interest
Atlantic Ocean

scale: 1/123 million
projection: Vertical Perspective
central point: 0° longitude, 9° latitude
elevation: 2 million meters

5.6 NEWPORT COAST ARCHAEOLOGICAL MAP

Here, soil types are combined with buffer calculations weighted by slope around archaeological sites for a study of resource procurement areas used by prehistoric peoples in the Newport Coast area in Southern California. Hues are used to categorize soil types. The map also overlays elevation, using dark shades for lower elevations progressing to white for higher elevations. The map does not incorporate hillshading, since elevation uses the full lightness range. Elevation grays are removed from the resource procurement areas, bounded by thin black lines, making the soil hues more saturated in the areas of interest. A dark gray fill marks the archaeological sites within the procurement areas. The overall effect provides a context of elevation and soil combinations outside the procurement areas, but emphasizes soil types inside these small areas of greatest relevance to the analysis. As a soil polygon moves from inside to outside a procurement area, the color becomes desaturated but the shared hue makes its continuation easy to interpret.

Courtesy of Nicole Pletka.

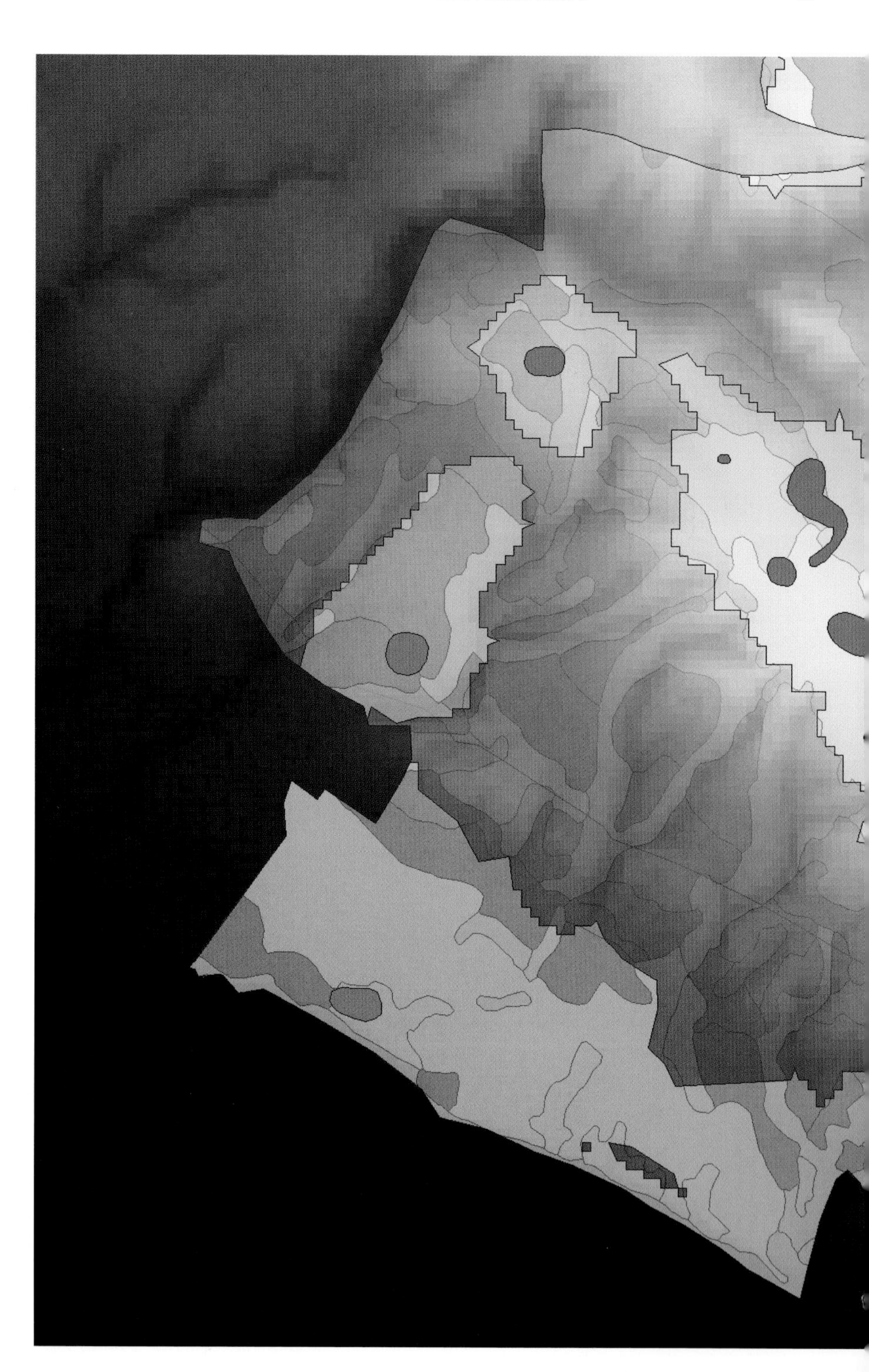

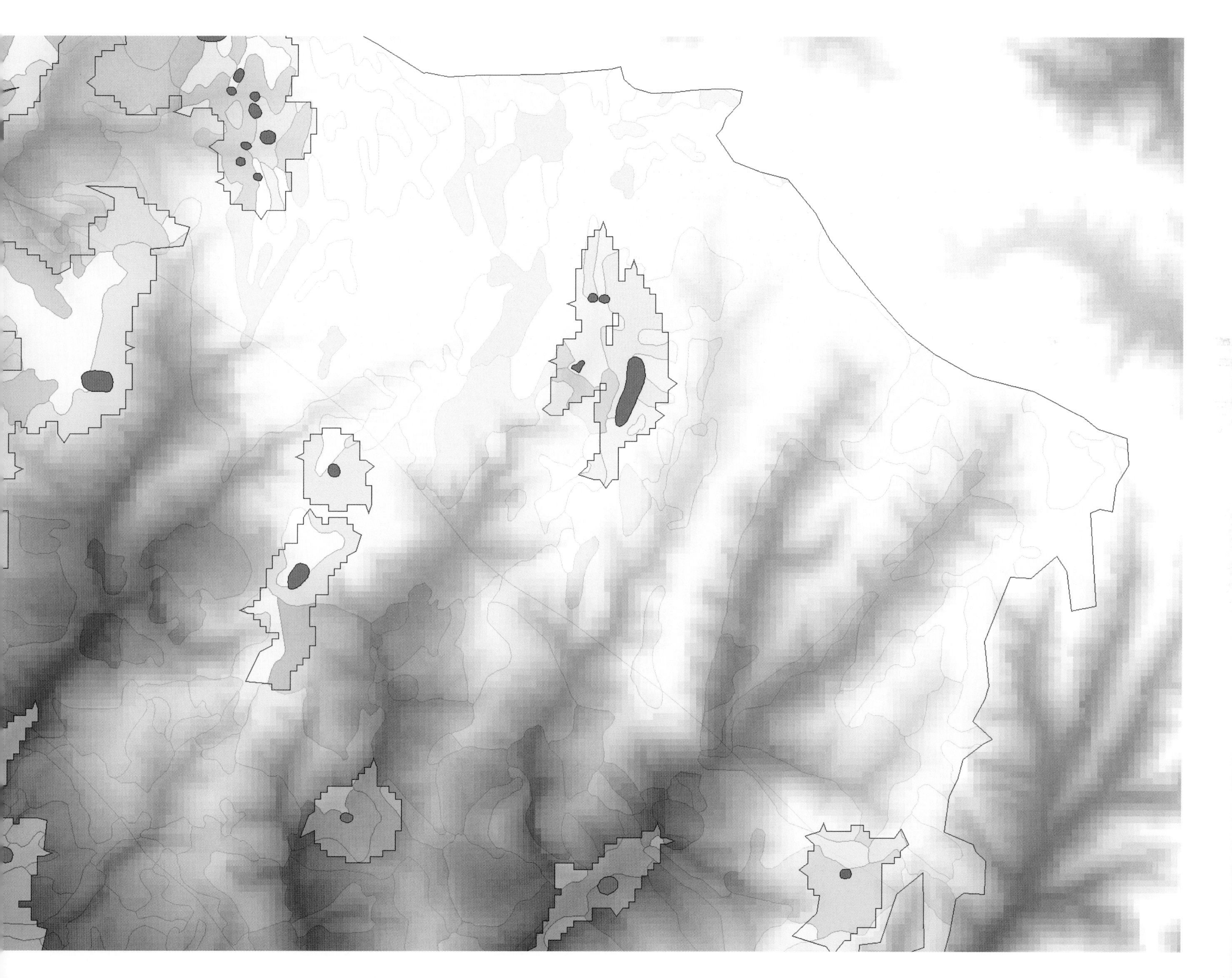

5.7 GLOBAL SOIL REGIONS MAP

World soil categories are aggregated into twelve orders in this U.S. Department of Agriculture (USDA) map. Colors are bold and contrast well with each other to aid map interpretation. Related soil orders have related colors. For example, the two dark greens for Alfisols and Mollisols represent the majority of crop areas in the world. Colors contrast in hue, lightness, and saturation to provide a visual message that soils around the world vary widely in their properties. Regionalization occurs as like colors cross country boundaries and even land masses at similar latitudes, creating a clear visual link to climate (soil moisture content is a significant factor in devising the soil taxonomy). Major river lines, highly generalized country boundaries, and white country labels with a fine black shadow provide basic location information without interfering substantially with the soil order colors. The white vignette along coasts is a design touch that provides maximum contrast for small coastal soil polygons.

Courtesy of Natural Resources Conservation Service.

Soil Orders

Alfisols

Most Alfisols were or are forested, with moderate to high base saturation. Typically they have a light-colored surface layer over a horizon of silicate clay accumulation (argillic). The cooler Alfisols tend to form a belt between the grassland Mollisols and the Spodosols of the more humid climates. Where temperatures are warmer they form a belt between the Aridisols and the older Ultisols and Oxisols. Along with Mollisols, Alfisols account for a major portion of soils that are used to grow crops in the world.

AREA - 13,159,000 km² - 10.1%

India

Entisols

The Entisols have little or no evidence of soil formation. They are most extensive on sub-recent alluvial plains and valleys or on steep slopes where erosion is rapid. The rate of soil formation is reduced for several reasons. Generally time has not elapsed since deposition of the material for soil forming processes to act. In some of these soils, peraquic conditions prevail where the soil is saturated with water during the whole year. The soil is permanently reduced, preventing cambic horizon formation. On steep slopes, rapid erosion results in shallow soils where weathered parent materials rest on hard rock.

AREA - 23,432,000 km² - 17.9%

Germany

Inceptisols

The Latin word 'inceptum' means "beginning" and the central concept of Inceptisols is that of soils in the early stages of soil formation. The initial stage of soil formation is exemplified by several attributes, which are the result of the presence or absence of certain processes. Soil formation on rocks consists of weathering of the rock which is essentially a geochemical process accompanied by soil forming processes acting on the weathered products. In cool humid climates, the soil forming process may be the accumulation of organic matter to give rise to a mollic or umbric epipedon. In warmer climates, cambic horizon formation takes place, which is expressed by clay or structure formation or release of iron to form a 'color B' horizon.

AREA - 19,854,000 km² - 15.2%

South Korea

Spodosols

A black, reddish brown to strong brown subsoil (spodic) horizon is the primary identifying characteristic of a Spodosol. It is often overlain by a gray to light gray eluvial horizon. These distinctive and contrasting colors make Spodosols easily identifiable, although there are always exceptions. The simple explanation for this horizon sequence holds that under cool, humid or perhumid climates, organic acids from a litter layer leach amorphous mixtures of organic matter and aluminum with or without iron from the eluvial horizon and deposit them in the illuvial spodic horizon. Most Spodosols have formed under such conditions and thus are common in the northern latitudes.

AREA - 4,596,000 km² - 3.5%

France

Andisols

Soils formed on volcanic ash and cinders and having andic properties are distributed along the circum-Pacific belts and occur sporadically elsewhere. The Andisols have mineralogical composition ranging from volcanic glass, short-range order minerals such as allophane and imogolite, and variable amounts of halloysite. This mineralogical association gives unique properties to such soils including a high phosphate fixing capacity, low cation retention, and a high water holding capacity. Many of these soils are found on volcanic slopes or are developed through the weathering of plateau basalts. These soils support a high human population density due to their general ease of cultivation and also because of the cool environment of the volcanic mountains, which is generally free of pests and diseases.

AREA - 975,000 km² - 0.8%

Japan

Gelisols

In areas where the mean annual soil temperature is less than 0°C, the soils are frozen for long periods of the year and thaw during the short warmer spells. The freezing and thawing processes promote physical changes in the soil. If there is sufficient water and the warm period is long enough, vegetation establishes and organic matter accumulates on the soil. Organic-rich soil develops. Due to low temperatures, these soils have unique features such as ice-lenses or permafrost may underlie the soil.

AREA - 11,869,000 km² - 9.1%

Alaska

Mollisols

Mollisols are typically base-rich with a dark colored surface horizon. To a large extent Mollisols are the breadbasket of the world—the prairies in the United States, the steppes of Russia, and the pampas of Argentina. Most Mollisols are cultivated, in fact there are only limited areas in the world where they have not been cultivated. Mollisols may initially be farmed with no additions of fertilizers. However to sustain the high yields of corn, soybeans, sorghum, and small grains of today, fertilizers must be used.

AREA - 9,161,000 km² - 7.0%

Missouri

Ultisols

Ultisols are morphologically similar to Alfisols in having a subhorizon of clay accumulation but have few extractable bases especially at depth. Most Ultisols are acid although some may have a high pH in the surface horizons due to vegetation recycling or from aerosolic additions. From a process point of view, the ideal Ultisol has a subsurface horizon of clay enrichment by clay translocation from the surface horizons. If the surface horizons have more than 40% clay, for practical purposes, these soils with textural change with depth are considered as Ultisols. If there is less than 40% clay, then they are classified as Oxisols.

AREA - 10,550,000 km² - 8.1%

Taiwan

Aridisols

Aridisols, as their name implies, are soils that do not have water available to mesophytic plants for long periods. During most of the time when the soil is warm enough for plants to grow, soil water is held at potentials less than permanent wilting point or it is salty, or both. There is no period of 90 consecutive days when moisture is continuously available for plant growth. The concept of Aridisols is based on the low availability of soil moisture for sustained plant performance. In areas bordering deserts, the absolute precipitation may be high but due to runoff or a very low storage capacity of the soil or both, the actual soil moisture regime is aridic.

AREA - 15,464,000 km² - 11.8%

Texas

Histosols

Most soil classifications including Soil Taxonomy separate mineral soils from organic soils. Histosols are soils that consist of dominantly organic soil materials. They develop where the rates of organic matter accumulation exceed decomposition and turnover. Most of these soils formed under saturated conditions where the soil was saturated or nearly saturated with water most of the year. These soils have been referred to as bogs, moors, peat or mucks. To be farmed, most Histosols must be drained. Management of the water table depth is critical to their use. When drained, Histosols oxidize and subside, and requiring further drainage.

AREA - 1,526,000 km² - 1.2%

Florida

Oxisols

Oxisols are reddish, yellowish, or grayish soils with a mineral composition dominated by intensive weathering. They are most common on the gentle undulating areas of geologically old surfaces in tropical and subtropical regions. The most extensive areas of Oxisols are on the interior plateaus of South America, the lower portion of the Amazon Basin, significant portions of the Central African Basin, and important areas in Asia, northern Australia, and several tropical islands of the Pacific. Their profiles are distinctive because of the lack of obvious horizons. Their surface horizons are usually somewhat darker in color than the subsoil, but the transition of subsoil layers is diffuse.

AREA - 9,811,000 km² - 7.5%

Thailand

Vertisols

Vertisols are clayey soils dominated by high shrink-swell, smectic clays, which have deep, wide cracks on some occasions during the year and slickensides within 100 cm of the soil surface. They shrink when dry and swell when moistened. Vertisols make up a relatively homogeneous order of soils because of the amount and kind of clay that is common to them. In many countries where Vertisols are common, they are known by their local names. Examples of local names include, Gilgai soils (Australia), Adobe (Philippines), Sha Chiang (China), Black Cotton Soils (India), Smonitza (Bulgaria), Tirs (Morocco), Makande (Malawi), Vleigrond (South Africa), and Sonsosuile (Nicaragua).

AREA - 3,160,000 km² - 2.4%

Texas

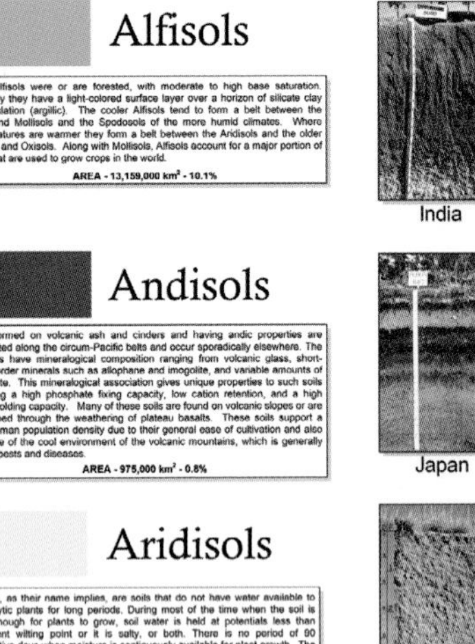

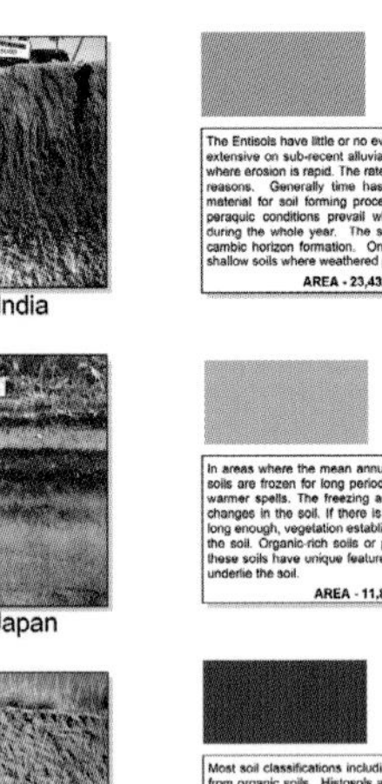

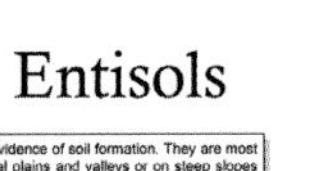

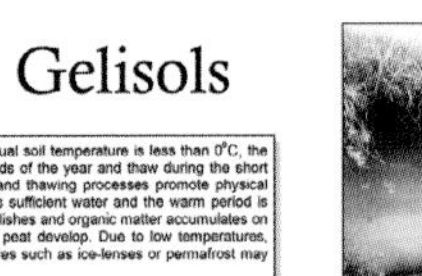

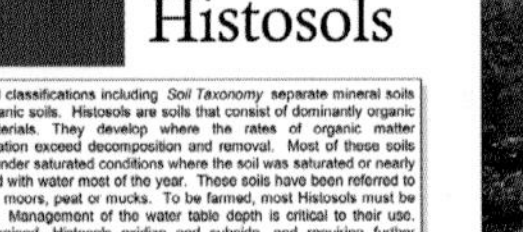

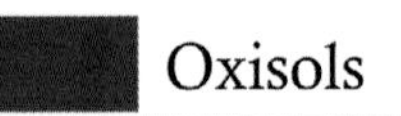

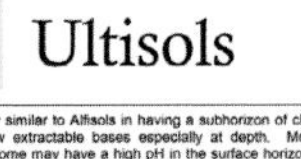

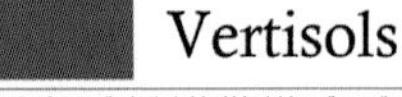

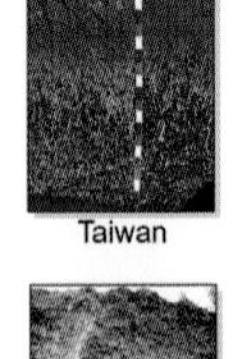

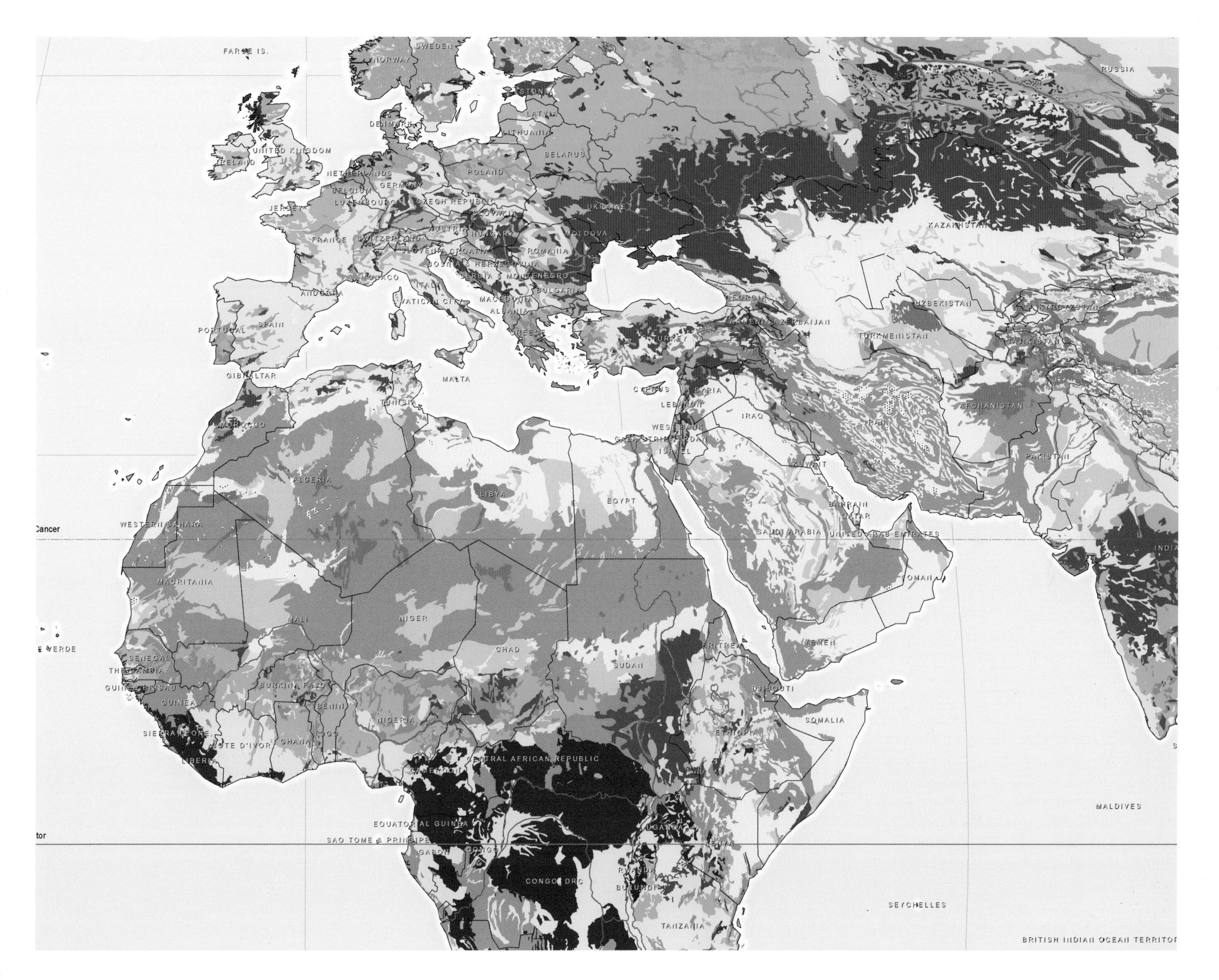
FAROE IS.
SWEDEN
NORWAY
RUSSIA
ESTONIA
LATVIA
DENMARK
LITHUANIA
UNITED KINGDOM
IRELAND
BELARUS
NETHERLANDS
POLAND
GERMANY
BELGIUM
LUXEMBOURG
CZECH REPUBLIC
JERSEY
SLOVAKIA
UKRAINE
AUSTRIA
HUNGARY
MOLDOVA
KAZAKHSTAN
FRANCE
SWITZERLAND
SLOVENIA
CROATIA
ROMANIA
BOSNIA & HERZEGOVINA
MONACO
SERBIA & MONTENEGRO
ANDORRA
ITALY
BULGARIA
GEORGIA
VATICAN CITY
MACEDONIA
UZBEKISTAN
KYRGYZSTAN
ALBANIA
PORTUGAL
SPAIN
ARMENIA
AZERBAIJAN
GREECE
TURKMENISTAN
TURKEY
TAJIKISTAN
GIBRALTAR
MALTA
CYPRUS
SYRIA
TUNISIA
LEBANON
AFGHANISTAN
IRAQ
IRAN
MOROCCO
WEST BANK
GAZA STRIP
JORDAN
ISRAEL
KUWAIT
PAKISTAN
ALGERIA
LIBYA
EGYPT
BAHRAIN
QATAR
Cancer
WESTERN SAHARA
SAUDI ARABIA
UNITED ARAB EMIRATES
INDIA
OMAN
MAURITANIA
MALI
NIGER
CHAD
ERITREA
YEMEN
SENEGAL
THE GAMBIA
SUDAN
GUINEA-BISSAU
BURKINA FASO
DJIBOUTI
GUINEA
BENIN
NIGERIA
SOMALIA
SIERRA LEONE
GHANA
TOGO
ETHIOPIA
LIBERIA
CENTRAL AFRICAN REPUBLIC
CAMEROON
MALDIVES
EQUATORIAL GUINEA
UGANDA
SAO TOME & PRINCIPE
KENYA
GABON
CONGO
RWANDA
CONGO DRC
BURUNDI
SEYCHELLES
TANZANIA

5.8 LITTLE BELT MOUNTAINS GEOLOGIC MAP

The Canyon Ferry Dam 30-by-60 minute quadrangle shows complicated geologic structures in an area of the Rocky Mountain thrust belt in Montana. Colors vary widely in hue but are fairly light to allow contours and other location information to show through the symbols. Hues are overlaid by patterns that vary in shape and hue to further differentiate rock types. For example, the scatter of ticks representing igneous rocks are rendered in different colors and overlaid on a range of other colors for a wide variety of classes. Related hues are used for related rock types, and different hues establish general differences in ages. For example, Mississipian blues are near older Devonian purples which are near still older Cambrian oranges. Letter labels, common on geologic maps, specify geologic characteristics of the areas. These labels are small so they do not compete with the visual patterns. They signify the rock age and group or formation name (for example, Mm for a Mississipian Madison group). The map legend contains approximately 150 symbols, so relying on matching color and pattern alone to identify geologic features would burden the map user—while they may seem redundant with the symbolization scheme, the labels are a helpful addition. At the same time, the general ages and types of rocks are evident from the logical use of color and pattern, so the savvy map user can comprehend the information presented without having to read the small labels.

Courtesy of U.S. Geological Survey.

Ebd	Biotite dacite (Eocene)
Ebhd	Biotite hornblende dacite (Eocene)
Eql	Quartz latite (Eocene)
Ebqd	Biotite quartz diorite (Eocene)
TKd	Garnet peridotite in diatreme (Tertiary or Upper Cretaceous)
Kau	Adel Mountain Volcanics, undivided (Upper Cretaceous)
Kad	Dacite
Kaql	Quartz latite
Kal	Latite
Kr	Rhyodacite flows or shallow intrusive bodies (Late Cretaceous)
Kbg	Biotite granite (Late Cretaceous)
Kqm	Quartz monzonite (Late Cretaceous)
Kbqm	Butte Quartz Monzonite (Late Cretaceous)
Kmd	Monzodiorite (Late Cretaceous)
Kmo	Monzonite (Late Cretaceous)
Kgd	Biotite granodiorite (Late Cretaceous)
Kda	Dacite (Late Cretaceous)
Kla	Latite (Late Cretaceous)
Ki	Basaltic and andesitic sills, dikes, and irregular-shaped bodies related to the Elkhorn Mountains Volcanics (Late Cretaceous)
Kt	Trachybasalt and syenogabbro (Late Cretaceous)
Kqd	Quartz diorite (Late Cretaceous)
Kd	Diorite (Late Cretaceous)
Kbd	Biotite diorite (Late Cretaceous)
Khd	Hornblende diorite (Late Cretaceous)
	Pyroxenite (Late Cretaceous?)
Kog	Olivine gabbro (Late Cretaceous)
Ktm	Two Medicine Formation, undivided (Upper Cretaceous)
Ktf	Volcanic member
	Elkhorn Mountains Volcanics (Upper Cretaceous)
Keva	Ash-flow tuff member
Kevm	Middle member
Kevl	Lower member
Ks	Slim Sam Formation (Upper Cretaceous)
Ket	Eagle Sandstone and Telegraph Creek Formation, undivided (Upper Cretaceous)
Kvt	Virgelle Sandstone and Telegraph Creek Formation, undivided (Upper Cretaceous)
Kvi	Intrusive igneous rocks, undivided (Late Cretaceous)
Ktc	Telegraph Creek Formation (Upper Cretaceous)
Kmu	Marias River Formation, undivided (Upper Cretaceous)
Kmk	Kevin Member
Kml	Lower part, undivided
Kb	Blackleaf Formation, undivided (Upper and Lower Cretaceous)
Kbv	Vaughn Member (Upper and Lower Cretaceous)
Kbt	Taft Hill Member (Lower Cretaceous)
Kbf	Flood Member (Lower Cretaceous)
Kc	Colorado Group, undivided (Upper and Lower Cretaceous)
Kck	Colorado Group (Upper and Lower Cretaceous) and Kootenai Formation (Lower Cretaceous), undivided
Kk	Kootenai Formation (Lower Cretaceous)
KJme	Morrison Formation (Lower Cretaceous and Upper Jurassic) and Ellis Group (Upper and Middle Jurassic), undivided
KJm	Morrison Formation (Lower Cretaceous and Upper Jurassic)
Je	Ellis Group, undivided (Upper and Middle Jurassic)
PPpq	Phosphoria (Permian) and Quadrant (Pennsylvanian) Formations, undivided
Pq	Quadrant Formation (Pennsylvanian)
Pa	Amsden Formation (Pennsylvanian)
Pal	Limestone succession at top
Par	Red-bed succession at base
PMab	Amsden Formation (Pennsylvanian) and Big Snowy Group (Upper Mississippian), undivided
Mt	Tyler Formation, undivided (Upper Mississippian)
Mtu	Upper part
Mts	Sandstone and conglomeratic sandstone in lower part
Mb	Big Snowy Group, undivided (Upper Mississippian)
Mho	Heath and Otter Formations, undivided
Mh	Heath Formation
Mo	Otter Formation
Mk	Kibbey Formation
Mmu	Madison Group, undivided (Upper and Lower Mississippian)
Mm	Mission Canyon Limestone (Upper and Lower Mississippian)
Ml	Lodgepole Limestone (Lower Mississippian)
MDt	Three Forks Formation (Lower Mississippian and Upper Devonian)
Du	Upper and Middle Devonian rocks, undivided
D€u	Upper and Middle Devonian and Upper and Middle Cambrian rocks, undivided
Dj	Jefferson Formation (Upper Devonian)
D€m	Maywood Formation (Upper and Middle Devonian) and local Upper Cambrian beds
€pp	Pilgrim Formation (Upper Cambrian) and Park Shale (Upper and Middle Cambrian), undivided
€pi	Pilgrim Formation (Upper Cambrian)
€p	Park Shale (Upper and Middle Cambrian)
€h	Hasmark Formation (Upper and Middle Cambrian)
€m	Meagher Limestone (Middle Cambrian)
€w	Wolsey Formation (Middle Cambrian)
€f	Flathead Sandstone (Middle Cambrian)
Zd	Diorite (Neoproterozoic)

Half of map legend shown.

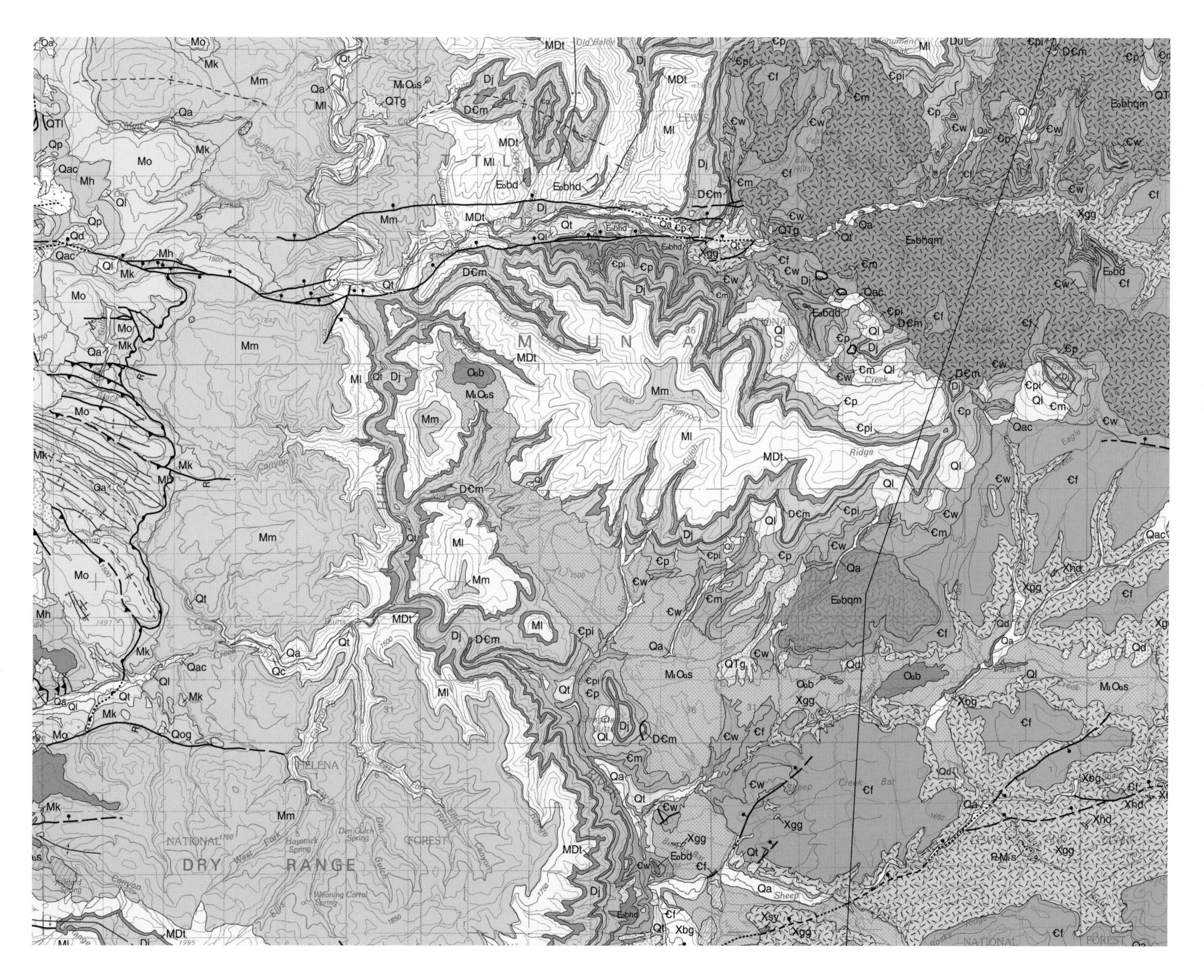
Old Baldy
Monument
LEWIS
NATIONAL
Rimrock
Ridge
Eagle Creek
HELENA
NATIONAL
DRY RANGE
FOREST
Haystack Spring
Den Gulch Spring
Weaning Corral Spring
West Fork
Ellis Canyon
Sheep Creek
LEWIS AND CLARK
NATIONAL FOREST

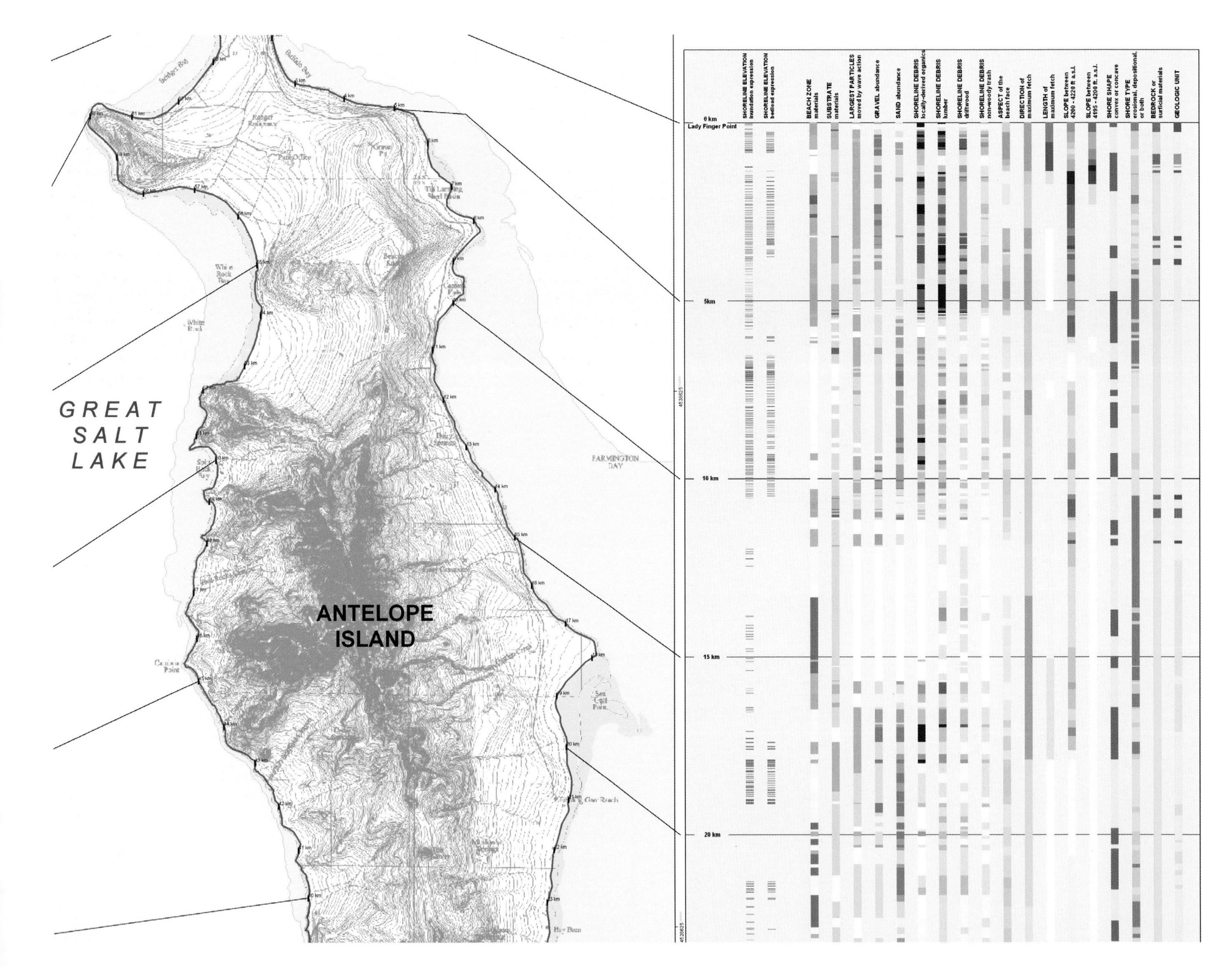
GREAT
SALT
LAKE
ANTELOPE
ISLAND
FARMINGTON
BAY
0 km
Lady Finger Point
5km
10 km
15 km
20 km
SHORELINE ELEVATION inundation expression
SHORELINE ELEVATION bedload expression
BEACH ZONE materials
SUBSTRATE materials
LARGEST PARTICLES moved by wave action
GRAVEL abundance
SAND abundance
SHORELINE DEBRIS locally-derived organics
SHORELINE DEBRIS lumber
SHORELINE DEBRIS driftwood
SHORELINE DEBRIS non-woody trash
ASPECT of the beach face
DIRECTION of maximum fetch
LENGTH of maximum fetch
SLOPE between 4200 - 4220 ft a.s.l.
SLOPE between 4195 - 4200 ft. a.s.l.
SHORE SHAPE convex or concave
SHORE TYPE erosional, depositional, or both
BEDROCK or surficial materials
GEOLOGIC UNIT

MATERIALS

Beach zone materials
Substrate materials
Largest particles moved by wave action

Mud
Sand
Sand with gravel
Fine gravel, gravel, and
coarse beach deposits
Cobbles
Boulders and fines
Colluvium
Small boulders
Boulders
Bedrock
Not Visible

ABUNDANCE of sand and gravel

None
Minor
Sheet of sand/gravel
Ridge of sand/gravel
Not visible

ABUNDANCE of shoreline debris

Locally derived organic material
such as twigs, branches, brinefly debris
Lumber including railroad ties
Driftwood and logs
Non-woody trash
such as tires, plastics, pottery

Rare
(unusual along a 100 ft stretch)
Uncommon
(occurs within a 30 - 50 ft stretch)
Common
(occurs within a 10 ft stretch)
Very common
(occurs within a 3 ft stretch)
Continuous
(relatively continuous line of debris)
Stacked
(piled up in stacks of debris)
Not Visible

COMPASS directions (in degrees from north)

Aspect (perpendicular to shore)
Direction of maximum fetch

15
30
45 Northeast
60
75
90 East
105

5.9 GREAT SALT LAKE SHORELINE MAP

Great Salt Lake, Utah, coastal characteristics are placed in context by a simple gray-scale scan of a reduced USGS topographic map. This map was not designed to clearly present the land features of Antelope Island; the map topic is the shoreline only. The coarse rendering retains just enough detail to see steep areas and some roads on the island interior. Besides a multicolor shoreline, the only color in the map is a light overlay of blues for water depths. Categories and measurements of shore characteristics are described in the columnar displays positioned to the left and right of the map. The data displays organize research data—they compare multiple attributes at each sampling point along the shore. Horizontal slices through these attribute columns are linked to specific locations along the shore at 5-kilometer intervals. Hue is partly used to categorize columns. For example, amounts of shoreline debris are represented in sequences of purple, and shore slope measures are in greens. The twenty columns shown here illustrate the varied characteristics of the shoreline in a simplified manner that allows comparison across attributes.

Courtesy of Genevieve Atwood, University of Utah.

5.10 PACIFIC TROPICAL CYCLONES MAP

The map of historical cyclone tracks is from Pacific Disaster Center's *Asia Pacific Natural Hazards and Vulnerabilities Atlas.* Cyclone track segments are coded by hues that follow a spectral sequence for ordered levels of severity, from red for super typhoon down to blue for tropical depression. All of these colors are dark for contrast with the light base of land and water. A single storm track may run through a selection of colors as the storm changes in strength. The individual lines in this representation are not intended to be discernable; rather, the dense tangle of lines tell us where, how frequent, and how strong storms tend to be across the Asia Pacific and Indian oceans. The glitter of orange and red line segments east of China show the area that suffered the strongest storms over more than sixty years (and thus the area most threatened by future cyclone hazards). Isolated lines curling away from the densest areas provide an understanding of the variation in storm tracks over the years.

Courtesy of Pacific Disaster Center.

Historical Tropical Cyclone Tracks

1945 - 2006

Super Typhoon

Hurricane

Tropical Storm

Tropical Depression

Country Boundaries

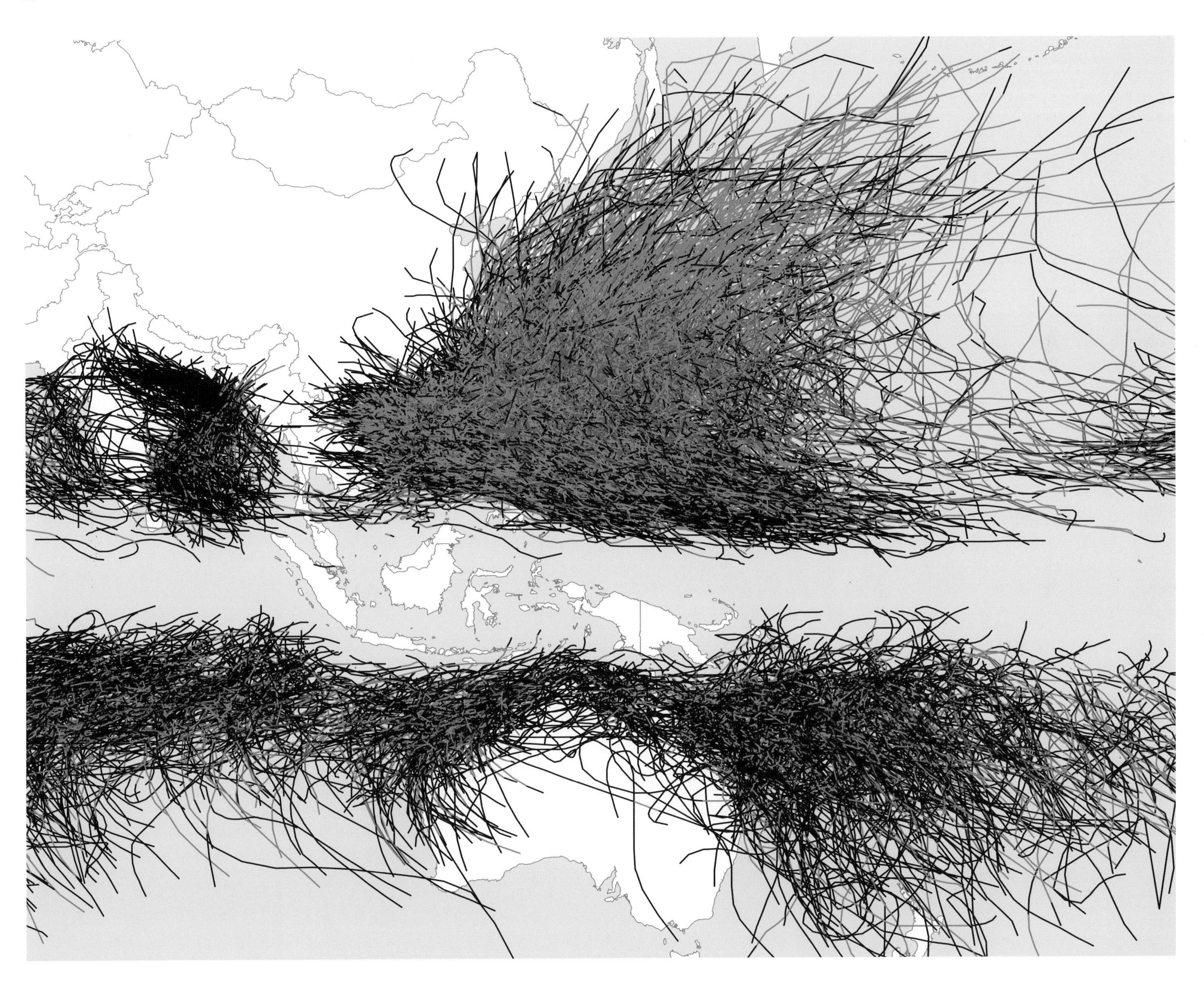

5.11 LAS VEGAS AIRPORT TRAFFIC MAP

Nevada's McCarran International Airport hosted over 41 million passengers the year this map was made. This map captures just a single day of flights in and out of McCarran and nearby smaller airports that serve the Las Vegas Valley, rendered as fine threads across the image base. All of the line colors are light so they contrast with the image background, and airport and airspace labels are readable over the maze of lines. Though lines are all the same weight, they cluster on popular flight paths, producing more solid colors and bolder line groups. The hues are pastel, but different enough that they are discernable from one another. Like Pacific Disaster Center's cyclone map, individual flight lines are not meant to be readable. Instead, the map gives an overview of the volumes and directions of traffic on one busy day. An image basemap showing terrain and landscape colors and textures, overlaid with a few roads and boundary lines, provides adequate base information and resembles the view experienced by the passengers in the sky.

Courtesy of Jeffrey S. Truby.

6

THEMATIC MAPS *QUANTITATIVE*

Thematic maps have a wide range of topics and objectives. Especially prevalent among thematic maps is the presentation of patterns in quantitative spatial data. The distributions shown in this chapter include data represented as points, lines, areas, and volumes. Some of the maps emphasize a single variable and others are multivariate maps that display the distributions of several related themes. Maps showing topics such as estimated hazards and underground water sources present themes that people cannot see on the land surface or in remotely sensed images. The maps organize crucial information in a visible form to allow people to understand environmental and social interrelationships.

Lightness and size differences are the primary tools used to present quantitative data on these maps. Gradations in hue between colors are used to add more contrast within a lightness sequence. Hue is also used to distinguish two lightness ranges, above and below a critical value or between time periods. In the multivariate maps, color and size are combined to moderate the visual impact of one variable using another. For example, small towns have smaller symbols, so the impact of colors representing their population changes is lessened by the small size of the town symbol. With good symbol choices, multivariate maps enable readers to make sense of relationships among three, four, or more distributions.

Quantitative data characteristics affect symbol choices. Continuous data that changes gradually across the surface may be represented with isolines, gradations in color, or a mesh of small arrows that cover an entire map area—snow loads, groundwater, and currents are examples of smooth and continuously changing data. Other continuous data may change abruptly at boundaries of areas, such as soil polygons or counties. Perceptual ordering by lightness in color sequences for these choropleth maps is more important than for smooth and continuous data representations. Discrete datasets are portrayed with points, lines, and discontinuous polygons representing features such as wells, buildings, earthquakes, towns, and linear paths or flows. These symbols may be graduated or proportioned in size to represent various populations, strengths, flows, or other measurements.

Multivariate maps often benefit from customized legends that pull symbols apart into the individual variables combined on the map. For example, the earthquake legend symbols are separated into magnitude circles of three sizes with no color and then single-size circles in colors representing depth. In other words, each combination of magnitude and depth seen on the map is not represented in the legend. The triangular legend for the soils map is perhaps the most customized of the selections in the chapter. These twelve colors could have been set in rectangles with long legend labels for each, but the balanced proportions of sand-silt-clay and variations in category ranges within these proportions would be much harder to visualize for the map user.

Base information on quantitative thematic maps may be minimal, such as administrative boundaries and coastlines, or more elaborate, such as hydrography and transportation networks. But the base is always symbolized as background, because the main messages of these maps are the geographic distributions of their themes.

6.0 COLORADO SNOW LOAD MAP

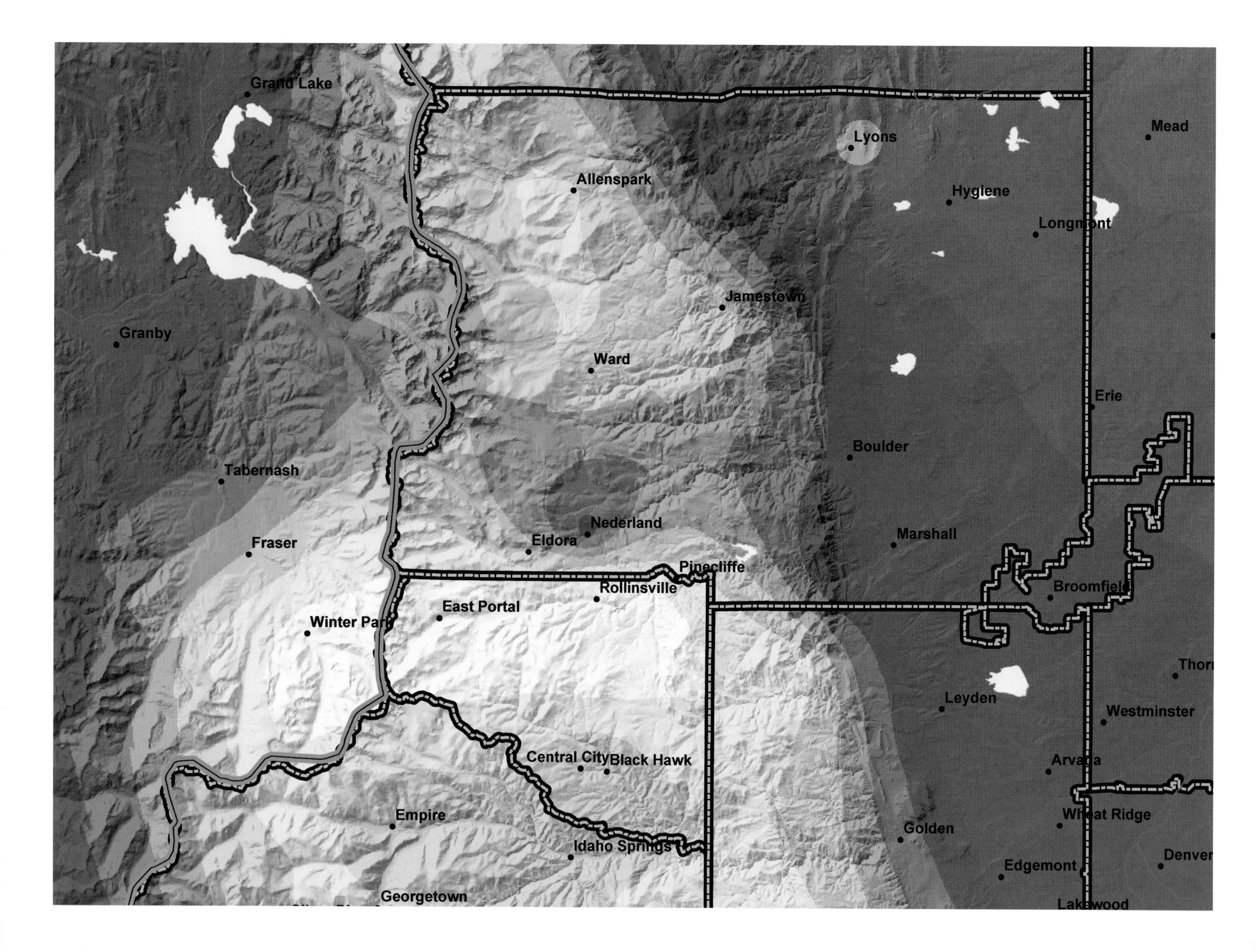

6.0 COLORADO SNOW LOAD MAP

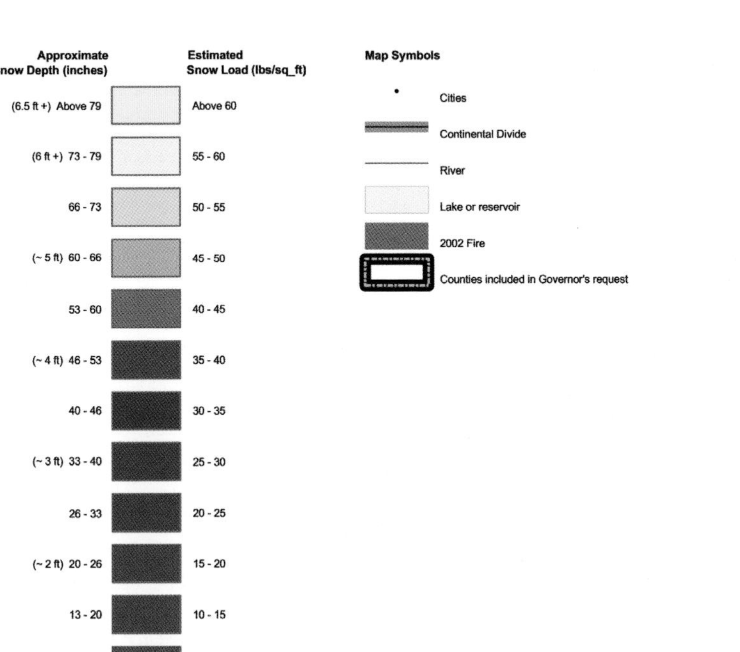

After record Colorado snowfall in March of 2002, USGS interpolated snow depths from NOAA/National Weather Service and weather spotter reports. Then snow depths, average water equivalents for snow, and the weight of water were used to calculate snow load, which is the estimated weight of snow in pounds per square foot, as a parallel measure sought by FEMA personnel.

Courtesy of U.S. Geological Survey.

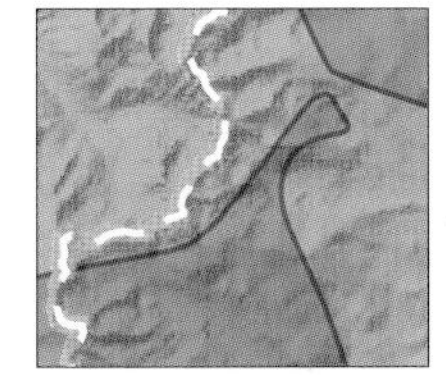

BLUE SEQUENTIAL SCHEME (Page 132)
Isolines in dark blue separate the gradual change in blues from light to dark for low to high snow loads. There is less contrast between these blues than the multiple hues of the original design, so the isolines assist in differentiating the bands of color. The isolines are light enough that they need not be broken where they run across labels. A light gray-scale hillshade sits below these partly transparent blues for locational reference.

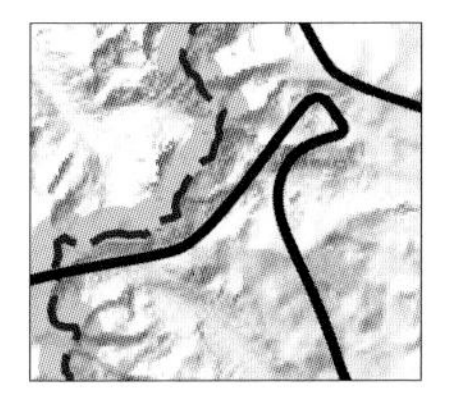

WEIGHTED ISOLINES (Page 133)
There are no color fills between isolines for this design, but the pattern from low to high is emphasized by a sequence of isoline weights. Low loads are represented by thin lines and higher loads are represented by progressively thicker lines. The light hillshade is more colorful, grading from brown shaded slopes to white illuminated slopes, with yellow for partly lit slopes and flat areas. A slight diagonal lightness gradient on water bodies gives them a glint that adds an additional touch of realism to this image. The continental divide is a wide band of transparent color running beneath the county boundaries.

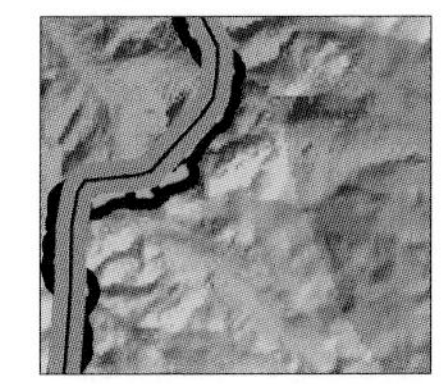

ORIGINAL DESIGN (Pages 130 and 132)
Snow loads are shown in a range of colors from magenta and purple for low weights, dark blues for medium weights, up to light cyan and yellow for the highest weights. Smooth lightness sequences between these hues signify the continuous nature of the data interpolation.

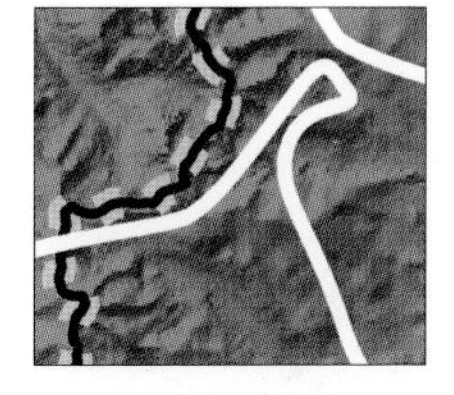

BLACK-AND-WHITE DESIGN (Page 133)
Contrast relationships are reversed on the black-and-white version of the map. Wide white isolines are prominent against the dark hillshade. These white lines also contrast in lightness with the black county boundaries and in pattern with the wide dash for the continental divide. Labels align with isolines, and lines are broken where labels are positioned along them. Town labels have halos similar in overall lightness to the terrain so they do not cut stark gaps in the landscape. They also prevent darker areas in the hillshading from obscuring the letters.

REDESIGNS

6A ORIGINAL DESIGN

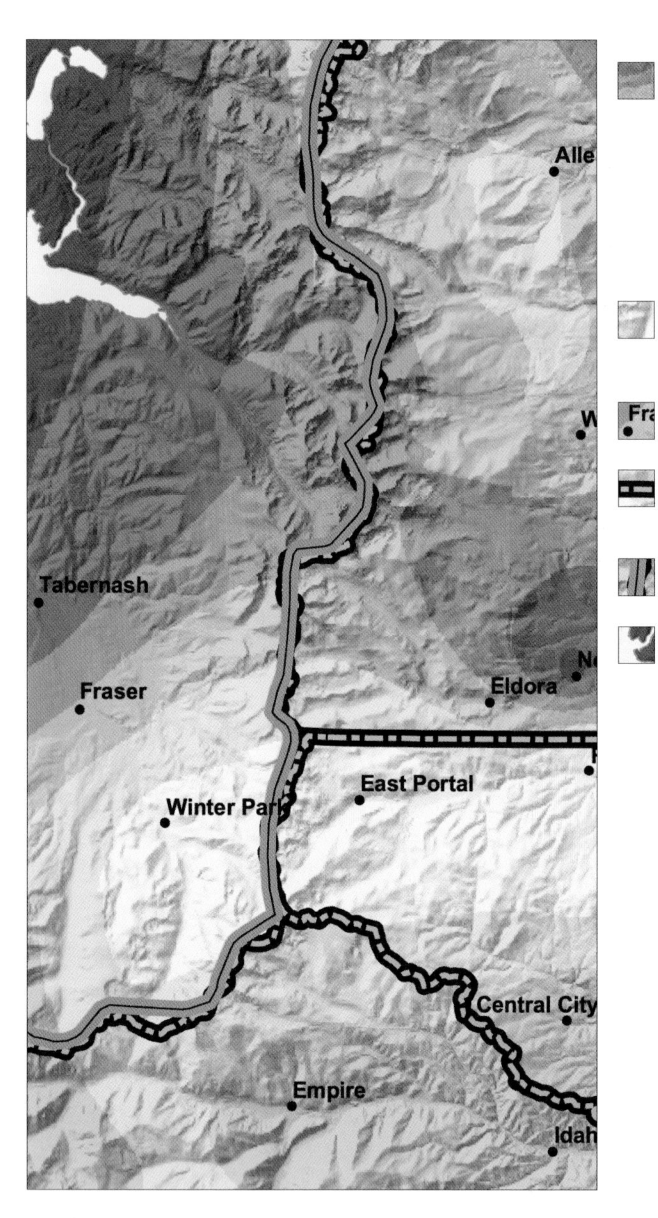

ISOLINES
Fills: All fills 35% transparent overlays:
60Y (yellow);
20C;
50C;
80C;
80C 30M;
80C 65M;
90C 90M;
60C 95M (purple)

HILLSHADE
Ramp from white to 100K

OTHER
All lines: round join
City
Symbol: 4 pt, 100K
Label: 8 pt, Arial Bold, 100K
County
Lines: 2 pt, 56C 100Y, dash pattern 7/2/3/2 *(dash/gap in pts)*; 6.5 pt, 100K
Divide
Lines: 0.75 pt, 100K, round join; 6 pt, 30C 38M 75Y
Water
Fill: 10C 10Y
Lines: 0.5 pt, 100C 40Y

ArcMap Tips (see pages 161 to 167)

1 Dashed line	22 Transparent fill

6B BLUE SEQUENTIAL SCHEME

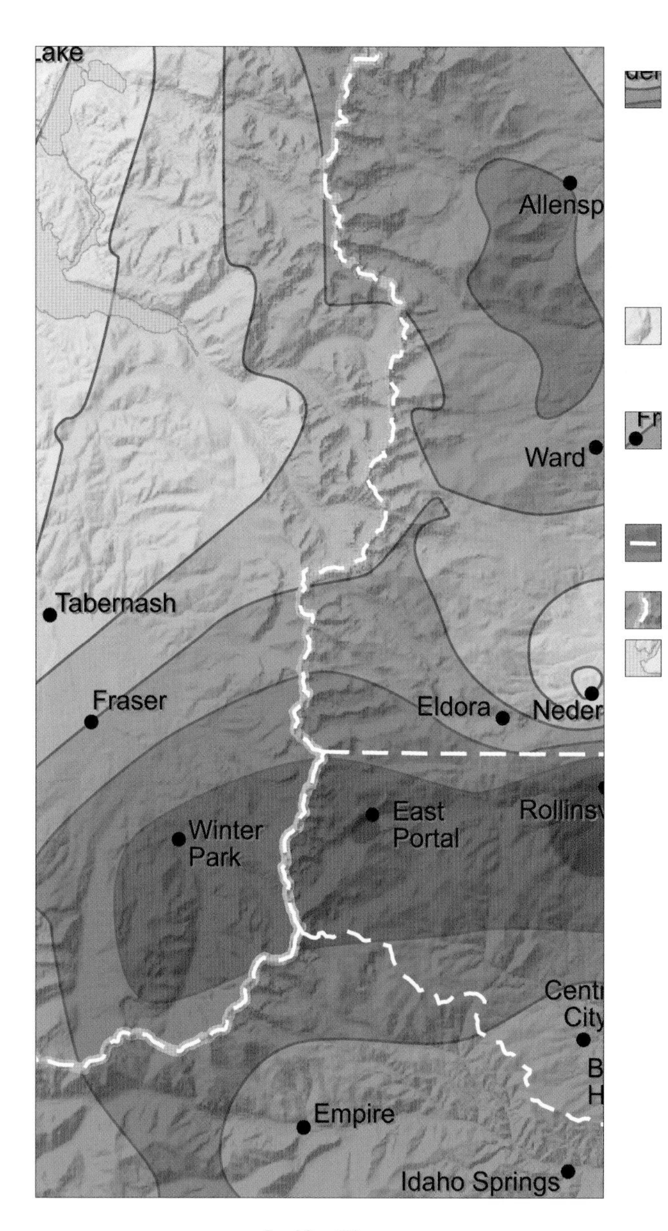

ISOLINES
Fills: 100C 55M 30K (dark blue);
100C 45M 7K;
90C 34M;
75C 22M;
57C 14M;
38C 8M;
24C 6M;
13C 3M (light blue)
Line: 1 pt, 100C 60M

HILLSHADE
Ramp from white to 100K, 80% transparent overlay

OTHER
City
Symbol: 5.5 pt, 100K
Label: 10 pt, Arial, 100K, Shadow: 50K, 0.75 pt X offset, –0.75 pt Y offset
County
Line: 1.5 pt, white, dash pattern 10/5, round join
Divide
Line: 4 pt, 30K, round join
Water
Fill: 20K, 40% transparent
Lines: 0.5 and 1 pt, 65K, 40% transparent, round join

ArcMap Tips

1 Dashed line	29 Label shadow
40 Layer order	

6C WEIGHTED ISOLINES

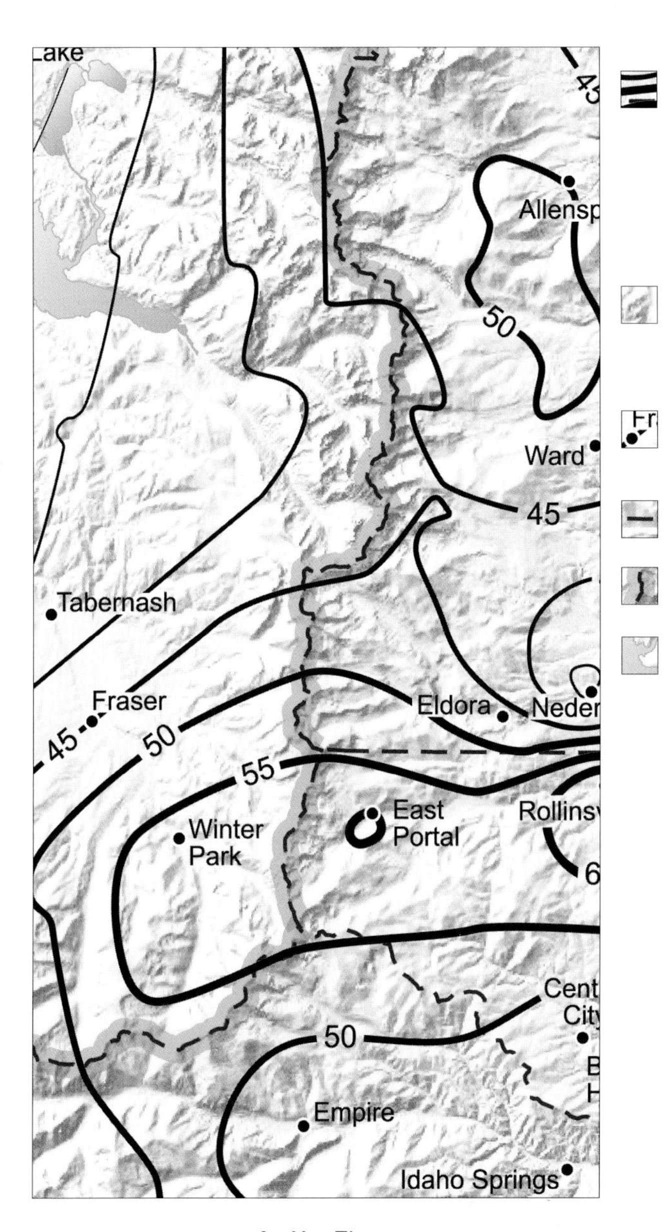

ISOLINES
Lines: 3.5 pt (thickest);
3 pt;
2.5 pt;
2 pt;
1.5 pt;
1 pt;
0.5 pt (thinnest);
all lines 100K
Label: 11.5 pt, Arial, 100K

HILLSHADE
Ramp from 29C 46M 82Y 7K (brown) to 25Y (yellow) to white

OTHER
City
Symbol: 5.5 pt, 100K
Casing: 1 pt, white
Label: 10 pt, Arial, 100K
County
Line: 1.5 pt, 40C 100M 100Y, dash pattern 10/5
Divide
Line: 10 pt, 75M, 80% transparent, round join
Water
Fill: -45° gradient from 11C 2Y to 42C 22M
Lines: 0.5 and 1 pt, 40C 20M, round join

ArcMap Tips

11 Polygon to line	13 Weighted isolines
19 Gradient fill	21 Hillshade colors

6D BLACK-AND-WHITE DESIGN

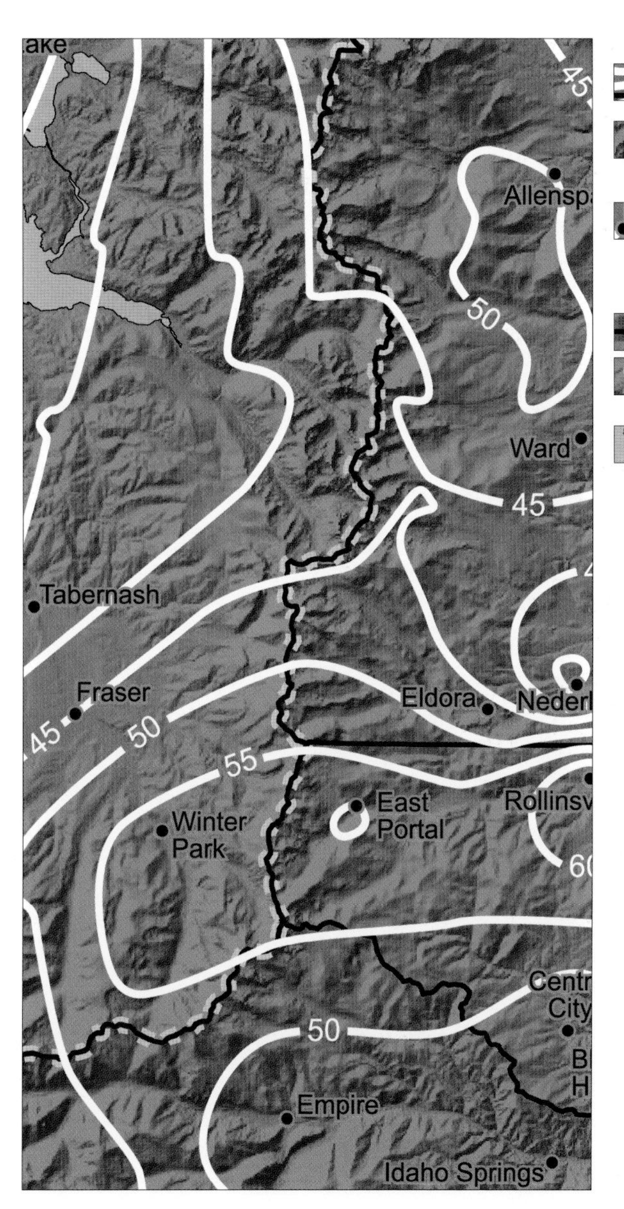

ISOLINES
Line: 3 pt, white
Label: 11.5 pt, Arial, white

HILLSHADE
Ramp from 45K to 100K

OTHER
City
Symbol: 5.5 pt, 100K
Casing: 1 pt, 45K
Label: 10 pt, Arial, 100K
Halo: 1 pt, 45K
County
Line: 2 pt, 100K
Divide
Line: 6 pt, 20K, dash pattern 5/6
Water
Fill: 20K
Lines: 0.5 and 1 pt, 100K, round join

ArcMap Tips

26 Halo	28 Variable-depth masking
38 Multiline labels	

6.1 U.S. SEISMIC HAZARD MAP

Filled isolines show the pattern of peak horizontal acceleration with 10 percent probability of exceedance in fifty years. Colors between isolines emphasize the data pattern, and the isoline interval increases at higher percent gravity accelerations. Grays, blues, and greens at the lower end of the data range color areas between isolines with a 1-percent difference. Isoline intervals progress from 5- to 10- to 20-percent differences at higher accelerations, and these ranges are represented by yellow to orange to red, with dark browns for the highest values. The colors help partition the map into low- and high-hazard zones, but do not interfere with point values and isolines that provide details of the hazard measurements. Light county lines as base data allow map readers to identify locations within the hazard pattern.

Courtesy of U.S. Geological Survey.

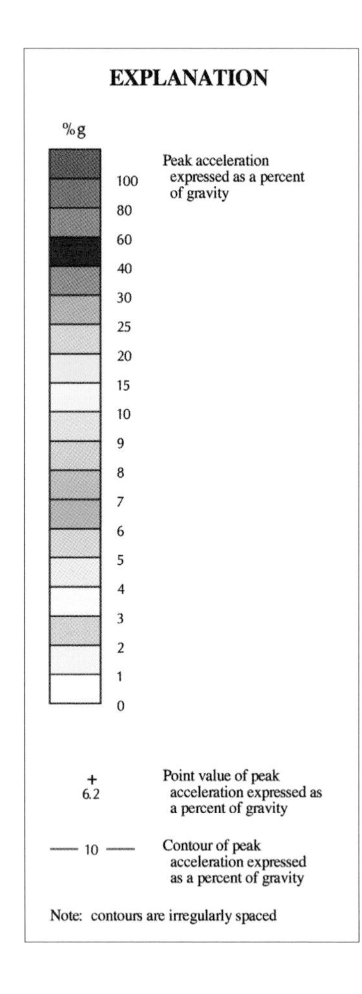

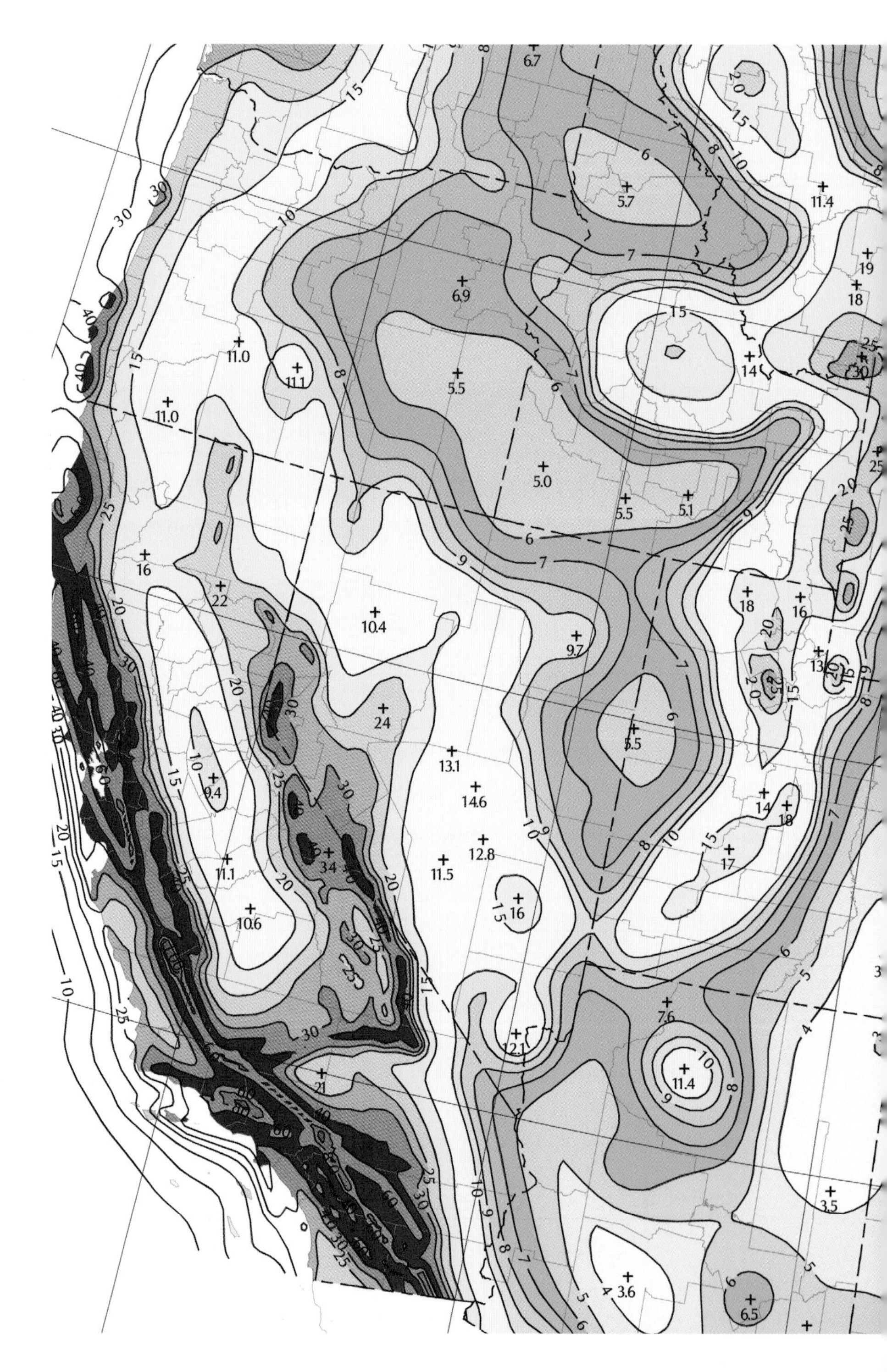

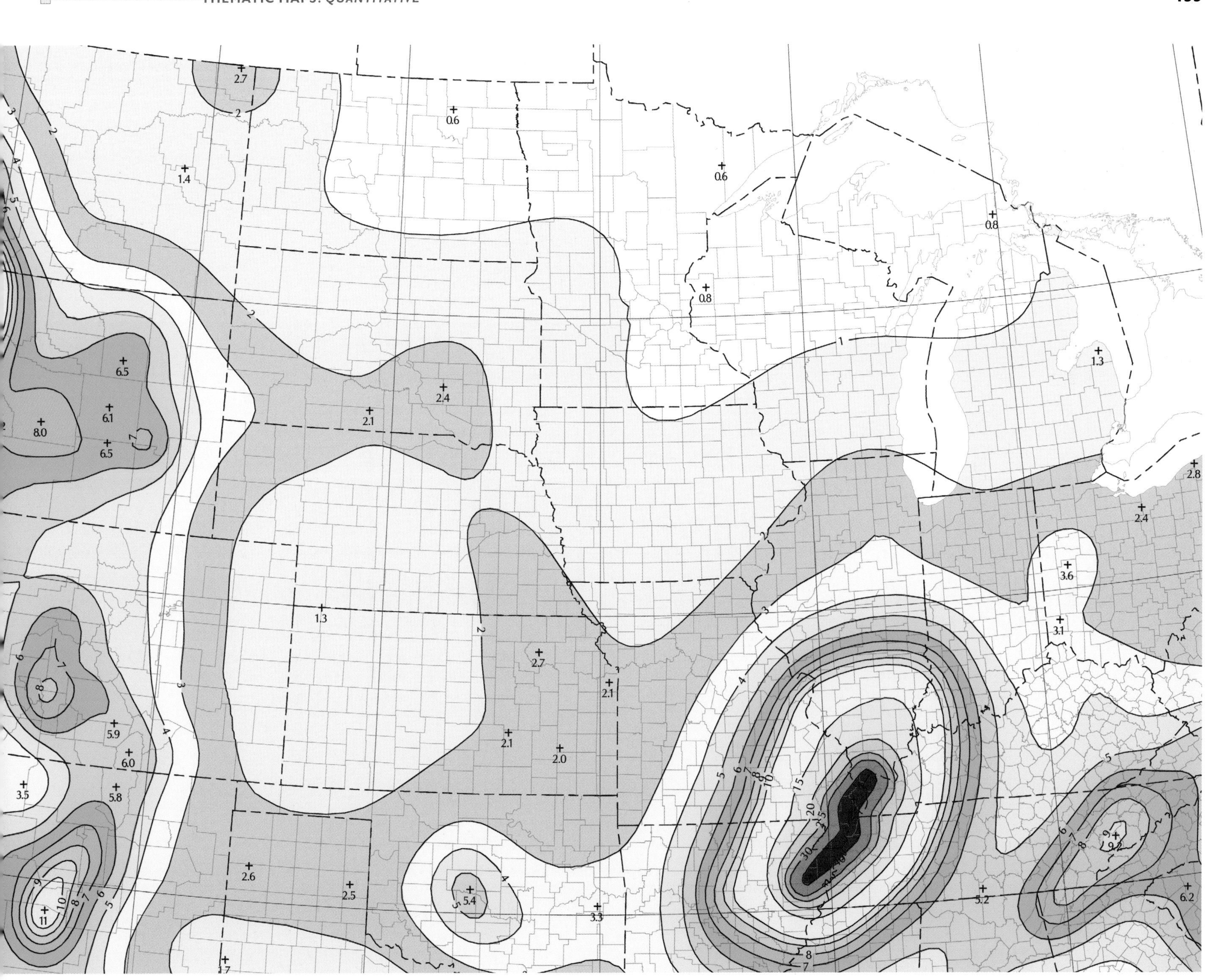
2.7
2
0.6
1.4
0.6
0.8
0.8
1
1.3
6.5
2.4
6.1
2.1
8.0
6.5
7
2.8
2.4
3.6
1.3
3.1
2.7
2.1
5.9
2.1
6.0
2.0
3.5
5.8
9.2
2.6
2.5
5.4
3.3
5.2
6.2
11
1.7

6.2 SAN DIEGO AREA WATER SOURCES MAP

Sweetwater Authority customers receive 70 percent of their water from local water supplies. The percentage of water supplied by each of three primary sources is represented by area colors in light-to-dark sequences of yellow, cyan, and red. Where percentages from multiple sources overlap, transparency in these colors indicates the combination of sources. For example, the orange northeast of the demineralization plant is the combination of red and yellow, signifying similar proportions of water supplied by the R.A. Reynolds and R.A. Perdue plants. The combination of lightness sequences with transparency over a lightly shaded terrain image prevents users from looking up exact supply values, but provides an overview of varying supply proportions across National City, Chula Vista, and Bonita, California.

Courtesy of Sweetwater Authority.

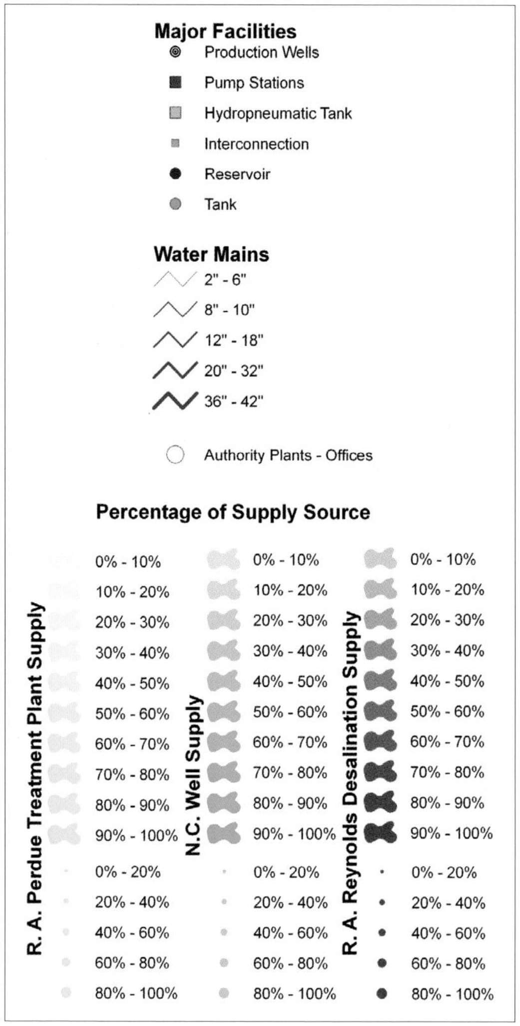

Major Facilities
Production Wells
Pump Stations
Hydropneumatic Tank
Interconnection
Reservoir
Tank

Water Mains
2" - 6"
8" - 10"
12" - 18"
20" - 32"
36" - 42"

Authority Plants - Offices

Percentage of Supply Source

R. A. Perdue Treatment Plant Supply
0% - 10%
10% - 20%
20% - 30%
30% - 40%
40% - 50%
50% - 60%
60% - 70%
70% - 80%
80% - 90%
90% - 100%
0% - 20%
20% - 40%
40% - 60%
60% - 80%
80% - 100%

N.C. Well Supply
0% - 10%
10% - 20%
20% - 30%
30% - 40%
40% - 50%
50% - 60%
60% - 70%
70% - 80%
80% - 90%
90% - 100%
0% - 20%
20% - 40%
40% - 60%
60% - 80%
80% - 100%

R. A. Reynolds Desalination Supply
0% - 10%
10% - 20%
20% - 30%
30% - 40%
40% - 50%
50% - 60%
60% - 70%
70% - 80%
80% - 90%
90% - 100%
0% - 20%
20% - 40%
40% - 60%
60% - 80%
80% - 100%

LINCOLN ACRES #1
National City
Richard A. Reynolds
Groundwater Demineralization
Facility
PLAZA
BONITA
SHOPPING
CENTER
RICE CANYON #22
HALECREST #21
Operations Center
Administrative Office
CLAIRE VISTA #10

6.3 MEXICO DOMINANT SOIL TEXTURE MAP

The North American Soil Properties (NOAM-SOIL) project uses color classes to summarize surface soil texture in this preliminary map of Mexico. The triangular legend indicates the percentages of sand, silt, and clay for each soil polygon. The relative weights in each texture class equal 100 percent, allowing the legend to collapse from a three-dimensional space to a two-dimensional triangle. Cyan for sand, yellow for silt, and magenta for clay distinguish three basic soil textures. The full logic of the color scheme assumes map users understand mixtures of these three primary colors. The potential number of legend colors is simplified by grouping percentage combinations into color areas, outlined in a bolder black line and labeled. Silty Clay, represented by orange, is a mixture of magenta and yellow, and Sandy Clay in purple is a mixture of cyan and magenta.

Courtesy of Sharon W. Waltman, USDA, Natural Resources Conservation Services, National Soil Survey Center, Lincoln, Nebraska; and David James, USDA, Agricultural Research Service, National Soil Tilth Laboratory, Ames, Iowa.

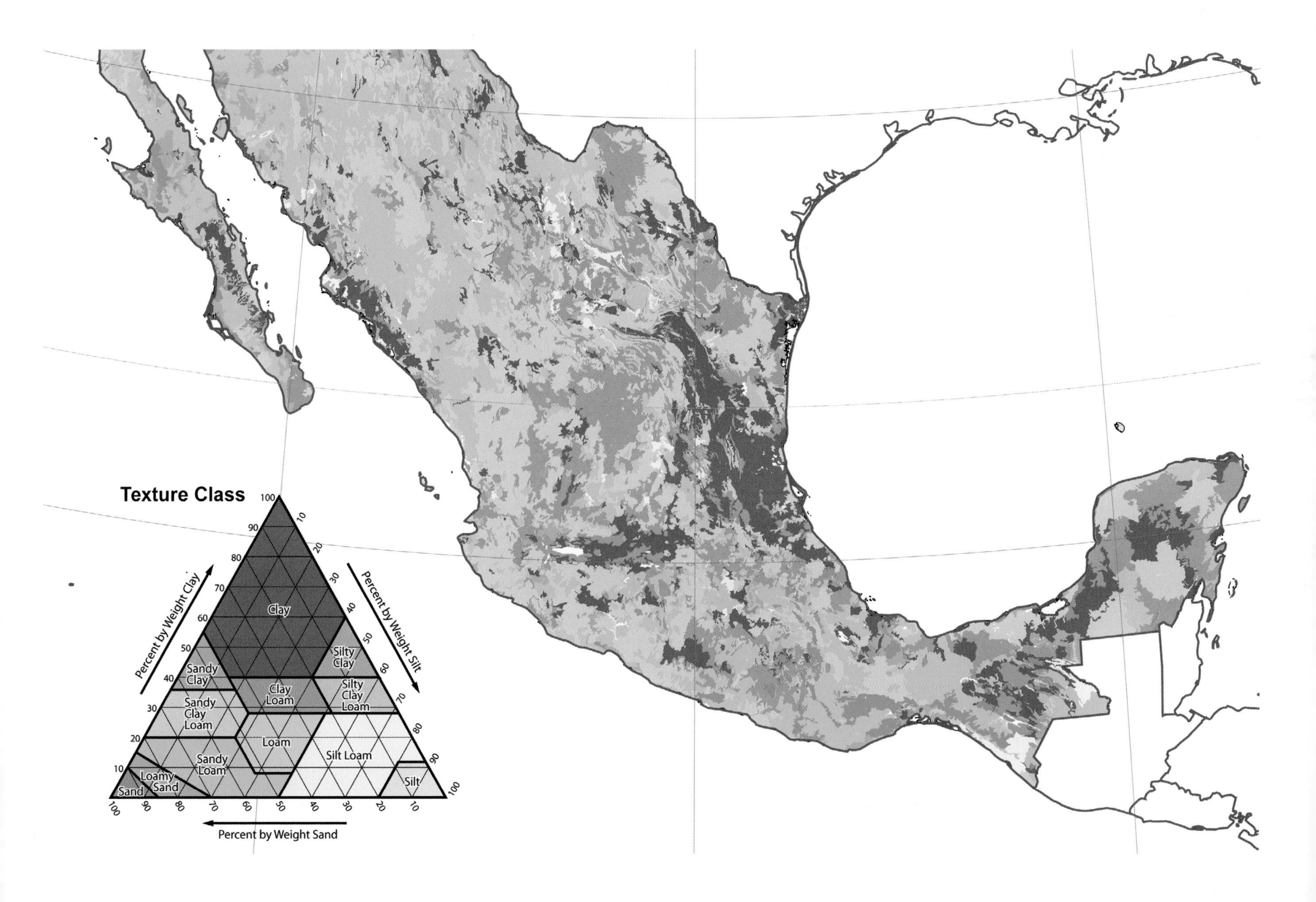
Texture Class
Clay
Silty Clay
Sandy Clay
Clay Loam
Silty Clay Loam
Sandy Clay Loam
Loam
Silt Loam
Sandy Loam
Silt
Loamy Sand
Sand
Percent by Weight Clay
Percent by Weight Silt
Percent by Weight Sand

6.4 COLUMBIA RIVER BIRD SPECIES RICHNESS MAP

This map uses a diverging scheme to emphasize watersheds along the Columbia River (Washington and Oregon), symbolized according to bird species richness. Predicted numbers of bird species are scored relative to high and low species numbers within ecoregions (bounded by wide white lines). This scaling procedure yields a species richness score from zero to one hundred for each watershed. Dark teal represents high species richness within each ecoregion, and dark brown represents low values. The lightest color in the scheme straddles the 50-percent level for watersheds in each ecoregion, and lightness diverges from this midpoint. Light hillshading and detailed river lines are background information that supports the watershed outlines, and major roads and selected towns overlay the theme to provide key location information.

Courtesy of CommEn Space.

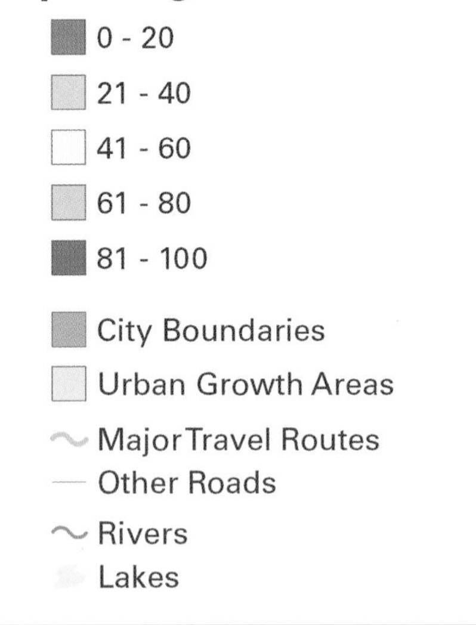

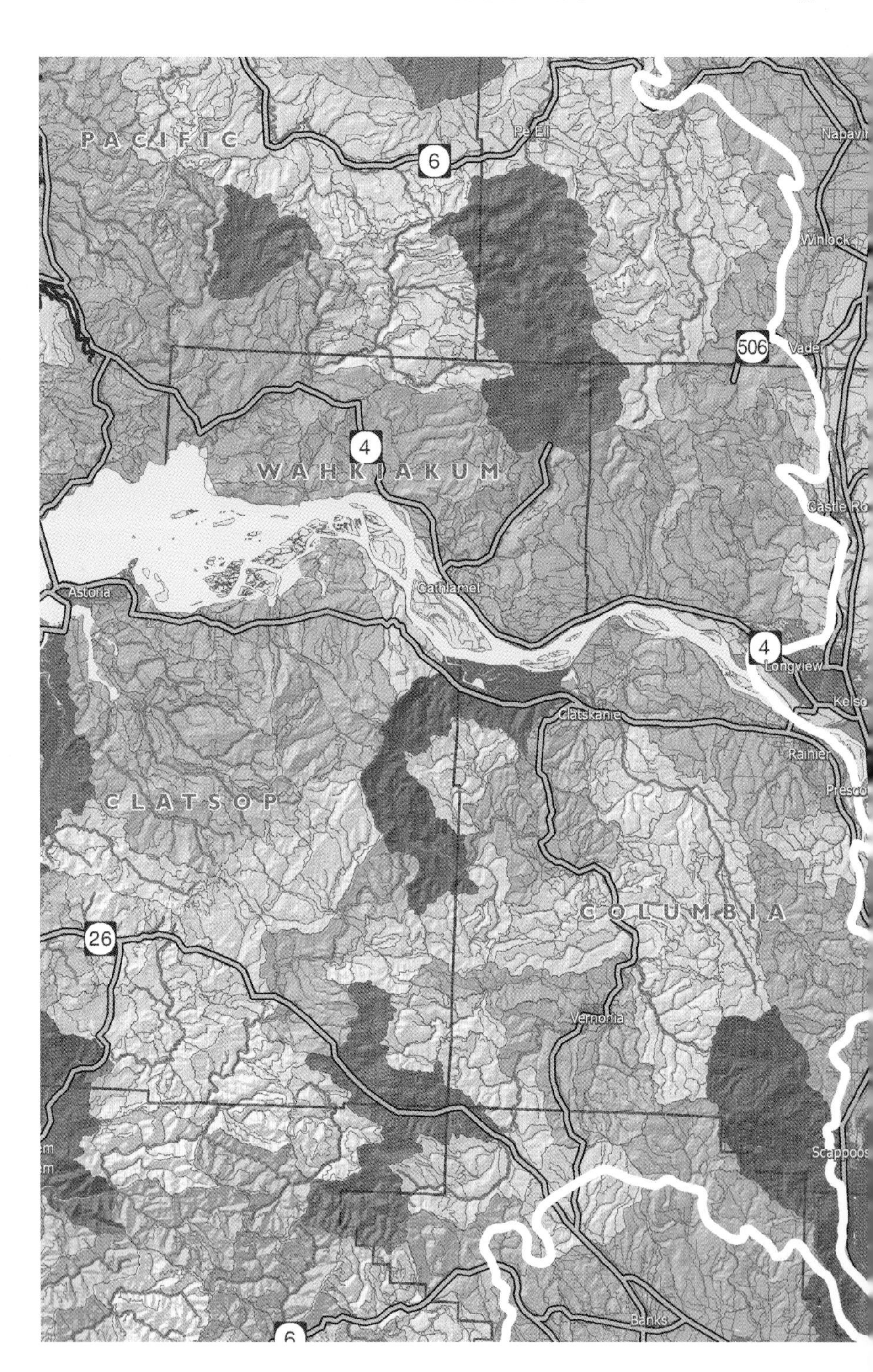

508
LEWIS
Morton
12
Mossyrock
505
504
COWLITZ
SKAMANIA
Woodland
La Center
503
Yacolt
Ridgefield
Battle Ground
502
CLARK
5
500
Vancouver
Stevenson
Cascade Locks
North Bonneville
White Salmon
Hood River
Bingen
Mosier
141
142
97

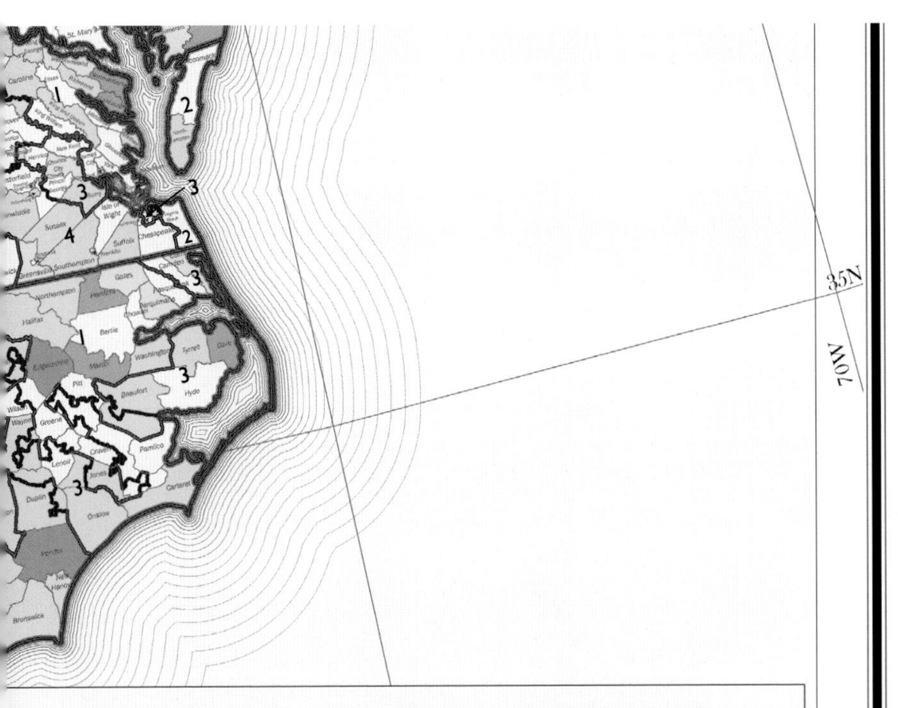

**CHANGE IN ANNUAL AVERAGE
UNEMPLOYMENT RATE - 2001 TO 2002**

- -13.70% TO -5.80%
- -5.79% TO -2.30%
- -2.29% TO -0.59%
- -0.58% TO 0.00% (-0.58% MEAN)
- +0.01% TO +0.60%
- +0.61% TO +2.30%
- +2.31% TO +5.90%

30N

6.5 U.S. CHANGE IN UNEMPLOYMENT MAP

This choropleth map uses a diverging color scheme to represent percentage changes in unemployment by county with an overlay of congressional districts to serve the map's intended audience of members and committees of the U.S. Congress and the Congressional Research Service. Congressional districts of the 108th Congress are numbered by state, and additional inset maps (for boxed areas such as Atlanta) offer enlarged views of complex districting in major cities. The map colors diverge from a central white class that ranges from the mean national decrease to zero unemployment change. Purples represent decreases in unemployment and oranges represent increases, with the darkest colors representing the largest changes. The mapmaker included historical touches in the design of the map, selecting elegant fonts and using GIS tools to produce a coastline vignette reminiscent of steel-engraving methods used by government cartographers in the 1800s.

Courtesy of Ginny Mason.

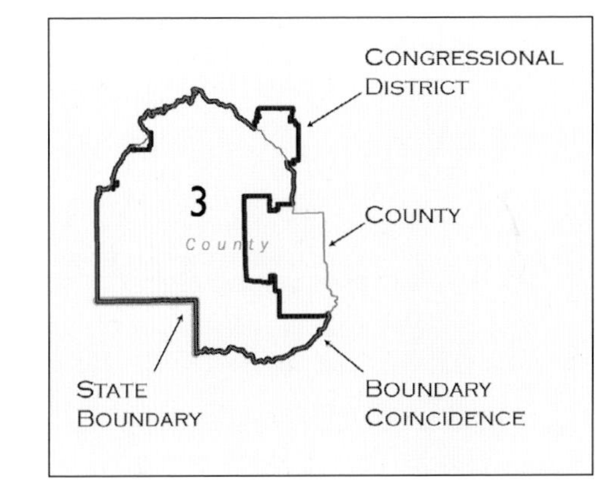

6.6 ALASKA FOREST FIRE PROGRESSION MAP

The Taylor Complex and Chicken Complex fires burned large areas north of the Alaska Range during the summer of 2004. This map shows the progression of the fire from June 24 to July 24 with cumulative acreage burned during each week noted in the legend. The history of the burn is represented with five colors ranging from light yellow through orange to dark red. The sequential scheme of these saturated colors carries the primary message of the map. Additional categorical data on vegetation and base data for trails, rivers, and terrain support the main map topic.

Courtesy of U.S. Department of Agriculture, Forest Service.

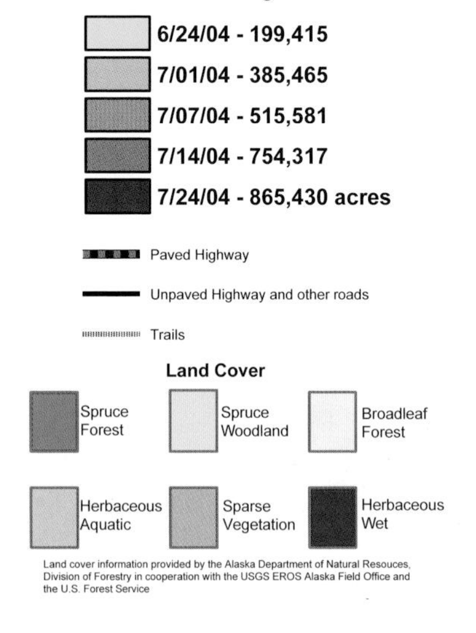

Logging Cabin Creek
Chicken
TAYLOR HIGHWAY
Cement Creek
Big Timber Creek
N/A
WINTER TRAIL
Porcupine

6.7 GETTYSBURG GROWTH MAP

Color is used to class structures in the borough of Gettysburg, Pennsylvania, by the era in which a building was first built at each location. Black outlines mark each site where a building stood in 1863, when the Battle of Gettysburg was fought during the American Civil War. The boldest buildings, in red crosshatching, are 1863 buildings still present today. Sites where 1863 buildings have been replaced by modern structures are represented in a lighter red. Orange identifies sites where buildings were present in 1924. Modern buildings, present in 1998, are represented with a light blue hue to contrast with the reds and oranges that emphasize the historical focus of the map. The road, walkway, and contour lines form the base for the building information.

GIS for Gettysburg Battlefield Planning and Rehabilitation map courtesy of National Park Service.

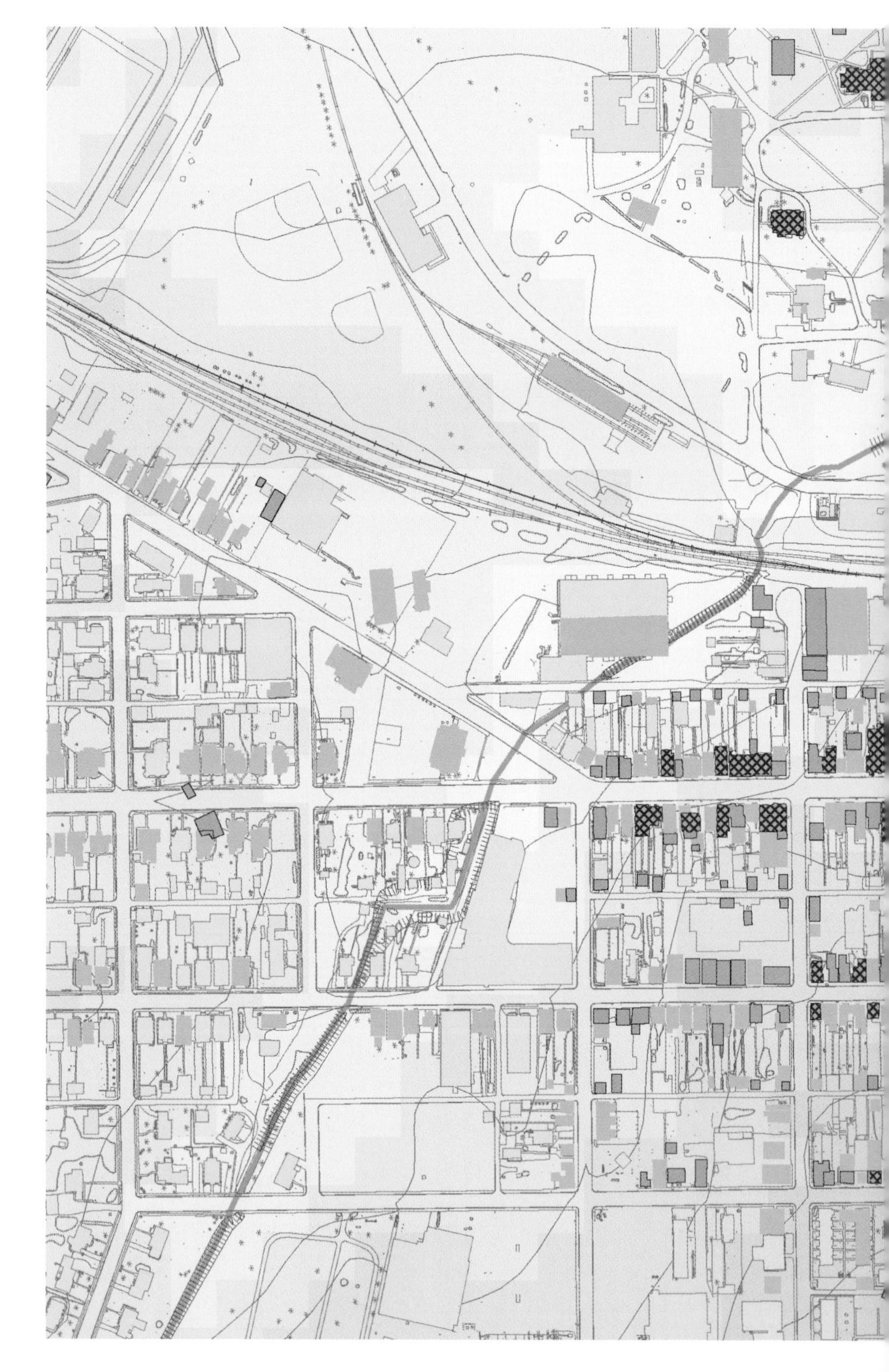

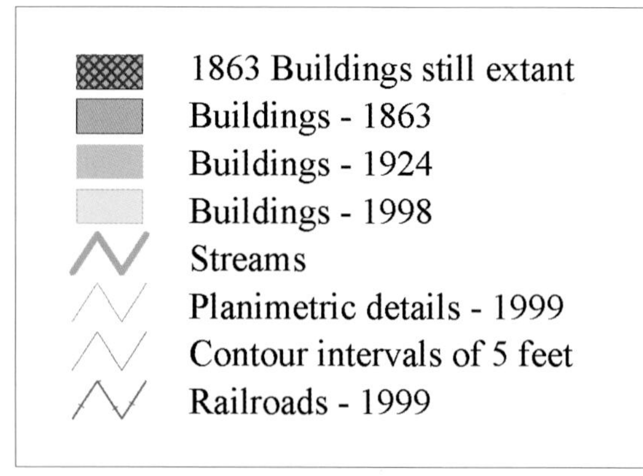

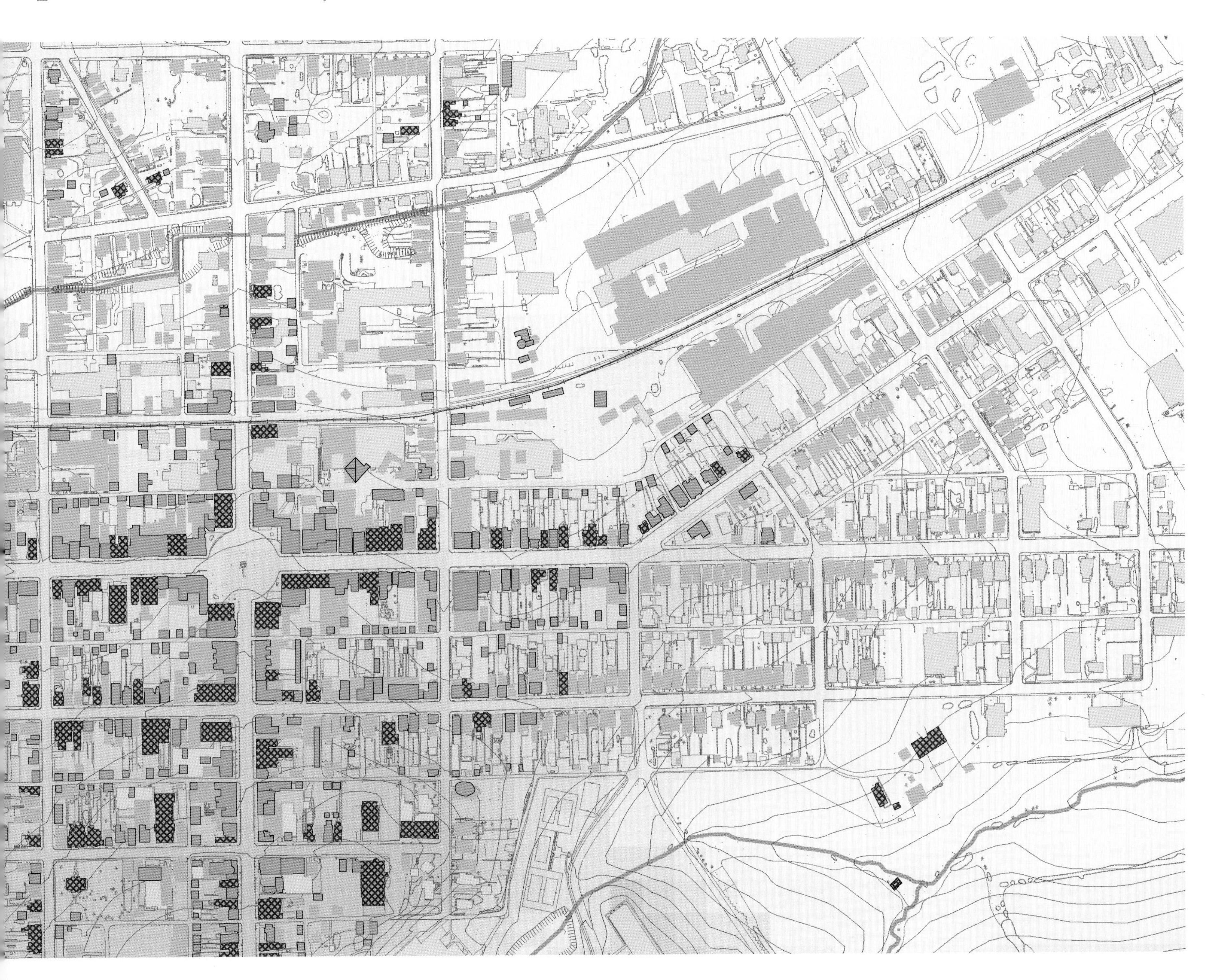

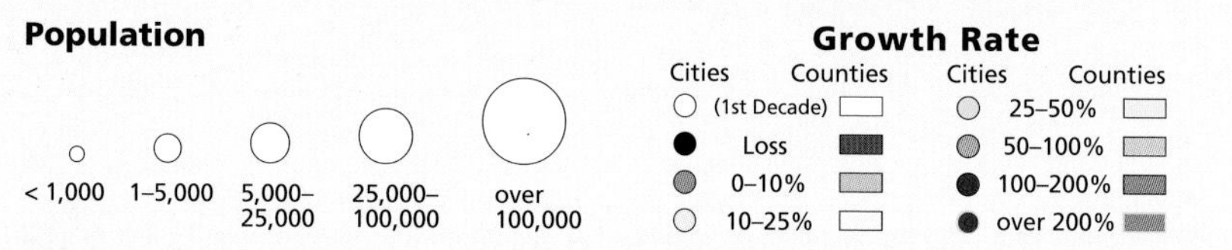

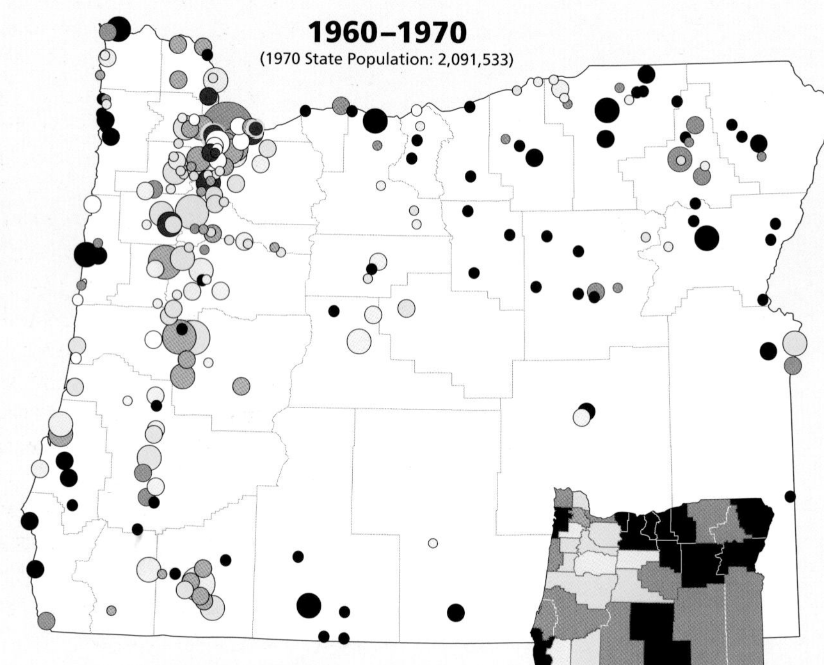

1970 Populations 75 largest incorporated places

Portland	379,967
Eugene	79,028
Salem	68,725
Corvallis	35,056
Medford	28,454
Springfield	26,874
Beaverton	18,577
Albany	18,181
Milwaukie	16,444
Klamath Falls	15,775
Hillsboro	14,675
Oswego	14,615
Roseburg	14,461
Bend	13,710
Coos Bay	13,466
Pendleton	13,197
Grants Pass	12,455
Ashland	12,342
The Dalles	10,423
Astoria	10,244
McMinnville	10,125
Gresham	10,030
La Grande	9,645
Baker	9,354
Oregon City	9,176
North Bend	8,553
Forest Grove	8,175
Woodburn	7,495
Lebanon	7,277
West Linn	7,091
Ontario	6,523
Newberg	6,507
Tigard	6,499
Dallas	6,361
Gladstone	6,254
St Helens	6,212
Cottage Grove	6,004
Monmouth	5,237
Newport	5,188
Hermiston	4,893
Coquille	4,437
Seaside	4,402
Silverton	4,301
Lincoln City	4,196
Milton–Freewater	4,105
Prineville	4,101
Reedsport	4,039
Central Point	4,004
Hood River	3,991
Tillamook	3,968
Canby	3,813
Sweet Home	3,799
Redmond	3,721
Oakridge	3,422
Burns	3,293
Stayton	3,170
Sutherlin	3,070
Toledo	2,818
Myrtle Creek	2,733
Brookings	2,720
Lakeview	2,705
Nyssa	2,620
Independence	2,594
Myrtle Point	2,511
Winston	2,468
Junction City	2,373
Florence	2,246
Molalla	2,005
Mount Angel	1,973
Cornelius	1,903
Sheridan	1,881
Scappoose	1,859
Bandon	1,832
Warrenton	1,825
Rainier	1,731

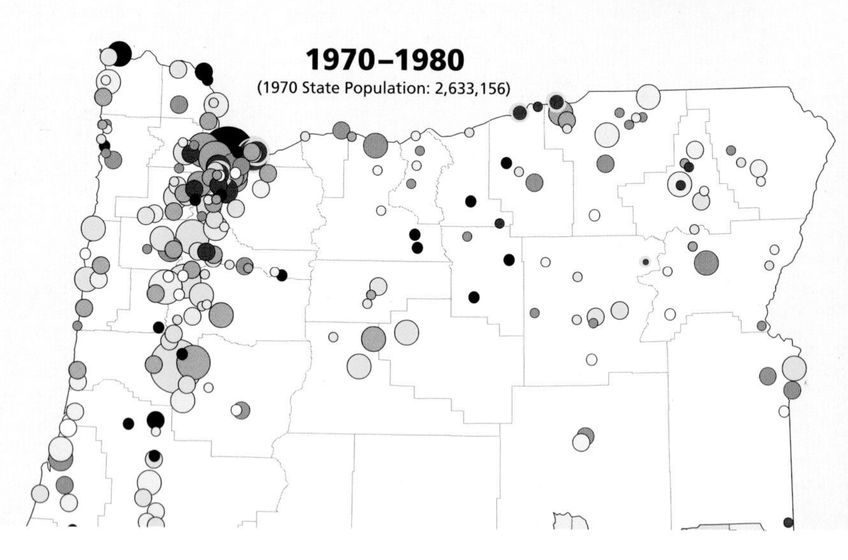

1980 Populations

Portland	366,383
Eugene	105,664
Salem	89,091
Springfield	41,621
Corvallis	40,960
Medford	39,746
Gresham	33,005
Beaverton	31,962
Hillsboro	27,664
Albany	26,511
Oswego	22,527
Milwaukie	17,931
Bend	17,260
Klamath Falls	16,661
Roseburg	16,644
Grants Pass	15,032
Ashland	14,943
Tigard	14,799
Oregon City	14,673
Woodburn	11,196
The Dalles	10,820
Lebanon	10,413
Newberg	10,394
Astoria	9,996
North Bend	9,779
Gladstone	9,500
Baker	9,471
Ontario	8,814
Dallas	8,530
Hermiston	8,408
Canby	7,659
Newport	7,519
Tualatin	7,483
Cottage Grove	7,148
St Helens	7,064
Sweet Home	6,921
Redmond	6,452
Central Point	6,357
Milton–Freewater	5,086
Reedsport	4,984
Sutherlin	4,560
Coquille	4,481
Florence	4,411
Cornelius	4,402
Stayton	4,396
Hood River	4,329
Independence	4,024
Tillamook	3,991
Oakridge	3,680
Burns	3,579
Brookings	3,384
Myrtle Creek	3,365
Winston	3,359
Junction City	3,320
Scappoose	3,213
Umatilla	3,199
Toledo	3,151

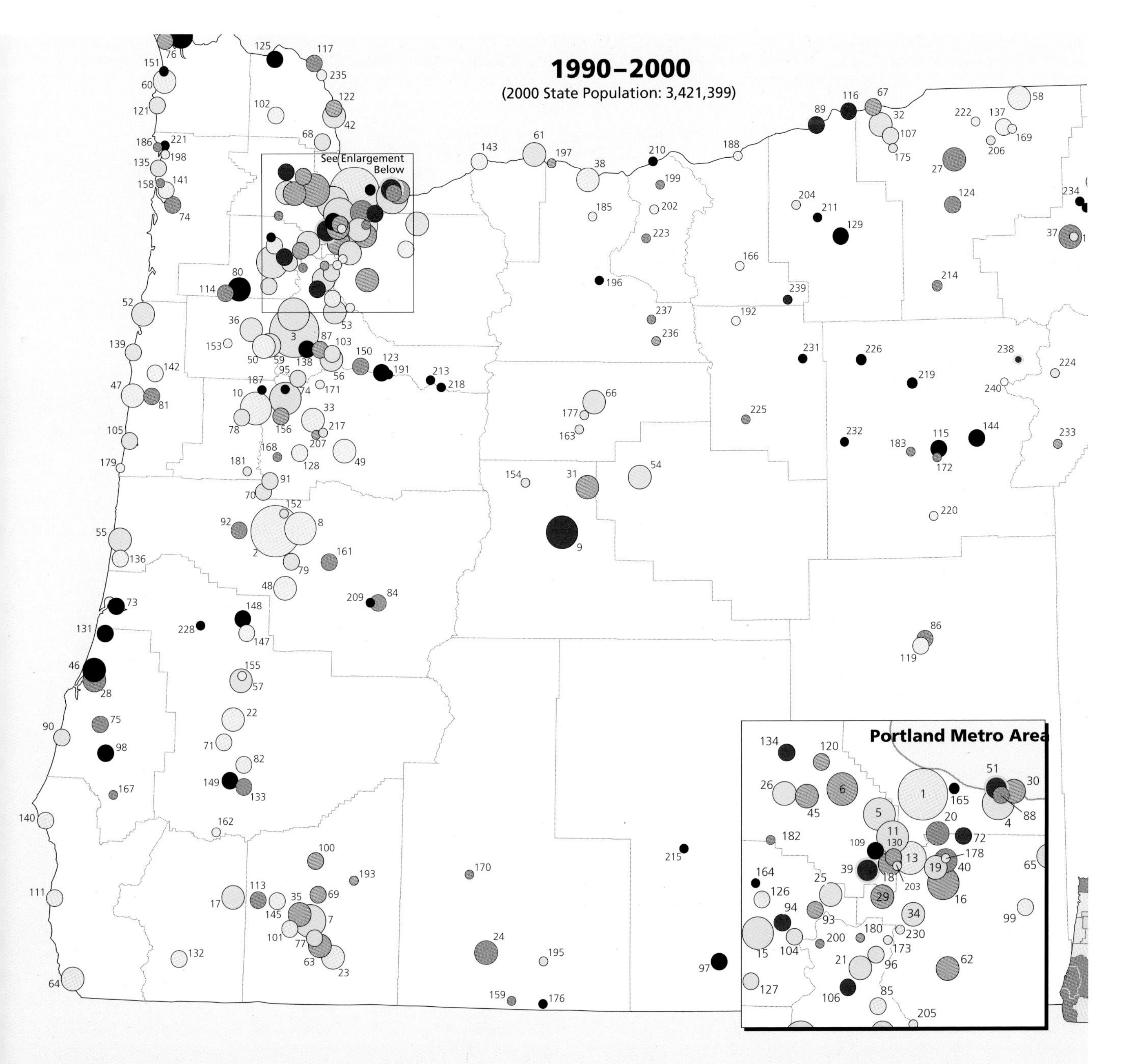

2000 Populations All Incorporated Places

1 ○ Portland 529,121	41 ○ Ontario 10,985	81 ● Toledo 3,472	121 ○ Cannon Beach.....1,588	161 ● Lowell 857
2 ○ Eugene 137,893	42 ○ St. Helens............10,019	82 ○ Myrtle Creek 3,419	122 ● Columbia City 1,571	162 ○ Glendale 855

6.8 OREGON POPULATION AND GROWTH MAPS

The growth of incorporated places each decade, from 1870 to 2000, is captured on 14 maps in the *Atlas of Oregon* (three are shown here). The carefully composed symbols on these maps detail multiple aspects of population. City symbol sizes are classed by the size of the population. The diverging color scheme shows loss in black, low growth in blues, and high growth (above 25 percent) in a yellow-orange-red sequence. The highest city growth rates are further emphasized by yellow outlines around the circles. The combination of circle size for population and color fill for growth appropriately scales the impact of the colors, so small cities that grow a lot have an appropriately lessened visual impact because the colored circle is smaller. Circles for smaller cities are placed above the larger circles so they are not lost when symbols overlap. Very small county maps that use the same growth-rate colors and classes complement the city information. A table next to each map lists city populations, and small markers next to each name identify each city's growth-rate category.

Maps by Stuart Allan and James Meacham. Courtesy of University of Oregon.

6.9 GEYSERS AREA DEPTH-DEPENDENT SEISMICITY MAP

USGS records seismic activity in the Geysers geothermal area of Northern California, an active area where eight earthquakes a day are detected on average. The map shows depth dependence and focal mechanisms for earthquakes from 1975 to 1995. Circle size is used for three classes, with the largest circle size for larger earthquakes (magnitude 4.0 to the highest at 4.9). A saturated color scheme classes earthquake depth: yellow (for less than a kilometer deep)-orange-red-magenta-blue-green-brown (for seven or more kilometers deep). Though there are many hues in this scheme, it is not a rainbow scheme. It takes advantage of using light saturated yellow at the start of the scheme and ends in dark brown, with nonspectral magentas midway through. The schematic "beach ball" diagrams show the focal mechanisms for a selection of the larger earthquakes, showing fault types and orientations. Depth information for larger earthquakes stands out because the symbols are larger, and they are pulled above the clouds of small symbols.

Courtesy of U.S. Geological Survey.

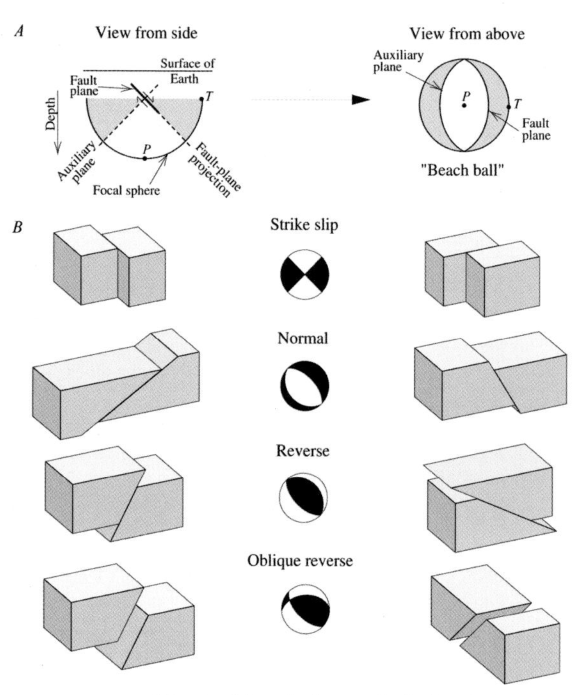

Figure 1. Schematic diagram of a focal mechanism.

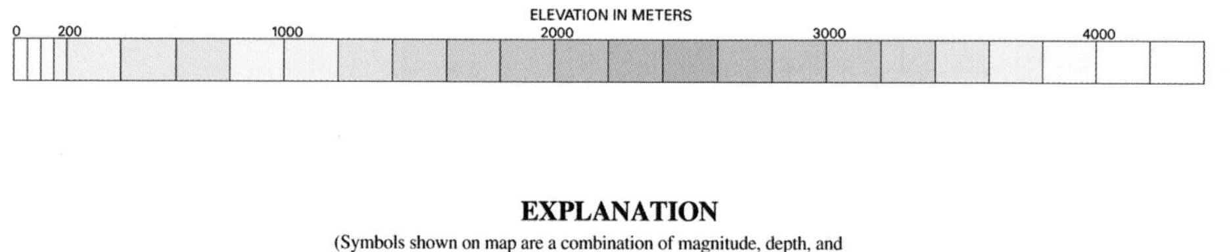

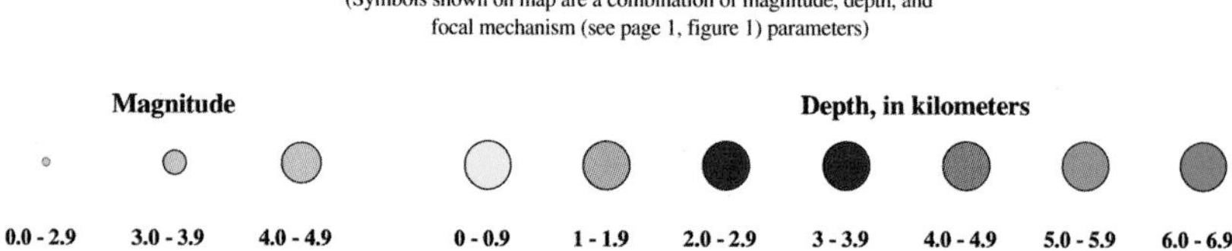

Faults

(Fault traces were digitized at 1:750,000 (Jennings, 1994) and may be mislocated by as much as 0.8 km (0.5 mi) when plotted on this map at 1:250,000. Fault traces are not intended for engineering or land-use purposes.)

Late Quaternary (displacement <700,000 yr)

Early Quaternary (displacement <1,600,000 yr)

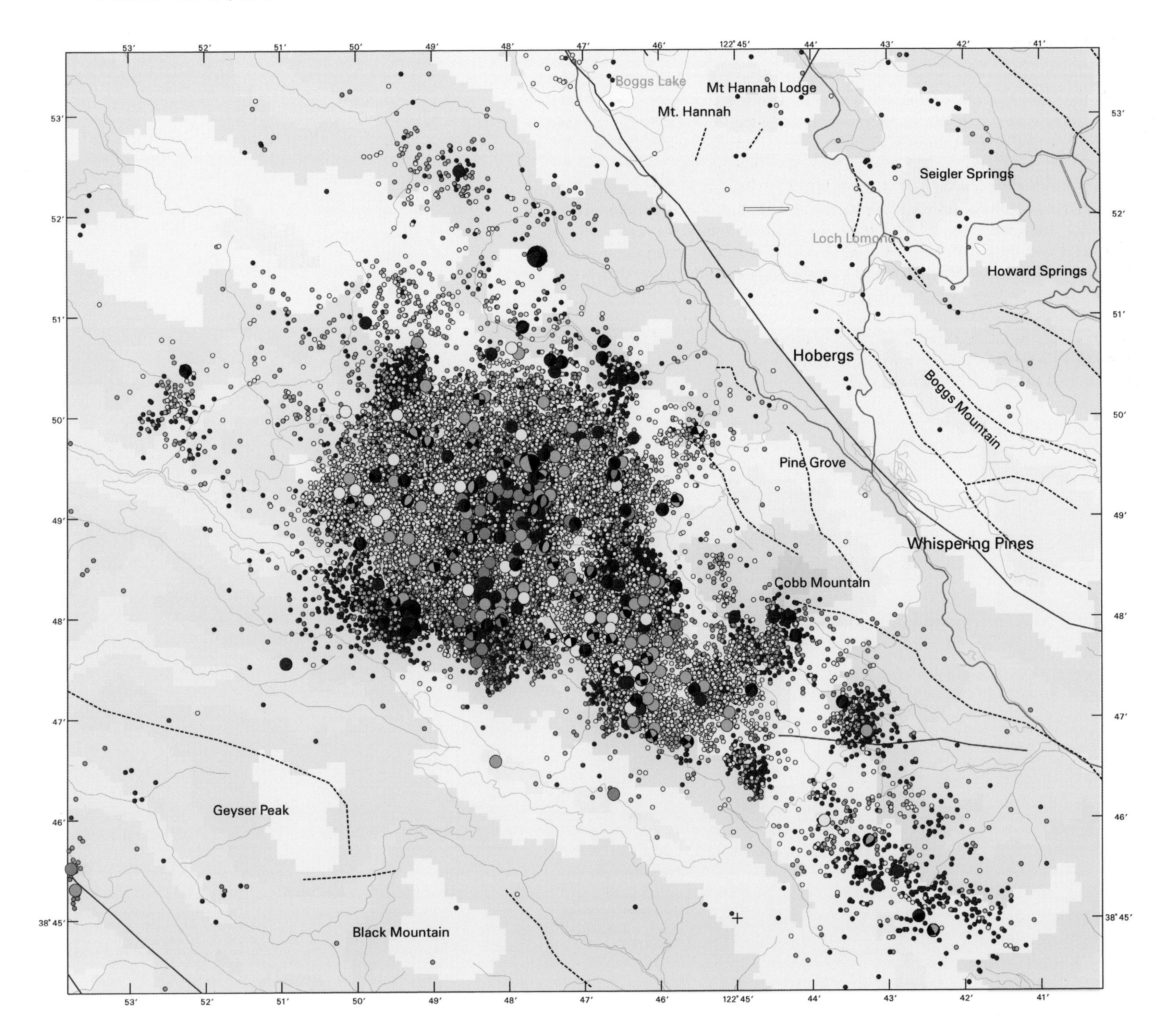
Boggs Lake
Mt Hannah Lodge
Mt. Hannah
Seigler Springs
Loch Lomond
Howard Springs
Hobergs
Boggs Mountain
Pine Grove
Whispering Pines
Cobb Mountain
Geyser Peak
Black Mountain
122° 45′
38° 45′

6.10 LOS ANGELES PERCHLORATE CONCENTRATION MAP

Dots that range in color from yellow, for low concentrations of perchlorate in drinking water wells, through orange, to dark red for high concentrations, are the primary symbols on this map. In contrast, small blue dots mark wells where perchlorate was not detected. The map is intended for analysis of the impact of this chemical on drinking water in EPA Region 9 of the Greater Los Angeles Basin, California. Perchlorate is a salt produced in the manufacture of rocket fuels and explosives, so additional information relevant to the map topic includes military and industrial sites. These feature categories are represented by hues and different point symbol shapes. Light roads, rivers, and hillshading provide base information.

Courtesy of U.S. Environmental Protection Agency and Titan Corp.

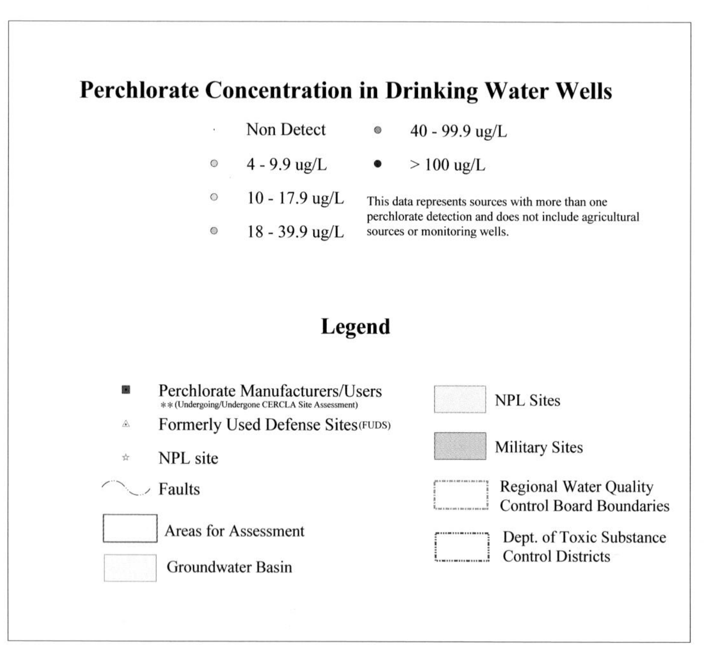

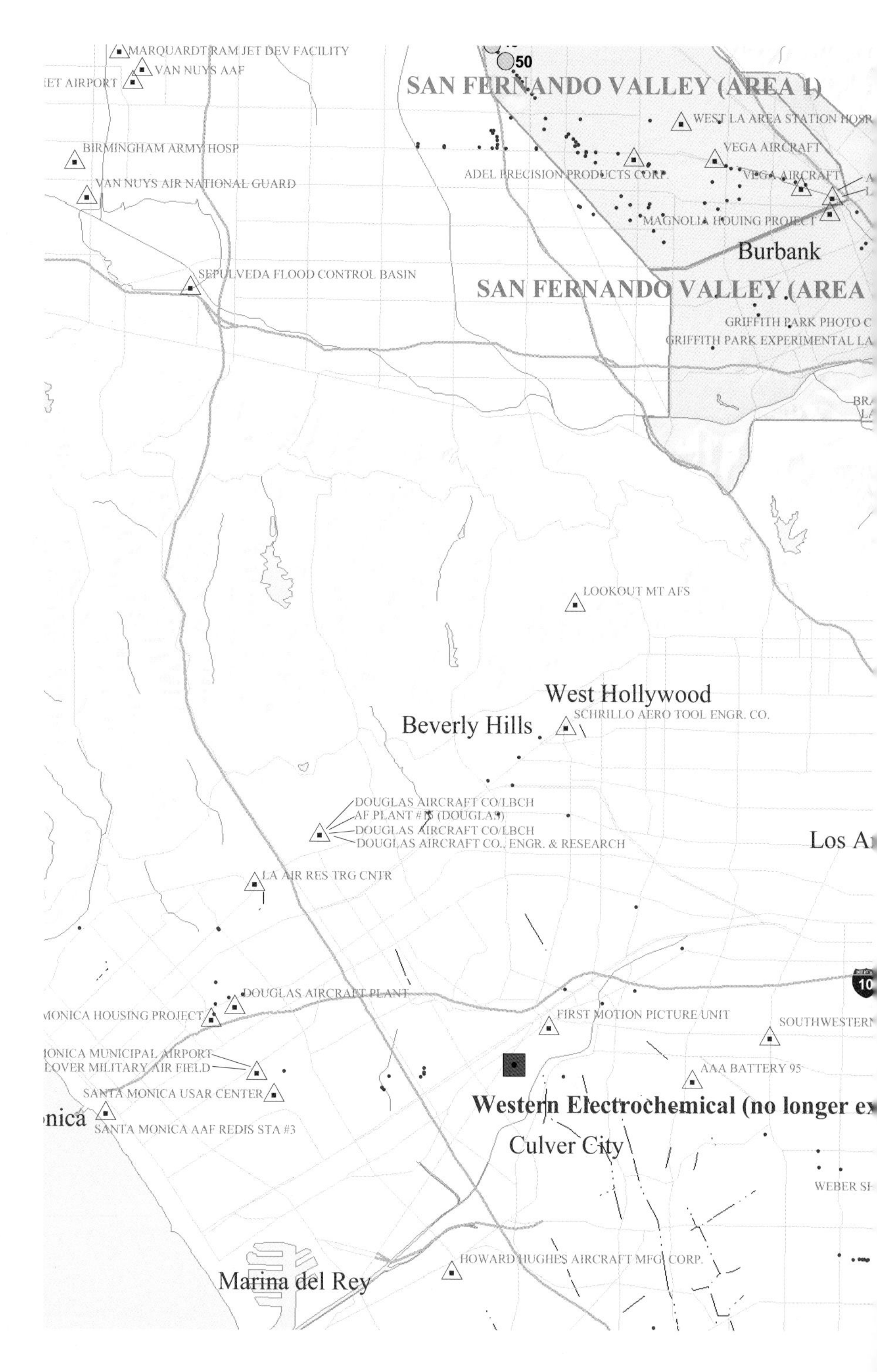

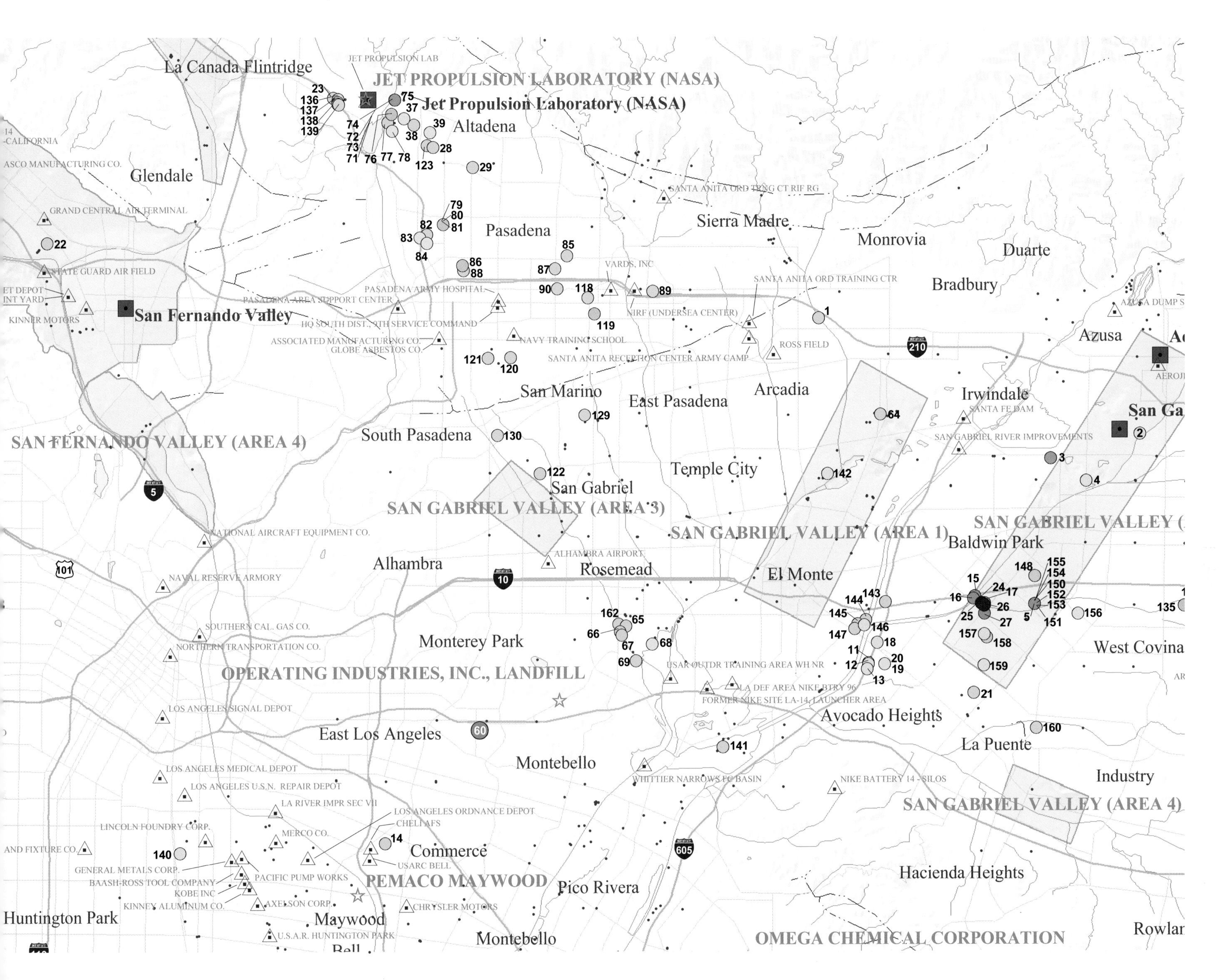
La Canada Flintridge
JET PROPULSION LAB
JET PROPULSION LABORATORY (NASA)
Jet Propulsion Laboratory (NASA)
Altadena
Glendale
Pasadena
Sierra Madre
Monrovia
Duarte
Bradbury
Azusa
San Fernando Valley
SAN FERNANDO VALLEY (AREA 4)
South Pasadena
San Marino
East Pasadena
Arcadia
Irwindale
Temple City
San Gabriel
SAN GABRIEL VALLEY (AREA 3)
SAN GABRIEL VALLEY (AREA 1)
Baldwin Park
Alhambra
Rosemead
El Monte
Monterey Park
West Covina
OPERATING INDUSTRIES, INC., LANDFILL
Avocado Heights
East Los Angeles
Montebello
La Puente
Industry
SAN GABRIEL VALLEY (AREA 4)
Commerce
PEMACO MAYWOOD
Pico Rivera
Hacienda Heights
Huntington Park
Maywood
OMEGA CHEMICAL CORPORATION
GRAND CENTRAL AIR TERMINAL
STATE GUARD AIR FIELD
KINNER MOTORS
PASADENA ARMY HOSPITAL
PASADENA AREA SUPPORT CENTER
HQ SOUTH DIST., 9TH SERVICE COMMAND
ASSOCIATED MANUFACTURING CO.
GLOBE ASBESTOS CO.
NAVY TRAINING SCHOOL
SANTA ANITA RECEPTION CENTER ARMY CAMP
VARDS, INC
NIRF (UNDERSEA CENTER)
SANTA ANITA ORD TRAINING CTR
SANTA ANITA ORD TRNG CT RIF RG
ROSS FIELD
SANTA FE DAM
SAN GABRIEL RIVER IMPROVEMENTS
AZUSA DUMP S
NATIONAL AIRCRAFT EQUIPMENT CO.
ALHAMBRA AIRPORT
NAVAL RESERVE ARMORY
SOUTHERN CAL. GAS CO.
NORTHERN TRANSPORTATION CO.
USAR OUTDR TRAINING AREA WH NR
LA DEF AREA NIKE BTRY 96
FORMER NIKE SITE LA-14, LAUNCHER AREA
LOS ANGELES SIGNAL DEPOT
WHITTIER NARROWS FC BASIN
NIKE BATTERY 14 - SILOS
LOS ANGELES MEDICAL DEPOT
LOS ANGELES U.S.N. REPAIR DEPOT
LA RIVER IMPR SEC VII
LOS ANGELES ORDNANCE DEPOT
CHELI AFS
LINCOLN FOUNDRY CORP.
MERCO CO.
USARC BELL
GENERAL METALS CORP.
BAASH-ROSS TOOL COMPANY
KOBE INC
PACIFIC PUMP WORKS
KINNEY ALUMINUM CO.
AXELSON CORP.
CHRYSLER MOTORS
U.S.A.R. HUNTINGTON PARK
Montebello
Rowlar

6.11 PUERTO RICO CRUISE SHIP WASTE DISCHARGE MAP

The data presented on this map of Puerto Rico and nearby islands was collected as part of a global analysis by the Ocean Conservation Tourism Alliance (OCTA) of waste discharge (food wastes, gray water, black water, and wastewater sludge). Dark blue line segments with direction arrows mark the beginning and estimated direction of travel during discharge events by individual cruise ships during October 2004. These are the most prominent symbols on the map, and the lightness of other symbols makes the abundance and pattern of these vectors the main map message. Conservation concerns are reflected in the chosen supporting information. Marine protected areas and coral reefs are symbolized with green and orange hues that stand out from the light land–water boundary. Light lines delineate 60- and 100-meter bathymetric contours to show general water depth information.

Courtesy of Conservation International.

Figure 13: Discharge Start Points with Vector

Puerto Rico and vicinity

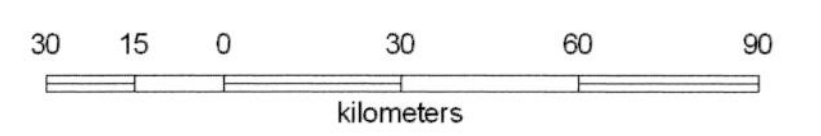

Scale: 1:2,220,000
Projection: Eckert IV

Data:
AIDA Cruises
Digital Chart of the World
International Council of Cruise Lines (Carnival Cruise Lines, Celebrity Cruises, Costa Cruise Lines N.V., Crystal Cruises, Cunard Line, Disney Cruise Line, Holland America Line, NCL America, Norwegian Cruise Line, Orient Lines, Princess Cruises, Radisson Seven Seas Cruises, Royal Caribbean International, Seabourn Cruise Line, Silversea Cruises, and Windstar Cruises)
P&O Cruises
reefbase.org
World Resources Institute, Reefs at Risk in the Caribbean, 2004
World Database of Protected Areas (WDPA) 2005

Cities
Discharge Start Points with Vector
Land
Marine Protected Area Polygons
Marine Protected Area Points
12 Nautical Miles from Coast
Coral Reefs
Bathymetric Contours (m)
60
100

This map was produced by the Conservation Mapping Program Center For Applied Biodiversity Science at Conservation International.

cartography: Erica Ashkenazi
July 2005

6.12 TAURANGA HARBOUR TIDAL MOVEMENTS MAP

Fields of small arrows model characteristics of spring high tide in Tauranga Harbour, New Zealand. The arrows provide insight on currents and flushing of the estuary that may be used for planning and environmental assessments. The arrows show multiple variables: direction, speed, and depth. The color sequence of the arrows, from white to dark blue, shows depth. The length of the arrows shows speed, with longer arrows for faster currents. Arrow azimuths show direction. A small, white arrow pointing north shows slow, shallow, north-flowing current, and a long, dark blue arrow pointing west shows fast, deep, west-flowing current. Patterns of flow are apparent in the fields of arrows as they shift direction in unison and form bold groups of darker and longer arrows through the harbor. The map's image base is subdued, with gray-scale land imagery and a blue overlay for the water body.

Courtesy of Environment Bay of Plenty (The Bay of Plenty Regional Council).

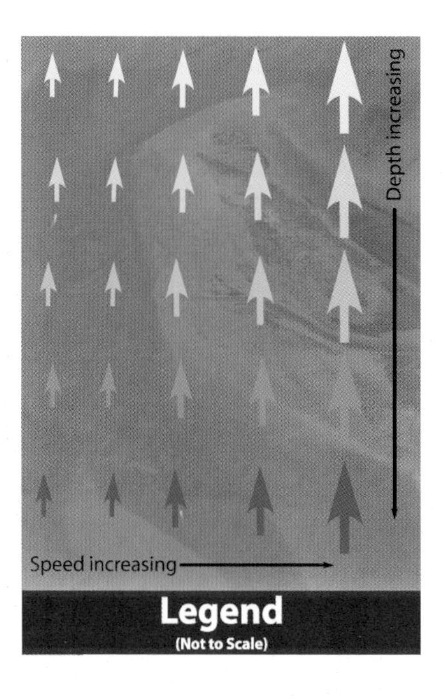

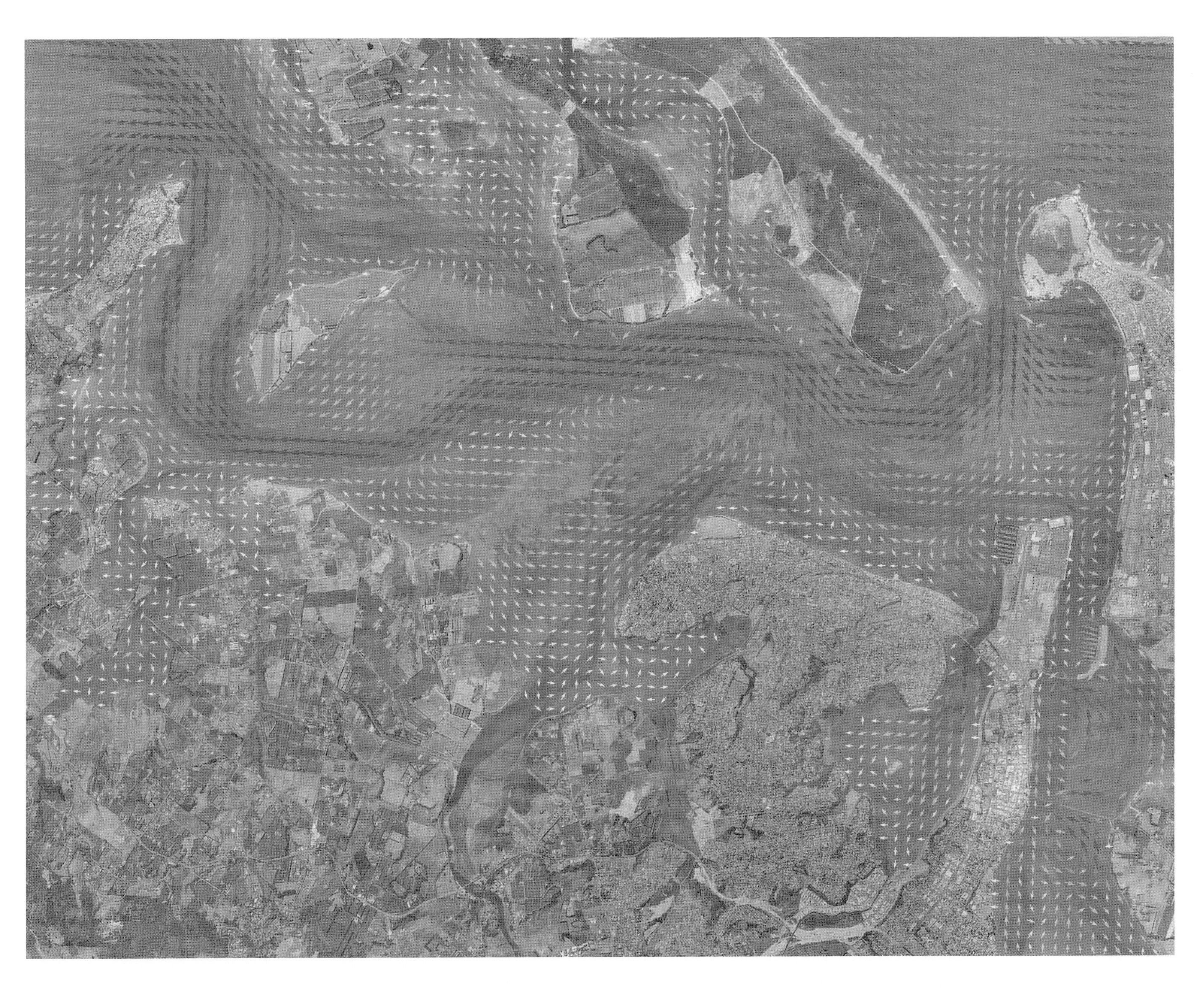

6.13 CALIFORNIA NET MIGRATION MAP

In this map from the *Census Atlas of the United States,* the largest net migrations between California and other states are shown for 1955 to 1960 in red and for 1995 to 2000 in blue. Widths of the flow lines are proportioned according to net numbers of people moving in and out of California. The arrow symbols make the trends—abruptly different in the two five-year periods—obvious to map readers at first glance. Readers gain a general understanding that similar numbers of people are moving out who had moved in forty years earlier by comparing the widths of the large red arrow pointing to California to the base widths of the blue arrows that originate in California. Small numbers, which do not interfere with the overall symbol pattern on the map, are placed next to the start and end points to note net migration numbers. Lightness gradation along the length of the arrows directs readers' attention to the final destinations of movers. Smooth arcs of smaller arrows coalesce toward a single direction, evoking a sense of movement that peaks as it reaches its destination (or leaves its source). Line hue, lightness, width, and direction carry this map's message.

U.S. Census Bureau by Alex Tait, International Mapping Associates.

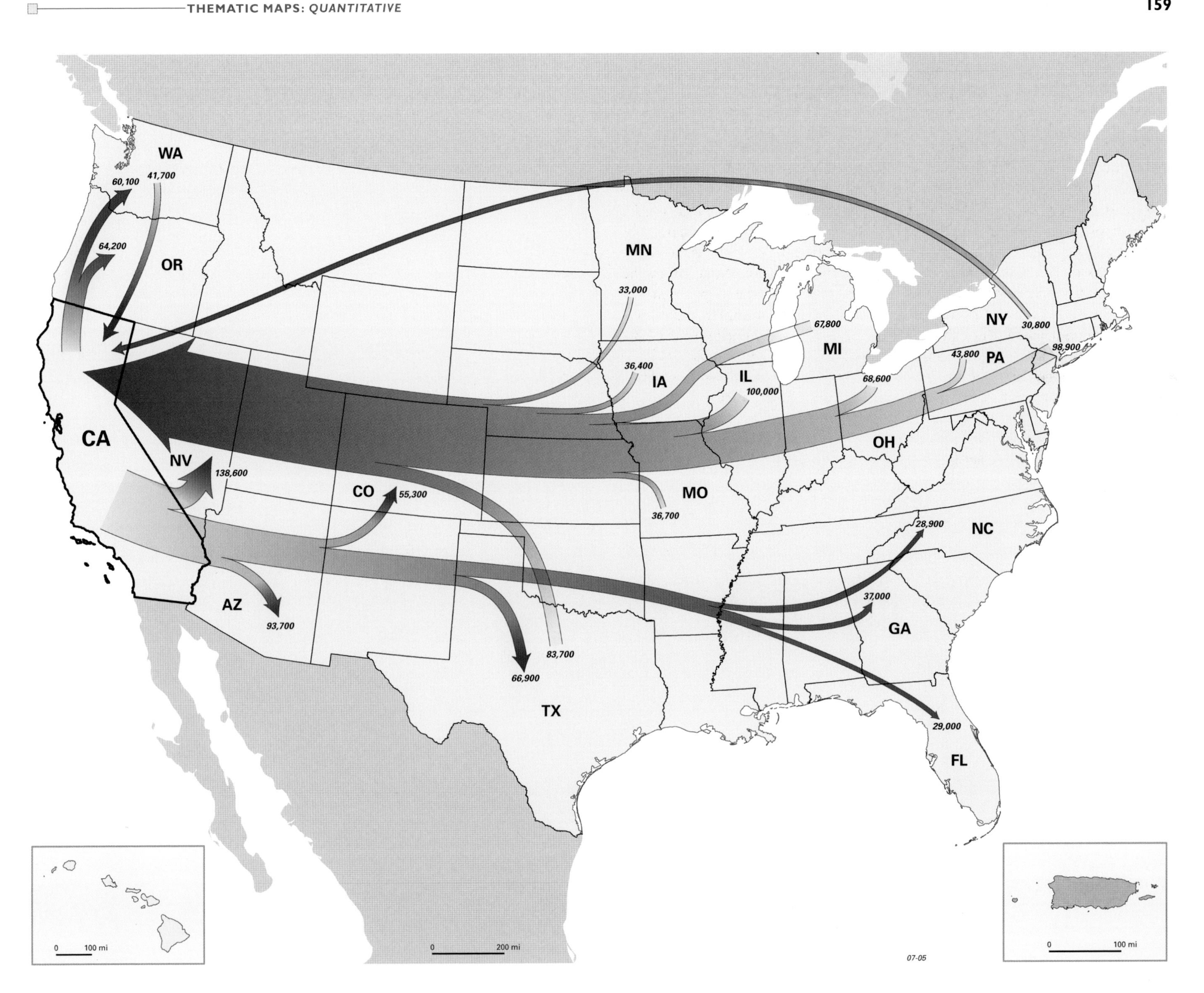
WA
60,100
41,700
64,200
OR
MN
33,000
NY
30,800
67,800
MI
98,900
43,800
PA
36,400
IA
IL
100,000
68,600
CA
OH
NV
138,600
CO
55,300
MO
36,700
28,900
NC
37,000
AZ
93,700
GA
83,700
66,900
TX
29,000
FL
0
100 mi
0
200 mi
07-05
0
100 mi

ArcMap TIPS

These tips are suited for people who are experienced ArcGIS users and are comfortable navigating the ArcMap application in particular. They do not offer full instructions but rather pointers and keywords that help users discover parts of the software that can be used to accomplish cartographic goals. The instructions are for versions 9.2 or higher.

LINES				
1 Dashed line	• Symbol Selector > Properties > Symbol Property Editor > Type: Cartographic Line Symbol > Template tab > drag and click template to create dash pattern • ArcToolbox > Data management Tools > Generalization > Dissolve to combine line segments to avoid unnecessary stops and restarts in a continuing dash pattern (dissolve on the field being symbolized) • Use Representations* to better control dashed lines: Stroke tab > Endings to constrain dash behavior at line ends; Representations > Geometric effect (+) > to control dashes at line corners and ends. Use control points to ensure dashes are placed over corners for clarity where line changes direction			
	1C	5A	6A	6B

* Right-click the layer name in the table of contents and choose Convert Symbology to Representation. Only available with ArcEditor and ArcInfo license levels.

2 Multilayer line	• Symbol Property Editor > Type: Cartographic Line Symbol > add layers for multilayer line effect (+ at bottom left) > adjust layer order using arrow buttons and set properties for each layer • A multilayer line that includes a solid line below a dash line protects two dashed lines from combining with each other to create an inconsistent pattern (such as overlaid dashes from the boundaries of two adjacent cities)			
	1A Two layers: dash and solid thicker line	2B Two layers: hash and wider line	2D Three layers: center line, fill line, and wider case	
3 Cased line	A cased line combines a thinner "fill" line overlaying a thicker line in a contrasting color that forms the case (an "outline" around the fill line). • Symbol Property Editor > Type: Cartographic Line Symbol > create two layers > adjust order, width, and color • Lock the case line layer to allow quicker changes to just the fill line color			
	1C	5D		

LINES (CONTINUED)	
4 Merge/overpass	Visual relationships between roads that merge, overpass, or underpass can be managed automatically with suitably structured data. • Set layer order in the table of contents (TOC) for features on different layers (dissolve roads if they have many small segments) • Split tool on the Editor toolbar to separate lines into overpassing and/or underpassing segments if they are part of one feature • Attributes tool on the Editor toolbar may be used to set an attribute for ordering the separate segments (e.g., top-most segments might be attributed "overpass" or "3"; add field to the attribute table if needed) • Put layers together in a group layer in TOC, right-click data frame > choose New Group Layer > drag layers into group; Group Layer Properties > Group tab > Symbol Levels > check "Draw this layer" > set levels for fill and case on each line type (use About Symbol Levels and Switch to Advanced View buttons) • Mapping Center[†] > Blog tab > enter search keyword **cased lines**
	1D: Wider road fill above ramp fills with casings for both road types below 2B: Split bridge portion of road and set both fill and case above highway fill/case
5 Marker line	• Symbol Property Editor > Type: Marker Line Symbol > choose symbol from Marker Line tab and set spacing on Template tab • For markers along a solid or dashed line, create a multilayer line that combines Marker Line and Cartographic Line types • Use Representations to control marker positions in relationship to line corners and ends.
	1A: Widely spaced markers with thin line 2A: Markers with no line 4D: Large open marker

[†] Mapping Center resources are online at mappingcenter.esri.com.

6 Offset line	• Symbol Property Editor > Line Properties tab > adjust offset • In Representations, use Offset Curve to control offset line placement; with patterned lines, clean up individual placements of pattern elements in Representations • These offset methods work well for simple shapes; acute angles in lines and wide line symbols (over 6 points) may cause problems
	1A: Simple offset line 1B: Three-layer line of marker symbols with three different offsets
7 Hash line	• Symbol Property Editor > Type: Hash Line Symbol > adjust settings using Hash Line, Cartographic Line, and Template tabs • For hashes along a solid line, create a multilayer line that combines Hash Line and Cartographic Line types • In Representations, insert control points to get hash marks on parallel railroad tracks to line up; use Cut Curve to control where the hashes begin and end along the lines (synchronized hashes are shown for railroads on the *ESRI Lunch Specials* map on Mapping Center)
	2D: Hashes only 2A: Railroad effect; hashes with line
8 Picture line	• Draw small picture and save as EMF file type • Symbol Property Editor > Type: Picture Line Symbol > load EMF file > adjust settings • In Representations, you have more control over where the picture is first positioned and how it is spaced (use the Cut Curve effect)
	2C: In this example, drawing of one pair of footprints (left and right) is repeated along line

LINES (CONTINUED)

9 **Transparent line**	• Layer Properties > Display tab > adjust Transparent value • To access this quickly, use the Effects toolbar > Adjust Transparency slider for the layer selected in the drop-down menu on the toolbar 5B
10 **Tint bands**	• For simple polygon shapes, Symbol Property Editor > Type: Simple Fill Symbol > Outline > Properties > Type: Cartographic Line > create additional Layers > set color and width; Line Properties tab > Offset tint band by half its width (e.g., a 6-point band offset –3 points sits inside the polygon); Cartographic Line tab > Round joins • For well-fit tint bands on more complex polygons, see tips at Mapping Center > Blog tab > enter search keyword **tint bands** 1C: Offset transparent band same hue as thin line 1B: Three bands, each progressively wider and lighter 2C: Two bands and fill color
11 **Polygon to line**	To prevent dash patterns on shared boundaries of adjacent polygons from interfering with each other. • Convert polygons into lines and remove duplicate lines of adjacent polygons using ArcToolbox > Data Management Tools > Features > Polygon to Line tool • Retain a separate polygon layer for fills with no line color or width set 6C
12 **Index contours**	Create a field in attribute table that designates index contours at a regular interval. • Mapping Center > Blog tab > enter search keyword **contour lines** 1B
13 **Weighted contours**	Symbolize increasing data values with increasingly wider isolines. • Layer Properties > Symbology tab > Quantities > Graduated Symbols > Field Value set to elevation attribute > adjust symbol sizes for each class 6C
14 **Tapered lines**	Tapered stream networks use thin lines for small tributaries and wider lines for main channels. • Layer Properties > Symbology tab > Quantities > Graduated Symbols or Categories > Field Value set to attribute that designates stream order, flow, or similar cartographic stream rating > adjust symbol sizes for each value or class • Or, right-click layer > Convert Features to Graphics > set each line's width • Additional tools are Wave and Streamline effects in Representations for ArcGIS 9.3 1D

POLYGONS

15 **Pattern fill**	Pattern fills are particularly useful for black-and-white mapping. • Symbol Property Editor > Type: Marker Fill Symbol > Marker Fill tab > Grid option and set marker properties > Fill Properties tab > X,Y separation for marker pattern 4D: Circle marker fill 5C: Open square and triangle at same fill spacing 5D: Thin rectangle fill

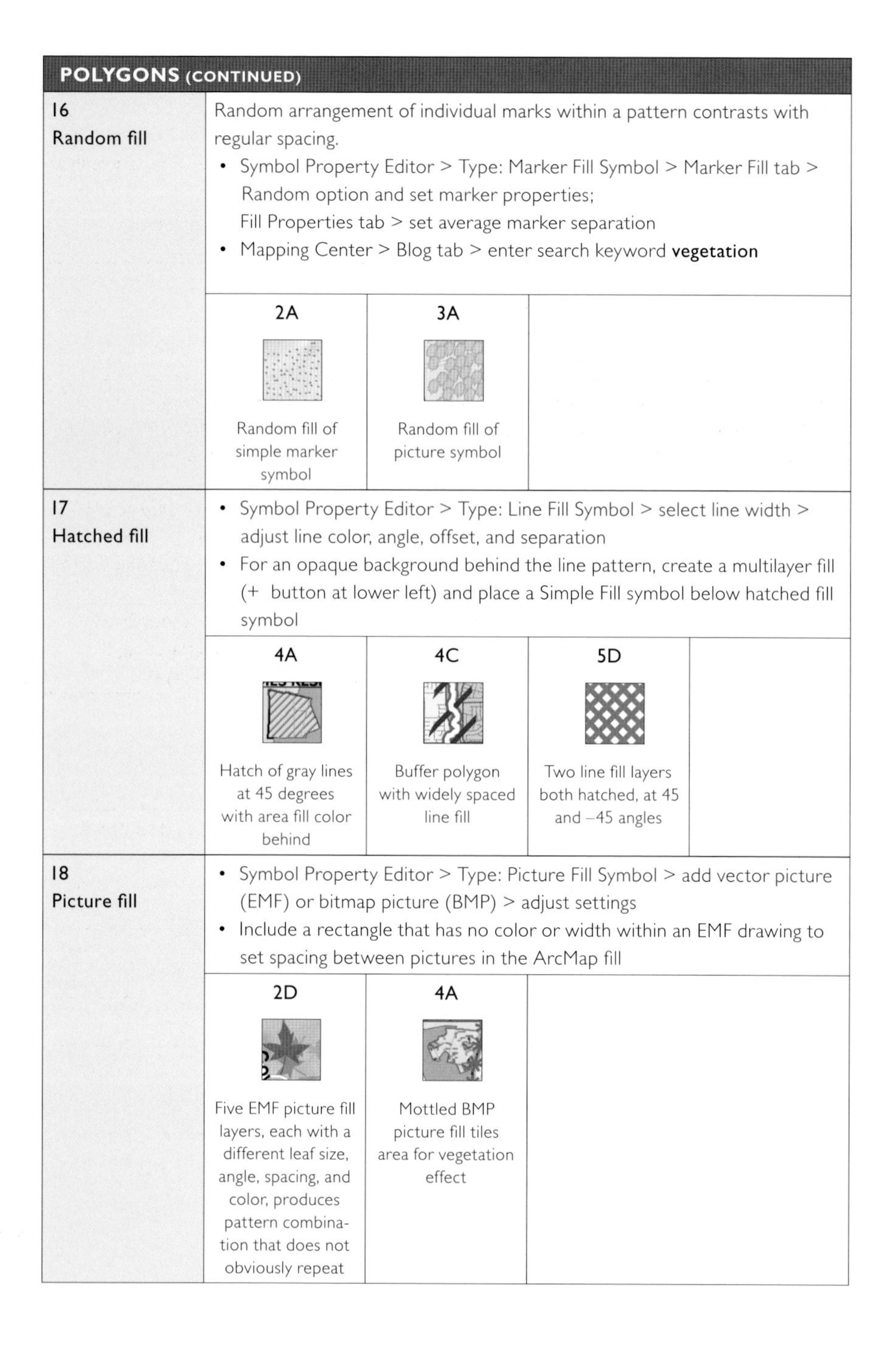

POLYGONS (CONTINUED)			
16 Random fill	Random arrangement of individual marks within a pattern contrasts with regular spacing. • Symbol Property Editor > Type: Marker Fill Symbol > Marker Fill tab > Random option and set marker properties; Fill Properties tab > set average marker separation • Mapping Center > Blog tab > enter search keyword **vegetation**		
	2A Random fill of simple marker symbol	3A Random fill of picture symbol	
17 Hatched fill	• Symbol Property Editor > Type: Line Fill Symbol > select line width > adjust line color, angle, offset, and separation • For an opaque background behind the line pattern, create a multilayer fill (+ button at lower left) and place a Simple Fill symbol below hatched fill symbol		
	4A Hatch of gray lines at 45 degrees with area fill color behind	4C Buffer polygon with widely spaced line fill	5D Two line fill layers both hatched, at 45 and –45 angles
18 Picture fill	• Symbol Property Editor > Type: Picture Fill Symbol > add vector picture (EMF) or bitmap picture (BMP) > adjust settings • Include a rectangle that has no color or width within an EMF drawing to set spacing between pictures in the ArcMap fill		
	2D Five EMF picture fill layers, each with a different leaf size, angle, spacing, and color, produces pattern combination that does not obviously repeat	4A Mottled BMP picture fill tiles area for vegetation effect	

19 Gradient fill	• Symbol Property Editor > Type: Gradient Fill Symbol > adjust settings • Mapping Center > Blog tab > enter search keyword **gradient**		
	4A Very subtle circular or buffered gradient for each area fill	6C Water glint is a light-to-medium blue linear gradient at angle	
20 Four-color fill	Four colors are the minimum needed to distinguish adjacent polygons (if an attribute is not being mapped). • Convert to Graphics > individually select polygons and apply fill color to each • Or, use the script at Mapping Center > ArcGIS Resources tab > Tools, Models & Scripts tab > **Four Color a Map**		
	4B		
21 Hillshade colors	• Layer Properties > Symbology tab > click on Color Ramp and select from drop-down menu of choices • Prepare custom ramps from Tools menu > Styles > Style Manager > Color Ramps folder > right-click in list area to create new two-color Algorithmic ramp or Multi-part series of algorithmic ramps • Adjust position of ramp colors in hillshade using Layer Properties > Symbology tab > Histogram button > move position of diagonal line in graph • Mapping Center > Blog tab > enter search keyword **hillshade**		
	2C Warmer hillshade with ramp from light to brown rather than to black or to gray	6C Multi-part ramp of white to yellow ramp continued by yellow to dark brown	
22 Transparent fill	Adjust layer order in TOC and set fill colors transparent over hillshade, or set hillshade transparent over fill color layers. • Effects toolbar > select Layer > Adjust Transparency tool • Or, Layer Properties > Display tab > set transparency		
	5A	6A	

POLYGONS (CONTINUED)	
23 **Transparent legend**	Create legend boxes that show the same combinations of transparency and color fills as seen on the map. • Mapping Center > Blog tab > enter search keyword **hypsometric legend** 5A
24 **Centerline**	Well-structured cartographic data has centerlines for linear polygons (e.g., polygons with lines along river banks or street curbs), which may be used with no line symbol to position labels or may replace the polygon at smaller scales. • Mapping Center > Blog tab > enter search keyword **centerline** 2C
25 **Eliminate by size**	Remove the clutter of small features by removing those below an area threshold. • Toolbox > Data Management Tools > Generalization > set minimum area within either Aggregate Polygons or Simplify Polygon • To better use generalization tools, structure hydrographic data with islands as a separate class, rather than using interior polygons 2A 2B

LABELS AND POINT SYMBOLS	
26 **Halo**	*Halo on label* • Layer Properties > Labels tab > Symbol > Properties > Mask tab > choose Halo > adjust settings • Use Label Manager on the Labeling toolbar for quick access to all the layers in the data frame *Halo on symbol* • Symbol Property Editor > Mask tab > choose Halo > adjust settings, • Or, set thin outline of marker to a contrasting color 5C 6D 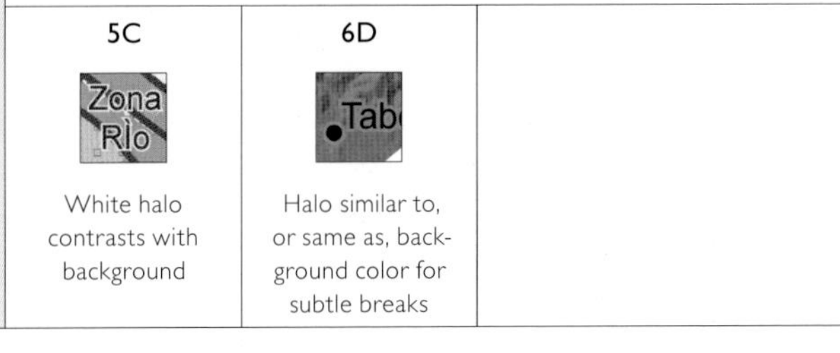 White halo contrasts with background — Halo similar to, or same as, background color for subtle breaks
27 **Label hull**	Label hulls are an alternative to halos. • Right-click frame > Convert Labels to Annotation; Toolbox > Cartography Tools > Masking Tools > Feature Outline Masks > set input layer and parameters > for Mask Kind choose Exact Simplified for a close fitting hull > set fill color for mask • A transparency setting for a mask fill improves contrast without eliminating background features 4B
28 **Variable-depth masking**	A variable-depth mask that blocks some layers is transparent so other underlying layers (e.g., terrain or land cover) are visible through the gaps in the masked layers (e.g., where contours break at elevation labels). • Right-click frame > Convert Labels to Annotation; Toolbox > Cartography Tools > Masking Tools > Feature Outline Masks > set input layer and parameters; Layer Properties for new feature outline mask layer > Symbol Selector > Fill color > No Color; Data Frame properties > Advanced Drawing Options > check Draw using masking options specified below > choose masking layer (e.g., feature outline mask layer) and masked layer (e.g., isoline layer) 1C 6D
29 **Label shadow**	Use small shadow offsets, such as 0.4 points, to raise a label slightly above the map surface and provide contrast. • Layer Properties > Labels tab > Symbol > Properties > Advanced Text tab > adjust shadow offsets and color 6B

LABELS AND POINT SYMBOLS (CONTINUED)

30 Custom symbols	• Layer Properties > Symbol Selector > Properties > Type: Character Marker, Picture Marker or Simple Marker Symbol > create multilayer marker that combines single characters in varied colors (e.g., outline, background, and letter) • For a shield behind each highway's number, Layer Properties > Symbol Selector > Properties > Advanced Text > check Text Background > Properties > Type: Marker Text Background > Symbol > create multi-layer marker with coordinated pieces of shield from font (outline, top color, and body color) • Select an existing pictogram or shield style and edit the many elements to build a custom symbol more quickly—copy-paste-edit rather than repetitively setting the same properties on each symbol			
	3A Picture marker (BMP or EMF)	3B Three-layer character marker	3D Character in custom font	4C Three-layer marker as background for highway number
31 Embed font	Prevent replacement of fonts and font-based symbols (such as north arrows and highway shields) with nonsense characters when displayed on computers without the same fonts installed. • File menu > Export map > Options (at bottom) > Format tab > check Embed All Document Fonts			
	3A	4B		
32 Open characters	Large labels have less visual prominence if character fill does not contrast with the background. • Layer properties > Symbol > Properties > Advanced Text tab > check Text fill pattern > Properties > Fill Color > No Color (or set text fill to color of background)			
	4C Open fill	4D Background fill		

33 Leader lines	Leader lines are thin short lines that clarify the association between a map feature and its label. • For a new label (one not generated using an attribute), Draw toolbar > Callout tool > label feature and adjust callout settings • Adjust leader line style from label's Properties > Text tab > Change Symbol > Properties > Advanced Text tab > check Text Background > Properties > Type: Line Callout (or Simple Line Callout) > adjust settings • Or, for a leader line without a label callout, draw a line using Draw toolbar > New Line tool			
	3C			
34 Maplex settings	Maplex makes more careful label position decisions than the standard labeling engine and provides refinement choices (e.g., allow labels to be placed outside or overrun features, stack long labels, remove duplicate labels or repeat at set distances, create curved labels, and spread characters). • Start Maplex from Tools menu > Extensions > check Maplex; Labeling toolbar > Labeling drop-down > Use Maplex Label Engine • Labeling toolbar > Label Manager tool > set label placement properties from Position and Properties buttons • See Maplex section (right-side links) on Mapping Center			
	1D Curved street labels	2B Centered street labels	1B Best position outside polygon	5B Overrun polygon
35 Annotation	A new label added to a map as annotation stores its own position, text string, and display properties in a geodatabase or map document; anchored to geography but can also be moved manually. • Activate Frame if in Layout View > Draw toolbar > New Text tool > click location on map > enter text and adjust properties			
	3B			

LABELS AND POINT SYMBOLS (CONTINUED)

36 Convert to annotation	Customize individual label positions and styles to improve automatically placed map labels. • Right-click data frame > Convert Labels to Annotation > adjust settings • Convert labels for different layers at the same time to retain relative positioning • Use Annotation toolbar in conjunction with the Editor toolbar to set target, edit, and reposition annotation: Start Editing > select an individual annotation > right-click and select Attributes > set detailed style and position edits under Annotation and Attributes tabs (change Value entries by clicking on them) • Annotation toolbar > Unplaced Annotation Window tool > displays unplaced annotation in a table to view omitted labels and return them to the map as needed 2A: Customize line spacing and indents 3C: Repositioned for improved placement 4C: Variable character spacing from area to area
37 Curved labels	*Custom curved labels* • Drawing toolbar > New Splined Text tool *Automated curved labels* • Label Manager > Placement Properties button > Orientation or Position button > choose a Curved option *Change annotation from straight to curved* • Editor toolbar > Start Editing > set Target to anno layer; Annotation toolbar > Edit Annotation tool > right-click on label and choose Curvature > Curved; right-click label again > Edit Baseline Sketch > move anchor points and handles for splined text 1D
38 Multiline labels	• Layer Properties > Labels tab > Expression > build multiline labels using label fields and the Visual Basic function vbCrLf • See ArcGIS Online Help search keyword **label expressions** • Or, use Stack Label setting in Maplex Placement Properties • Or, convert single-part annotation to multiple parts and reposition 4A 6D

OVERALL EFFECTS

39 Spot color	Use a graphics program (such as Adobe Illustrator) to change from RGB or CMYK to print an image using custom ink colors. • Mapping Center > Blog tab > enter search keyword **CMYK Adobe** 5C
40 Layer order	Click and drag layers to reorder them in the table of contents for desired design effects (layers in the examples below are named in order from top to bottom as seen in the maps). 1A: Road/ River/ Boundary line1/ Boundary line2 (with pink band)/ Contours 3B: Building/ Parking line/ Road/ Parking fill/ Walkway/ Campus fill 3C: Label/ Campus building/ Boundary line/ City building/ City fill/ Campus fill/ Road line/ Road fill 4B: Highway line/ Highway case/ Wire center line (brown)/ Rate center line (yellow)/ Area code line (blue)/ Local roads/ Wire center fill 4D: Wire center line (black dash)/ Rate center line (gray line)/ Rate center case (black)/ Wire center line (white mask)/ Road lines 6B: County line (white dash)/ Transparent hillshade/ Continental divide (gray line)/ Snowload fill and line (blues)
41 Convert grid	Convert grid lines to data (not to graphic) so they can be positioned behind other features (e.g., roads) in the TOC. • Mapping Center > Blog tab > enter search keywords **grid polygon** 2D
42 3D buildings	Use ArcGIS 3D Analyst, ArcScene, or the SketchUp ArcGIS plug-in to create 3D buildings, or insert them as picture markers. 3D

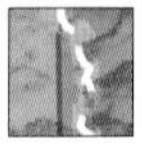

RESOURCES

BOOKS

Bertin, Jacques. 1983. *Semiology of graphics: Diagrams, networks, maps.* Madison, Wisc.: University of Wisconsin Press (translation of 1967 edition).

Brewer, Cynthia A. 2005. *Designing better maps: A guide for GIS users.* Redlands, Calif.: ESRI Press.

Brown, Allan, and Wim Feringa. 2003. *Colour basics for GIS users.* Harlow, England; New York: Prentice Hall.

Campbell, John. 2001. *Map use and analysis.* 4th ed. Boston: McGraw-Hill.

Dent, Borden D. 1999. *Cartography: Thematic map design.* 5th ed. Boston: WCB/ McGraw-Hill.

Dorling, Daniel, and David Fairbairn. 1997. *Mapping: Ways of representing the world.* Harlow, England: Prentice Hall.

Fairchild, Mark D. 2005. *Color appearance models.* 2nd ed. Chichester, West Sussex, England; Hoboken N.J.: John Wiley and Sons, Inc.

Flacke, Werner, and Birgit Kraus. 2005. *Working with projections and datum transformations in ArcGIS.* Norden. Netherlands: Points Verlag Norden Halmstad.

Kraak, Menno-Jan, and Ferjan Ormeling. 2003. *Cartography: Visualization of geospatial data.* 2nd ed. Harlow, England; New York: Prentice Hall.

Krygier, John, and Denis Wood. 2005. *Making maps: A visual guide to map design for GIS.* New York: Guilford Press.

MacEachren, Alan M. 1995. *How maps work: Representation, visualization, and design.* New York: Guilford Press.

Monmonier, Mark S., 1993. *Mapping it out: Expository cartography for the humanities and social sciences.* Chicago: University of Chicago Press.

——. 1996. *How to lie with maps.* 2nd ed. Chicago: University of Chicago Press.

Muehrcke, Phillip C., Juliana O. Muehrcke, and A. Jon Kimerling. 2001. *Map use: Reading, analysis, interpretation.* 4th ed. (revised). Madison, Wisc.: J.P. Publications.

Robinson, Arthur H., Joel L. Morrison, Phillip C. Muehrcke, A. Jon Kimerling, and Stephen C. Guptill. 1995. *Elements of cartography.* 6th ed. New York: John Wiley and Sons, Inc.

Slocum, Terry A., Robert B. McMaster, Fritz C. Kessler, and Hugh H. Howard. 2004. *Thematic cartography and geographic visualization.* 2nd ed. Upper Saddle River, N.J.: Prentice Hall.

JOURNALS

Cartographica, published by the Canadian Cartographic Association. *www.cca-acc.org.*

Cartographic Perspectives, published by the North American Cartographic Information Society. *www.nacis.org.*

Cartography and Geographic Information Science, published by the Cartography and Geographic Information Society. *www.cartogis.org.*

The Cartographic Journal, published by the British Cartographic Society. *www.cartography.org.uk.*

ESRI MAP BOOKS

Many example maps in this book were showcased in one of four *Map Books* published by ESRI. *Map Books* contain a variety of maps presented by GIS users in the Map Gallery at the annual International Users Conference in San Diego.

Law, Michael, ed. 2007. *ESRI Map Book Volume 22.* Redlands, Calif.: ESRI.

Sappington, Nancy, ed. 2004. *ESRI Map Book Volume 19.* Redlands, Calif.: ESRI.

——. 2005. *ESRI Map Book Volume 20.* Redlands, Calif.: ESRI.

——. 2006. *ESRI Map Book Volume 21.* Redlands, Calif.: ESRI.

ESRI MAPPING CENTER

http://mappingcenter.esri.com

ESRI Mapping Center is a Web site that helps you make great looking maps with ArcGIS by using the same cartographic concepts that professional cartographers use. Visit the Mapping Center to learn a variety of cartographic techniques and find information on best practices for mapping and cartography with GIS.

ESRI Mapping Center features include the following:

Blog—Offers the opportunity to read about and participate in the latest topics of conversation.

Maps—Highlights cartographic techniques that you can try out on your own.

ArcGIS Resources—Provides downloads you can use for creating your own cartographic effects.

Ask a Cartographer—Presents solutions to your particular mapping challenges.

Other Resources—Contains links to ESRI publications and presentations, additional mapping resources on the ESRI Web site, and a special collection of Cartographers' Favorites.

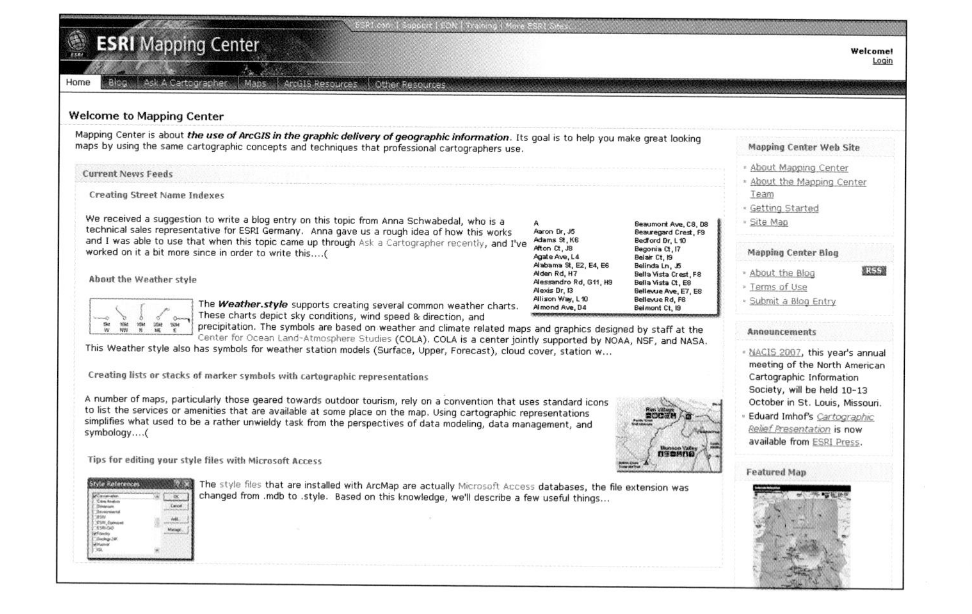